Aircraft Systems

This book is dedicated to Sheena Moir and Susan Seabridge for their support, forbearance, and encouragement.

Aircraft Systems

Mechanical, electrical, and avionics subsystems integration

Ian Moir

Allan Seabridge

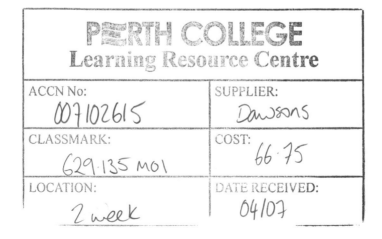

Professional Engineering Publishing

Professional Engineering Publishing Limited,
London and Bury St Edmunds, UK

First edition published 1991 by Longman Group UK Limited.

This edition published 2001, reprinted 2002 by Professional Engineering Publishing, UK.
Published in USA by American Institute of Aeronautics and Astronautics, Inc.

ISBN 1 86058 289 3

A CIP catalogue record for this book is available from the British Library.

Printed and bound by CPI Antony Rowe, Eastbourne

List of Plates

About the Authors

Ian Moir BSc, CEng, FRAeS, FIEE served twenty years in the Royal Air Force as an engineering cadet/officer, retiring with the rank of squadron leader. He then went on to work for eighteen years at Smiths Industries, Cheltenham, UK. Here he had responsibilities for the introduction of avionics technology into aircraft utilities systems on both military and civil aircraft. He was programme manager for the integrated Utilities Management System on the UK Experimental Aircraft Programme (EAP) and technology demonstrator for the European Fighter Aircraft. Ian's principal successes at Smiths Industries included the selection and development of new integrated systems for the McDonnell Douglas/Boeing AH-64C/D Longbow Apache attack helicopter and Boeing 777 (Queen's Award for Technology – 1998), both of which are major production programmes.

Ian has over 40 years' experience in the aerospace industry. He is currently an International Aerospace consultant, operating in the areas of aircraft electrical and utilities systems and avionics.

Allan Seabridge BA, MPhil is currently the chief flight systems engineer at BAE SYSTEMS, a position held since 1998. Before that he was the avionics integrated product team leader on the Nimrod MRA4 programme for five years. He has worked in the aerospace industry for over 30 years in flight systems and avionics systems engineering, business development, and project management. He has been involved in a wide range of military fast jet, trainer, and ground or maritime surveillance aircraft projects.

Allan has worked in many international collaborative programmes in Europe and the United States, and he has led a number of national and international engineering teams. This has resulted in an interest in all aspects of system engineering capability – the practice of engineering, the processes and tools employed, and the people and skills required.

Related Titles

Title	Author	ISBN
Aircraft Conceptual Design Synthesis	D Howe	1 86058 301 6
IMechE Engineers' Data Book – Second Edition	C Matthews	1 86058 248 6
Optimizing the super-turbocharged aeroengine	J Panting	1 86058 080 7
Military Aerospace Technologies – FITEC '98	IMechE Conference	1 86058 168 4
Civil Aerospace Technologies – FITEC '98	IMechE Conference	1 86058 168 4
Ground Support Equipment in the 21st Century	IMechE Seminar	1 86058 333 4
Journal of Aerospace Engineering	Proceedings of the IMechE, Part G	ISSN: 0954/4100

For the full range of titles published by Professional Engineering Publishing (publishers to the Institution of Mechanical Engineers) contact:

Sales Department
Professional Engineering Publishing Limited
Northgate Avenue
Bury St Edmunds
Suffolk IP32 6BW
UK
Tel: +44 (0) 1284 724384; Fax: +44 (0) 1284 718692
E-mail: sales@pepublishing.com
www.pepublishing.com

Contents

Foreword from First Edition

This publication sets in context the relationship between the various systems of aircraft management and concentrates on those 'Systems' that are fundamental to the best performance of the primary aircraft tasks.

Modern computers provide us with enormously enhanced power to solve the problems associated with advanced performance and create the opportunity to expand the envelope of capability of the aircraft as never before. The unstable aircraft is here to stay, and this is a contradiction in terms for all of us who have a long history in aviation! The understanding and solution of these problems is fundamental to the safe and sustained development of more and better aircraft at affordable cost. Systems Engineering holds the key to many of the improvements we seek, particularly in reliability.

The book corrects any impressions there may be that 'Systems' concern solely avionics or the more esoteric onboard equipment which is not particularly concerned with the basic flight profile of the aircraft. First we have to design an aircraft that is as advanced aerodynamically as we can make it before we start fitting the extra equipment on board.

The development of basic aircraft systems has not stood still. We can check this simple fact by looking at the wing size of a modern passenger aircraft and see that its size is reducing while the lifting power of the wing is still increasing. This is a measure of improvements now capable of being made in wing design which in turn are dependent on 'Systems' capable of developing the maximum performance from the minimum of weight, hence fly by wire. There is nothing new in all this: aircraft performance has ever been about the relationship between power and weight. Indeed, until adequate power at the right weight was available, sustained manned flight was not possible.

This is a straightforward textbook, which is written in an extremely readable style and enables the reader to acquire a comprehensive knowledge of the principles of aircraft systems, without overwhelming them with too much jargon. As such, it makes an important contribution to our reference library of knowledge.

Admiral Sir Raymond Lygo KCB, Hon. FRAeS, FPCL, FRSA, CBIM
Chief Executive, British Aerospace, 1978–1989

Foreword

Aircraft design begins with dreams and design requirements, and eventually proceeds to detailed drawings of every part of the aircraft being fabricated. To the outside world, the disciplines of aerodynamics and structures often seem most important – they lead to the overall shaping of the aircraft and to the design of the parts that, when fabricated and assembled, comprise the physical geometry of the aircraft. These are obviously important, but without some other things inside, the aircraft could never fly.

These 'other things' – more properly known as 'aircraft subsystems' or just 'systems' – play a crucial role in aircraft design and operation. Systems turn an aerodynamically shaped structure into a living, breathing, flying machine. Systems include flight control, hydraulics, electrical, pneumatic, fuel, environmental control, landing gear, and the ever-more-capable avionics. In the early stages of conceptual or preliminary design the systems must be initially defined, and their impacts must be incorporated into design layouts, weight analyses, and performance calculations. Anyone seeking to become a good aircraft conceptual designer must learn about all types of systems. During detail design the systems are fully defined, including system architecture, functional analysis, component design, and safety and failure analysis. This is done by highly experienced systems specialists, and any of these systems areas can be a rewarding career.

Ian Moir and Allan Seabridge have written a wonderful and urgently needed book on the subject of aircraft systems that can serve as both an introduction and a reference for engineers who specialize in such areas. It covers all the systems areas listed previously, and more, and includes overall concepts, systems architectures, design approaches, and both proven and emerging technologies. These systems are briefly introduced in aircraft conceptual design textbooks, so this book also serves as a follow-up to them and as a reference for aircraft conceptual designers. *Aircraft Systems – mechanical, electrical, and avionics subsystems integration* is sure to become a standard industry reference seen in industry design offices and student design labs everywhere.

AIAA are pleased to have joined with Professional Engineering Publishing, (publishers to the IMechE), in the co-publication of this book and to include it as part of their Education Series. *Aircraft Systems* offers a unique source of information on current design practice in the area of aircraft systems and provides a balanced approach to the more practical aspects of aircraft design and development. As an extensively revised second edition to a previously published title, *Aircraft Systems* benefits from use and reader feedback, as well as the peer review and editorial process of both Professional Engineering Publishing and AIAA.

Daniel P Raymer
President, Conceptual Research Corporation, USA

Acknowledgements

The authors have been practising engineers in aerospace for a combined total of well over 70 years, much of that time spent in the specification, design, and development of avionics and ultilities systems for high-performance combat aircraft. Both authors were heavily involved in the conception and development of the Utilities Management System equipment that was demonstrated on the British Aerospace Experimental Aircraft Programme, better known as EAP. The EAP – a technology demonstrator for Eurofighter – was an aircraft programme, which generated commitment, loyalty, and enthusiasm among all those in the industry who are fortunate enough to play a part.

It was this project that led the authors and others to collaborate on a number of technical papers describing the EAP USM system. It also brought a realization that many of the aircraft utilities systems had failed to achieve the exposure lavished upon the more glamorous avionics systems, despite their equal importance in assuring mission success.

The writing of *Aircraft Systems* would not have been possible without the generous assistance and support of many colleagues, individuals, and companies within the aerospace industry on both sides of the Atlantic. The range of topics covered would have been beyond the competence of the authors without such support. In particular: Chapter 4 *Hydraulic Systems* was written with the great assistance of Fred Greenwood of BAE SYSTEMS (British Aerospace); Chapter 7 *Environmental Control Systems* was written with the extensive assistance of Carole Todd, Assistant Project Manager at BAE SYSTEMS (British Aerospace); and Geoff Worral also of BAE SYSTEMS (British Aerospace) rendered significant support with Chapter 8 *Emergency Systems*.

Invaluable support and assistance has been given by the following individuals, companies, and organizations:

Airbus UK Limited
BAE SYSTEMS (formerly British Aerospace)
Bell/Boeing V-22 Tiltrotor Team
Boeing (including the former McDonnell Douglas)
Boeing Vertol
Claverham/FHL
Dunlop Aerospace International
Flight Refuelling Limited
GKN Westland Helicopters
Gordon G Bartley
Handley Page Association
High Temp Engineers
Honeywell
Honeywell Normalair-Garret Limited

Interavia
JSF Program Office
Kidde-Graviner
Leland Electrosystems
Martin Baker Engineering
Parker Aerospace
Raytheon (formerly Hawker Siddeley)
Rolls-Royce
Rolls-Royce/Turbomeca
Royal Aero Club
Smiths Industries
TI Group
TRW Lucas Aerospace
United Technologies
US Air Force
Vickers Systems

Authors' Preface

Some clarification of terminology is needed to establish the extent and compass of *Aircraft Systems* and the term 'avionics'. Avionics is now universally associated with those aeronautical and aviation electronics systems connected with flight deck systems, flight control, systems management, navigation, communications, radar, and electronic warfare. These are the systems that provide the aircraft with the capabilities in order to fulfil a particular operational role.

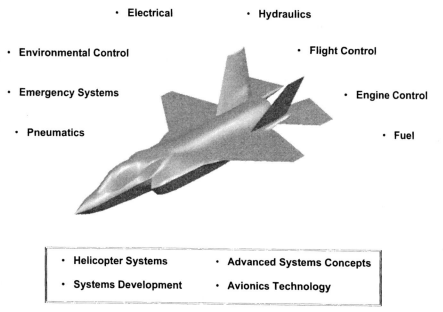

- **Electrical**   - **Hydraulics**
- **Environmental Control**   - **Flight Control**
- **Emergency Systems**   - **Engine Control**
- **Pneumatics**   - **Fuel**

- **Helicopter Systems**   - **Advanced Systems Concepts**
- **Systems Development**   - **Avionics Technology**

There is another world of aircraft systems that are required to enable the aircraft to fly and function – the 'general' or 'utilities' systems. These are less glamorous than the classical avionics systems, but are nevertheless essential for the aircraft to operate, since without them the aircraft will not leave the ground. They are associated with flight control; engine control; and the control of fuel, hydraulics, electrical, pneumatic, environmental, and emergency systems. These systems have, in recent years, increasingly adopted electronics technologies in order to improve system control and diagnostics. Therefore, without exception, these systems are today also 'avionic' in nature.

Aircraft Systems – mechanical, electrical, and avionics subsystems integration describes the nature of these systems in detail, giving both military and civil examples. In addition, the book describes the unique nature of helicopter systems and some of the more advanced systems concepts that are being developed or have recently reached fruition. Finally – given the magnitude and scope of the development of aircraft systems – the development methodologies and avionics technology typically used in the implementation of aircraft systems are also outlined.

During the ten years since the publication of first edition of this book, many developments and technological advances have pushed the subject of *Aircraft Systems* forward. This completely revised book is now longer, with more illustrations, and covers more ground but, we trust, still retains an immediate and straightforward handling of what can be a complex subject.

We hope that our work will educate, explain, and enlighten.

Ian Moir and Allan Seabridge
2001

Acronyms and Abbreviations

A429 ARINC 429 Data Bus
A629 ARINC 629 Data Bus
AAWWS Airborne Adverse Weather Weapons System (Apache)
AC Advisory Circular (FAA)
AC Alternating Current
ACE Actuator Control Electronics (B777)
ACMP AC Motor Pump
ACT Active Control Technology
A/D Analogue to Digital
ADC Air Data Computer
ADIRS Air Data & Inertial Reference System
ADM Air Data Module
ADP Air Driven Pump
ADU Actuator Drive Unit
ADV Air Defence Variant (Tornado)
AFCS Automatic Flight Control System
AFDC Autopilot Flight Director Computers (B777)
AFDS Autopilor Flight Director System
AFTI Advanced Fighter Technology Integration (F-16)
AIAA American Institute of Aeronautics & Astronautics
AIMS Airplane Information Management System (B777)
AIR Aerospace Infermaton Report (SAE)
Aj Jet Pipe Area
AMAD Airframe-Mounted Accessory Drive
Amp or A Ampere
APB Auxiliary Power Breaker
APU Auxiliary Power Unit
ARINC Air Radio Inc
ASCB Avionics Standard Communications Bus

ASI Airspeed Indicator
ASIC Application Specific Integrated-Circuit
AS/PCU Air Supply/Pressurisation Control Unit (B777)
ATA Air Transport Association
ATC Air Traffic Control
ATF Advanced Tactical Fighter
Atm atmosphere
ATM Air Transport Management
ATP Advanced Turbo-Prop
ATR Air Transport Radio
AUW All-Up Weight
AVM Airplane Vibration Monitoring

BAES BAE SYSTEMS
Batt Battery
BC Bus Controller (MIL-STD-1553B)
BCF Bromo-Chloro-diFluoro-Methane
BIT(E) Built-In Test (Equipment)
BOV Blow Off Valve
BPCU Bus Power Control Unit
BSCU Brake System Control Unit
BTB Bus Tie Breaker
BTMU Brake Temperature Monitoring Unit

C Centigrade
CAA Civil Aviation Authority
CASA Construcciones Aeronauticas Socieda Anonym
CBL™ Control-By-Light™ (Raytheon proprietary fibre optic bus)
CCA Common Cause Analysis
CCB Converter Control Breaker (B777)
CDA Concept Demonstration Aircraft
CDR Critical Design Review

CDU Cockpit Display Units
CF Constant Frequency
CFD Continuous Fire Detector
CG Centre of Gravity
CH Channel
CHRG Charger
CNS Communications, Navigation, Surveillance
COTS Commercial Off-The-Shelf
CPG Co-Pilot Gunner (AH-64 Apache)
CPT Combined Processor Totaliser
CSAS Control Stability Augmentation System
CSD Constant Speed Drive
CT Current Transformer
CTC Cabin Temperature Control
CTOL Conventional Take-Off & Landing
CV Carrier Vehicle

D/A Digital to Analogue
DATAC Digital Autonomous Terminal Access Communication (forerunner to ARINC 629)
DC Direct Current
DDP Declaration of Design & Performance
DECU Digital Engine Control Unit
Def Stan Defence Standard
Dem/Val Demonstration/Validation
DFCC Digital Flight Control Computer (AFTI F-16)
DTD Directorate of Technical Development
DTI Department of Trade & Industry
DVO Direct Vision Optics

E2PROM Electrically Erasable Programmable Read Only Memory
EAI Engine Anti-Ice

EAP Experimental Aircraft Programme
ECAM Electronic Crew Alerting & Monitoring
ECS Environmental Control System
EDP Engine Driven Pump
EEC Electronic Engine Controller
EFA European Fighter Aircraft
EFAB Extended Forward Avionics Bay
EFIS Electronic Flight Instrument System
EGT Exhaust Gas Temperature
EHA Electro-Hydrostatic Actuator
EHI European Helicopter Industries
EICAS Engine Indication & Crew Alerting System
ELAC Elevator Aileron Computer (A320)
ELCU Electronic Load Control Unit
ELMS Electrical Load Management System (B777)
EMA Electro-Mechanical Actuator
EMI Electro-Magnetic Interference
EPC External Power Contactor
EPMS Electrical Power Management System (AH-64C/D Apache)
EPROM Electrically Programmable Read Only Memory
EPS Emergency Power Supply
EPU Emergency Power Unit
ERA Electrical Research Agency
ESS Essential
ESS Environmental Stress Screening
ETOPS Extended Twin Operations
EU Electronics Unit
EXT or Ext External

FAA Federal Aviation Authority
FAC Flight Augmentation Computer
FADEC Full Authority Digital Engine Control
FAR Federal Aviation Regulations
FBW Fly-By-Wire
FCC Flight Control Computer
FCDC Flight Control Data Concentrators (A330/A340)
FCP Fuel Control Panel
FCPC Flight Control Primary Computer (A330/A340)
FCS Flight Control system

FCSC Flight Control Secondary Computer (A330/A340)
FCU Flight Control Unit (Autopilot)
FCU Fuel Control Unit (Engine)
FHA Functional Hazard Analysis
FITEC Farnborough International Technology Exploitation Conference (1998)
FLIR Forward Looking Infra Red
FMEA Failure Modes & Effects Analysis
FMGEC Flight Management Guidance & Envelope Computers (A330/A340)
FMQGS Fuel Management & Quantity Gauging System (Global Express)
FMS Flight Management System
FQIS Fuel Quantity Indication System
FQPU Fuel Quantity Processor Unit (B777)
FSD Full Scale Development
FSEU Flap Slats Electronics Unit (B777)
FTA Fault Tree Analysis

G or Gen Generator
GA General Aviation
Gal Gallon
GCB Generator Control Breaker
GCU Generator Control Unit
GE General Electric (US)
GEC General Electric Company
GLY Galley
GND Ground
gpm Gallons per minute
GPS Global Positioning System
GPU Ground Power Unit
GR Ground Reconnaissance

HP/hp High Pressure
hp Horse Power
HT Horizontal Tail
HUMS Health & Usage Management System
Hyd Hydraulic
Hz Hertz

IC Integrated Circuit
IDEA Integrated Digital Electric Airplane
IDG Integrated Drive Generator
IDS InterDictor Strike (Tornado)
IEE Institution of Electrical Engineers

IEEE Institute of Electrical & Electronic Engineers
IFE In-Flight Entertainment
IFPC Integrated Flight & Propulsion Control
IFSD In-Flight Shut Down
IFSR In-Flight Shutdown Rate
IGV Inlet Guide Vanes
IMA Integrated Modular Avionics
IMechE Institution of Mechanical Engineers
In Inch
INS Inertial Navigation System
INV Inverter
I/O Input/Output
IPN Iso-Propyl Nitrate
I/Press,ip Intermediate Pressure (of engine)
IPT Integrated Product Team
IPU Integrated Power Unit
IR Infra Red
IRS Inertial Reference System
ISA Instruction Set Architecture
ISA International Standard Atmosphere

JAA Joint Airworthiness Authority
JAR Joint Aviation Regulation
J/IST Joint Strike Fighter/Integrated Subsystems Technology
JSF Joint strike Fighter

K Kelvin
kg Kilogram
kN Kilo Newton
kPa Kilo Pascal
KT or kt Knot
kVA Kilo Volt-Ampere
kW Kilo Watt

L Left
L Lift
LAF Load Alleviation Function
LAN Local Area Network
LB or lb Pound
LH Left Hand
LHX or LH Light Helicopter
LIB Left Inboard
LOB Left Outboard
LOX Liquid Oxygen
LP Low Pressure
LRM Line Replaceable Module
LRU Line Replaceable Unit
LSI Large Scale Integration
LVDT Linear Variable Differential Transformer

M   Mach Number
m   Metre
MA   Markov Analysis
mA   Milli Ampere
MADMEL   Management and Distribution of More Electric Aircraft
MAU   Modular Avionics Unit (Honeywell EPIC system)
Max   Maximum
MBB   Messerschmit Bolkow Blohm
MCDU   Multipurpose Control & Display Unit
MCM   Multi Chip Module
MCU   Modular Concept Unit
MDC   Miniature Detonation Cord
MDHC   McDonnell Douglas Helicopter Company (now Boeing)
MEA   More-Electric Aircraft
MECU   Main Engine Control Unit
MDP   Motor Driven Pump
MFD   Multi-Function Display
MHz   Mega Hertz
MIL-H-   Military Handbook
MIL-STD-   Military Standard
Min   Minimum
Min   Minute
MK   Mark
ml   Millilitre
MLC   Main Line Contactor
MLI   Magnetic Level Indicator
mm   Millimetre
MN   Mega Newton
mph   miles per hour
MR   Maritime Reconnaissance
m/s   Metres/second
MSOC   Molecular Sieve Oxygen Concentrator
MSOV   Modulating Shut-Off Valve
MTBF   Mean Time Between Failures

N   North Pole
NADC   Naval Air Development Center
NASA   National Space & Aerospace Agency
NATO   North Atlantic Treaty Organisation
Nav   Navigation
NH or N2   Speed of rotation of engine HP shaft
Ni-Cd   Nickel-Cadmium
NL or N1   Speed of rotation of engine LP shaft

NOTAR   NO TAil Rotor
NRV   Non-Return Valve
Ny   Longitudinal Acceleration
Nz   Normal Acceleration

OBIGGS   On-Board Inert Gas Generation System
OBOGS   On-Board Oxygen Generating System
OEM   Original Equipment Manufacturer
OMS   On-board Maintenance System
Ox   Pitch Axis
Oy   Roll Axis
Oz   Yaw Axis

P   Pressure
PC   Pressure Capsule
PCU   Power Control Unit
PDC   Power Distribution Center
PDE   Power Drive Electronics (AFTI F-16)
PDR   Preliminary Design Review
PDU   Power Drive Unit
PFC   Primary Flight Computer (B777)
PFCS   Primary Flight Control System (B777)
PM   Permanent Magnet
PMA   Permanent Magnet Alternator
PMG   Permanent Magnet Generator
PNVS   Pilot Night Vision System (Apache)
Press   Pressure
PRSOV   Pressure Reducing Shut-Off Valve
PRV   Pressure Reducing Valve
Ps or Po   Ambient Static Pressure
PSA   Power Supply Assembly (B777)
PSEU   Proximity Switch Electronics Unit (B777)
psi   Pounds/Square Inch
PSSA   Preliminary System Safety Analysis
Pt   Dynamic Pressure
Pt   Total Pressure
PTFE   Poly-Tetra-Fluoro-Ethylene
PSU   Power Supply Unit
PTU   Power Transfer Unit
PWR   Power

'Q' feel   A pitch feel schedule used in aircraft flight control systems

R   Right
R & D   Research & Development
RAeS   Royal Aeronautical Society
RAF   Royal Air Force
RAM   Random Access Memory
RAT   Ram Air Turbine
RDCP   Refuel/Defuel Control Panel (Global Express)
RFP   Request For Proposal
RH   Right Hand
RI   Right Inboard
RJ   Regional Jet
ROB   Right Outboard
ROM   Read Only Memory
rpm   Revolutions Per Minute
RT   Remote Terminal (MIL-STD-1553B)
RTCA   Radio Technical Committee Association
RTZ   Return-To-Zero
RVDT   Rotary Variable Differential Transformer

S   South Pole
SAARU   Secondary Attitude Air data Reference Unit
SAE   Society of Automobile Engineers
SCR   Silicon Controlled Rectifier (Thyristor)
SDR   System Design Review
SEC   Spoiler Elevator Computer (A320)
SFCC   Slat/Flap Control Computers (A330/A340)
SFENA   Societe Francaise d'Equipments pour la Navigation Aerienne
shp   Shaft horse Power
SIM   Serial Interface Module (A629)
S/Ldr   Squadron Leader
SMP   Systems Management Processor (EAP)
SMTD   STOL Manoeuvre Technology Demonstrator (F-15)
SOL   Solenoid
SOV   Shut-Off Valve
sq m   Square Metre
SR   Switched Reluctance
SRR   System Requirements Review
SSA   System Safety Analysis
SSPC   Solid State Power Controller
SSR   Software Specification Review
STBY   Standby
STC   Supplementary Type Certificate

STOL Short Take-Off and Landing
STOVL Short Take-Off Vertical
   Landing
SVCE Service

T Temperature
T1 Intake Total Temperature
TADS Target acquisition &
   Designator System (Apache)
TBT Turbine Blade Temperature
TCD Total contents Display
T/EMM Thermal/Energy
   Management Module
T/F Transformer
TGT Turbine Gas Temperature
TPMU Tyre Pressure Monitoring
   Unit
TRU or TR Transformer Rectifier
   Unit

u/c Undercarriage
UCS Utilities Control System
UK United Kingdom
UMS Utilities Management
   System
US United States
USA United States of America
USG US Gallon (1 USG = 0.8
   Imperial Gallon)
USM Utility Systems Management
UTIL Utility
UV Ultra-Violet
U/V Under Voltage

V Velocity
V Volts
VDU Visual Display Unit
VIB Vibration
VIGV Variable Inlet Guide Vane

VF Variable Frequency
VLSI Very Large Scale Integrated-
   Circuit
VMS Vehicle Management System
VOR VHF Omni-Range
VSCF Variable Speed Constant
   Frequency
V/STOL Vertical/Short Take-Off
   & Landing
VSU Voltage Sense Unit
VSV Variable Stator Vane
VTOL Vertical Take-Off &
   Landing

W Watts
W Weight
WRDC Wright Research and
   Development Centre
WWII World War II

CHAPTER 1

Flight Control Systems

Introduction

Flight controls have advanced considerably throughout the years. In the earliest biplanes flown by the pioneers flight control was achieved by warping wings and control surfaces by means of wires attached to the flying controls in the cockpit. Figure 1.1 shows the multiplicity of rigging and control wires on an early monoplane. Such a

Fig. 1.1 Morane Saulnier Monoplane refuelling before the 1913 Aerial Derby (Royal Aero Club)

means of exercising control was clearly rudimentary and was usually barely adequate for the task in hand. The use of articulated flight control surfaces followed soon after but the use of wires and pulleys to connect the flight control surfaces to the pilot's controls persisted for many years until advances in aircraft performance rendered the technique inadequate for all but the simplest aircraft.

When top speeds advanced into the transonic region the need for more complex and more sophisticated methods became obvious. They were needed first for high-speed fighter aircraft and then with larger aircraft when jet propulsion became more widespread. The higher speeds resulted in higher loads on the flight control surfaces which made the aircraft very difficult to fly physically. The Spitfire experienced high control forces and a control reversal which was not initially understood. To overcome the higher loadings powered surfaces began to be used with hydraulically powered actuators boosting the efforts of the pilot to reduce the physical effort required. This brought another problem: that of 'feel'. By divorcing the pilot from the true effort required to fly the aircraft it became possible to undertake manoeuvres which could overstress the aircraft. Thereafter it was necessary to provide artificial feel so that the pilot was given feedback representative of the demands he was imposing on the aircraft. The need to provide artificial means of trimming the aircraft was required as Mach trim devices were developed.

A further complication of increasing top speeds was aerodynamically related effects. The tendency of many high-performance aircraft to experience roll/yaw coupled oscillations – commonly called 'dutch roll' – led to the introduction of yaw dampers and other auto-stabilization systems. For a transport aircraft these were required for passenger comfort whereas on military aircraft it became necessary for target tracking and weapon aiming reasons.

The implementation of yaw dampers and auto-stabilization systems introduced electronics into flight control. Autopilots had used both electrical and air-driven means to provide an automatic capability of flying the aircraft, thereby reducing crew workload. The electronics used to perform the control functions comprised analogue sensor and actuator devices which became capable of executing complex control laws and undertaking high-integrity control tasks with multiple lanes to guard against equipment failures. The crowning glory of this technology was the Category III autoland system manufactured by Smiths Industries and fitted to the Trident and Belfast aircraft.

The technology advanced to the point where it was possible to remove the mechanical linkage between the pilot and flight control actuators and rely totally on electrical and electronic means to control the aircraft. Early systems were hybrid: using analogue computing with discrete control logic. The Control and Stability Augmentation System (CSAS) fitted to the Tornado was an example of this type of system though the Tornado retained some mechanical reversion capability in the event of total system failure. However the rapid development and maturity of digital electronics soon led to digital 'fly-by-wire' systems. These developments placed a considerable demand on the primary flight control actuators which have to be able to accommodate multiple channel inputs and also possess the necessary failure logic to detect and isolate failures (see Fig. 1.2).

Most modern fighter aircraft of any sophistication now possess a fly-by-wire system due to the weight savings and considerable improvements in handling characteristics which may be achieved. Indeed many such aircraft are totally unstable and would not be able to fly otherwise. In recent years this technology has been applied to civil

transports: initially with the relaxed stability system fitted to the Airbus A320 family and A330/A340. The Boeing 777 airliner also has a digital fly-by-wire system, the first Boeing commercial aircraft to do so.

Fig. 1.2 Tornado ADV (F 3) Prototype (BAE SYSTEMS)

Principles of flight control

All aircraft are governed by the same basic principles of flight control, whether the vehicle is the most sophisticated high-performance fighter or the simplest model aircraft.

The motion of an aircraft is defined in relation to translational motion and rotational motion around a fixed set of defined axes. Translational motion is that by which a vehicle travels from one point to another in space. For an orthodox aircraft the direction in which translational motion occurs is the direction in which the aircraft is flying, which is also the direction in which it is pointing. The rotational motion relates to the motion of the aircraft around three defined axes: pitch, roll and yaw. See Fig. 1.3.

This figure shows the direction of the aircraft velocity in relation to the pitch, roll and yaw axes. For most of the flight an aircraft will be flying straight and level and the velocity vector will be parallel with the surface of the earth and proceeding upon a

Fig. 1.3 *Definition of flight control axes*

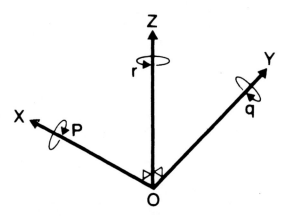

heading that the pilot has chosen. If the pilot wishes to climb the flight control system is required to rotate the aircraft around the pitch axis (Ox) in a nose-up sense to achieve a climb angle. Upon reaching the new desired altitude the aircraft will be rotated in a nose-down sense until the aircraft is once again straight and level.

In most fixed wing aircraft, if the pilot wishes to alter the aircraft heading then he will need to execute a turn to align the aircraft with the new heading. During a turn the aircraft wings are rotated around the roll axis (Oy) until a certain bank angle is attained. In a properly balanced turn the roll altitude will result in an accompanying change of heading while the roll angle (often called the bank angle) is maintained. This change in heading is actually a rotation around the yaw axis (Oz). The difference between the climb (or descent) and the turn is that the climb only involves rotation around one axis whereas the turn involves simultaneous co-ordination of two axes. In a properly co-ordinated turn, a component of aircraft lift acts in the direction of the turn, thereby reducing the vertical component of lift. If nothing were done to correct this situation, the aircraft would begin to descend; therefore in a prolonged turning manoeuvre the pilot has to raise the nose to compensate for this loss of lift. At certain times during flight the pilot may in fact be rotating the aircraft around all three axes, for example during a climbing or descending turning manoeuvre.

The aircraft flight control system enables the pilot to exercise control over the aircraft during all portions of flight. The system provides control surfaces that allow the aircraft to manoeuvre in pitch, roll and yaw. The system has also to be designed so that it provides stable control for all parts of the aircraft flight envelope; this requires a thorough understanding of the aerodynamics and dynamic motion of the aircraft. As will be seen, additional control surfaces are required for the specific purposes of controlling the high-lift devices required during approach and landing phases of flight. The flight control system has to give the pilot considerable physical assistance to overcome the enormous aerodynamic forces on the flight control surfaces. This in turn leads to the need to provide the aircraft controls with 'artificial feel' so that he does not inadvertently overstress the aircraft. These 'feel' systems need to provide the pilot with progressive and well harmonized controls that make the aircraft safe and pleasant to handle. A typical term that is commonly used today to describe this requirement is 'carefree handling'. Many aircraft embody automatic flight control systems to ease the burden of flying the aircraft and to reduce pilot workload.

Flight control surfaces

The requirements for flight control surfaces vary greatly between one aircraft and another, depending upon the role, range and agility needs of the vehicle. These varying requirements may best be summarized by giving examples of two differing types of aircraft: an agile fighter aircraft and a typical modern airliner.

The Experimental Aircraft Programme (EAP) aircraft is shown in Fig. 1.4 and represents the state-of-the-art fighter aircraft as defined by European manufacturers at the beginning of the 1990s. The EAP is similar to the European fighter aircraft (EFA) being developed by the four nation Eurofighter consortium comprising Alenia (Italy), BAE SYSTEMS (UK), CASA (Spain) and Daimler Chrysler (Germany).

Primary flight control

Primary flight control in pitch, roll and yaw is provided by the control surfaces described below.

Pitch control is provided by the moving canard surfaces, or foreplanes, as they are sometimes called, located either side of the cockpit. These surfaces provide the very powerful pitch control authority required by an agile high-performance aircraft. The position of the canards in relation to the wings renders the aircraft unstable. Without the benefit of an active computer driven control system the aircraft would be uncontrollable and would crash in a matter of seconds. While this may appear to be a fairly drastic implementation, the benefits in terms of improved manoeuvrability enjoyed by the pilot outweigh the engineering required to provide the computer controlled or 'active' flight control system.

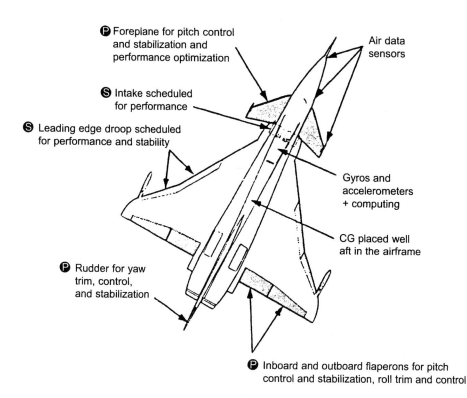

Ⓟ Foreplane for pitch control and stabilization and performance optimization

Air data sensors

Ⓢ Intake scheduled for performance

Ⓢ Leading edge droop scheduled for performance and stability

Gyros and accelerometers + computing

CG placed well aft in the airframe

Ⓟ Rudder for yaw trim, control, and stabilization

Ⓟ Inboard and outboard flaperons for pitch control and stabilization, roll trim and control

Ⓟ Primary controls
Ⓢ Secondary controls

Fig. 1.4 *Example of flight control surfaces – EAP (BAE SYSTEMS)*

Roll control is provided by the differential motion of the foreplanes, augmented to a degree by the flaperons. In order to roll to the right, the left foreplane leading edge is raised relative to the air flow generating greater lift than before. Conversely, the right foreplane moves downwards by a corresponding amount relative to the air flow thereby reducing the lift generated. The resulting differential forces cause the aircraft to roll rapidly to the right. To some extent roll control is also provided by differential action of the wing trailing-edge flaperons (sometimes called elevons). However, most of the roll control is provided by the foreplanes.

Yaw control is provided by the single rudder section. For high performance aircraft yaw control is generally less important than for conventional aircraft due to the high levels of excess power. There are nevertheless certain parts of the flight envelope where control of yaw (or sideslip) is vital to prevent roll–yaw divergence.

Secondary flight control

High-lift control is provided by a combination of flaperons and leading-edge slats. The flaperons may be lowered during the landing approach to increase the wing camber and improve the aerodynamic characteristics of the wing. The leading-edge slats are typically extended during combat to further increase wing camber and lift. The control of these high-lift devices during combat may occur automatically under the control of an active flight control system. The penalty for using these high-lift devices is increased drag, but the high levels of thrust generated by a fighter aircraft usually minimizes this drawback.

The EAP has airbrakes located on the upper rear fuselage. They extend to an angle of around 30 degrees, thereby quickly increasing the aircraft drag. The air brakes are deployed when the pilot needs to reduce speed quickly in the air; they are also often extended during the landing run to enhance the aerodynamic brake effect and reduce wheel brake wear.

Commercial aircraft

An example of flight control surfaces of a typical commercial airliner is shown in Fig. 1.5. Although the example is for the Airbus Industrie A320 it holds good for similar airliners produced by Boeing or other manufacturers. The controls used by this type of aircraft are described below.

Pitch control is exercised by four elevators located on the trailing edge of the tailplane or horizontal stabilizer. Each elevator section is independently powered by a dedicated flight control actuator, powered in turn by one of several aircraft hydraulic power systems. This arrangement is dictated by the high integrity requirements placed upon flight control systems. The entire tailplane section itself is powered by two or more actuators in order to trim the aircraft in pitch. In a dire emergency this facility could be used to control the aircraft, but the rates of movement and associated authority are insufficient for normal control purposes.

Roll control is provided by two aileron sections located on the outboard third of the trailing edge of each wing. Each aileron section is powered by a dedicated actuator powered in turn from one of the aircraft hydraulic systems. At low air speeds the roll control provided by the ailerons is augmented by differential use of the wing spoilers mounted on the upper surface of the wing. During a right turn the spoilers on the inside wing of the turn, that is the right wing, will be extended. This reduces the lift of the right

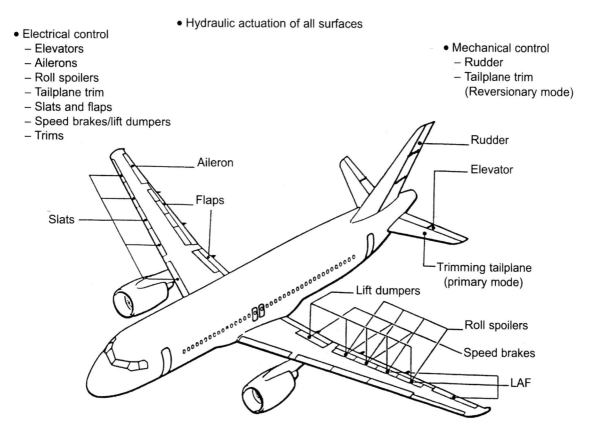

- Electrical control
  - Elevators
  - Ailerons
  - Roll spoilers
  - Tailplane trim
  - Slats and flaps
  - Speed brakes/lift dumpers
  - Trims

• Hydraulic actuation of all surfaces

• Mechanical control
  - Rudder
  - Tailplane trim
    (Reversionary mode)

Rudder

Elevator

Aileron

Flaps

Slats

Trimming tailplane
(primary mode)

Lift dumpers

Roll spoilers

Speed brakes

LAF

Fig. 1.5 *Example of flight control surfaces – commercial airliner (A320)*
(BAE SYSTEMS)

wing causing it to drop, hence enhancing the desired roll demand.

Yaw control is provided by three independent rudder sections located on the trailing edge of the fin (or vertical stabilizer). These sections are powered in a similar fashion to the elevator and ailerons. On a civil airliner these controls are associated with the aircraft yaw dampers. These damp out unpleasant 'dutch roll' oscillations which can occur during flight and which can be extremely uncomfortable for the passengers, particularly those seated at the rear of the aircraft.

Secondary flight control

Flap control is effected by several flap sections located on the inboard two-thirds of the wing trailing edges. Deployment of the flaps during take-off or landing extends the flap sections rearwards and downwards to increase wing area and camber, thereby greatly increasing lift for a given speed. The number of flap sections may vary from type to type; typically for this size of aircraft there would be about five per wing, giving a total of ten in all.

Slat control is provided by several leading-edge slats, which extend forwards and outwards from the wing leading edge. In a similar fashion to the flaps described above, this has the effect of increasing wing area and camber and therefore overall lift. A typical aircraft may have five slat sections per wing giving a total of ten in all.

Speed brakes are deployed when all of the overwing spoilers are extended together which has the effect of reducing lift as well as increasing drag. The effect is similar to the use of air brakes in the fighter, increasing drag so that the pilot may adjust his air speed rapidly; most air brakes are located on rear fuselage upper or lower sections and may have a pitch moment associated with their deployment. In most cases compensation for this pitch moment would be automatically applied within the flight control system.

While there are many identical features between the fighter and commercial airliner examples given above, there are also many key differences. The greatest difference relates to the size of the control surfaces in relation to the overall size of the vehicle. The fighter control surfaces are much greater than the corresponding control surfaces on an airliner. This reflects its prime requirements of manoeuvrability and high performance at virtually any cost. The commercial airliner has much more modest control requirements; it spends a far greater proportion of flying time in the cruise mode so fuel economy rather than ultimate performance is a prime target. Passenger comfort and safety are strong drivers that do not apply to the same degree for a military aircraft.

Flight control linkage systems

The pilot's manual inputs to the flight controls are made by moving the cockpit control column or rudder pedals in accordance with the universal convention:

- Pitch control is exercised by moving the control column fore and aft; pushing the column forward causes the aircraft to pitch down, and pulling the column aft results in a pitch up.
- Roll control is achieved by moving the control column from side to side or rotating the control yoke; pushing the stick to the right drops the right wing and vice versa.
- Yaw is controlled by the rudder pedals; pushing the left pedal will yaw the aircraft to the left while pushing the right pedal will have the reverse effect.

There are presently two main methods of connecting the pilot's controls to the rest of the flight control system. These are:

- Push-pull control rod systems.
- Cable and pulley systems.

An example of each of these types will be described and used as a means of introducing some of the major components which are essential for the flight control function. A typical high-lift control system for the actuation of slats and flaps will also be explained as this introduces differing control and actuation requirements.

Push–pull control rod system

The example chosen for the push–pull control rod system is the relatively simple yet high performance Hawk 200 aircraft. Figure 1.6 shows a simplified three-dimensional schematic of the Hawk 200 flight control which is typical of the technique widely used for combat aircraft. This example is taken from BAE SYSTEMS publicity information relating to the Hawk 200 (reference (1)). The system splits logically into pitch/yaw (tailplane and rudder) and roll (aileron) control runs respectively.

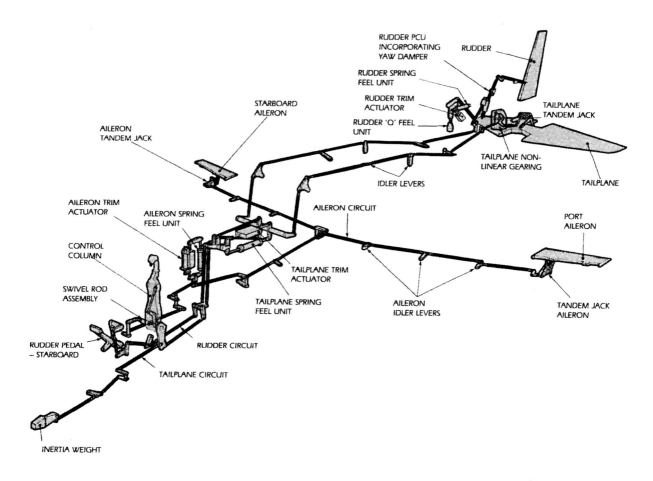

The pitch control input is fed from the left-hand or starboard side (looking forward) of the control column to a bell-crank lever behind the cockpit. This connects in turn via a near vertical control rod to another bell-crank lever which returns the control input to the horizontal. Bell-crank levers are used to alter the direction of the control runs as they are routed through a densely packed aircraft. The horizontal control rod runs parallel to a tailplane trim actuator/tailplane spring feel unit parallel combination The output from these units is fed upwards into the aircraft spine before once again being translated by another bell-crank lever. The control run passes down the left side of the fuselage to the rear of the aircraft via several idler levers before entering a non-linear gearing mechanism leading to the tandem jack tailplane Power Control Unit (PCU). The idler levers are simple lever mechanisms which help to support the control run at convenient points in the airframe. The hydraulically powered PCU drives the tailplane in response to the pilot inputs and the aircraft manoeuvres accordingly.

The yaw input from the rudder pedals is fed to a bell-crank lever using the same pivot points as the pitch control run and runs vertically to another bell-crank which translates the yaw control rod to run alongside the tailplane trim/feel units. A further two bell-cranks place the control linkage running down the right-hand side of the rear fuselage via a set of idler levers to the aircraft empennage. At this point the control linkage accommodates inputs from the rudder trim actuator, spring feel unit and 'Q' feel

Fig. 1.6 Hawk 200 push–pull control rod system (BAE SYSTEMS)

unit. The resulting control demand is fed to the rudder hydraulically powered PCU which in turn drives the rudder to the desired position. In this case the PCU has a yaw damper incorporated which damps out undesirable 'dutch roll' oscillations.

The roll demand is fed via a swivel rod assembly from the right-hand or port side (looking forward) of the control column and runs via a pair of bell-crank levers to a location behind the cockpit. At this point a linkage connects the aileron trim actuator and the aileron spring feel unit. The control rod runs aft via a further bell-crank lever and an idler lever to the centre fuselage. A further bell-crank lever splits the aileron demand to the left and right wings. The wing control runs are fed outboard by means of a series of idler levers to points in the outboard section of the wings adjacent to the ailerons. Further bell-cranks feed the left and right aileron demands into the tandem jacks and therefore provide the necessary aileron control surface actuation.

Although a simple example, this illustrates some of the considerations which need to be borne in mind when designing a flight control system. The interconnecting linkage needs to be strong, rigid and well supported; otherwise fuselage flexing could introduce 'nuisance' or unwanted control demands into the system. A further point is that there is no easy way or route through the airframe, therefore an extensive system of bell-cranks and idler levers is required to support the control rods. This example has also introduced some of the major components which are required to enable a flight control system to work while providing safe and pleasant handling characteristics to the pilot. These are:

- Trim actuators in tailplane (pitch), rudder (yaw) and aileron (roll) control systems.
- Spring feel units in tailplane (pitch), rudder (yaw) and aileron (roll) control systems.
- 'Q' feel unit in the rudder (yaw) control system.
- Power control units (PCUs) for tailplane, rudder and aileron actuation.

Cable and pulley system

The cable and pulley system is widely used for commercial aircraft; sometimes used in conjunction with push–pull control rods. It is not the intention to attempt to describe a complete aircraft system routing in this chapter. Specific examples will be outlined which make specific points in relation to the larger aircraft. Refer to Fig. 1.7.

Figure 1.7(a) shows a typical aileron control system. Manual control inputs are routed via cables and a set of pulleys from both captain's and first officer's control yokes to a consolidation area in the centre section of the aircraft. At this point aileron and spoiler runs are split both left/right and into separate aileron/spoiler control runs. Both control column/control yokes are synchronized. A breakout device is included which operates at a predetermined force in the event that one of the cable runs fails or becomes jammed. Control cable runs are fed through the aircraft by a series of pulleys, idler pulleys, quadrants and control linkages in a similar fashion to the push–pull rod system already described. Tensiometer/lost motion devices situated throughout the control system ensure that cable tensions are correctly maintained and lost motion eliminated. Differing sized pulleys and pivot/lever arrangements allow for the necessary gearing changes throughout the control runs. Figure 1.7(b) shows a typical arrangement for interconnecting wing spoiler and speed brake controls. Trim units, feel units and PCUs are connected at strategic points throughout the control runs as for the push–pull rod system.

Fig. 1.7 *Examples of wire and pulley aileron control system (Boeing)*

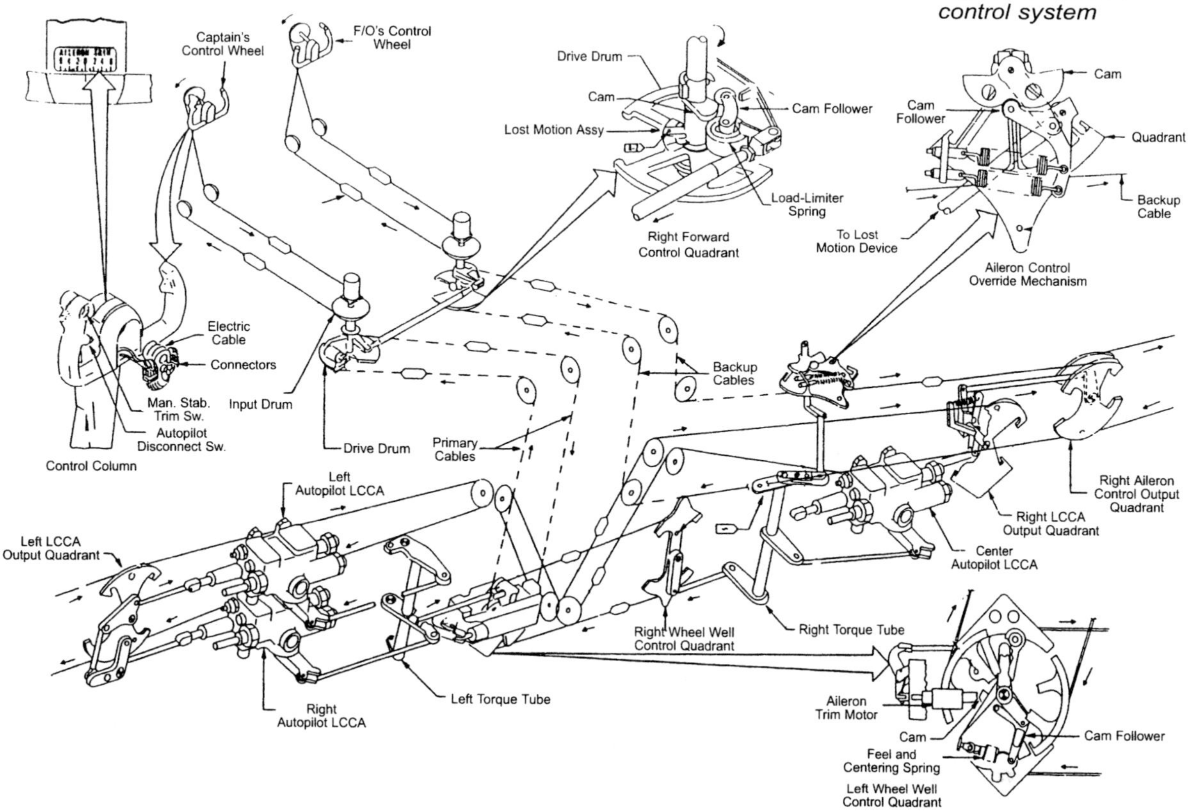

Fig. 1.7a *Aileron control system*

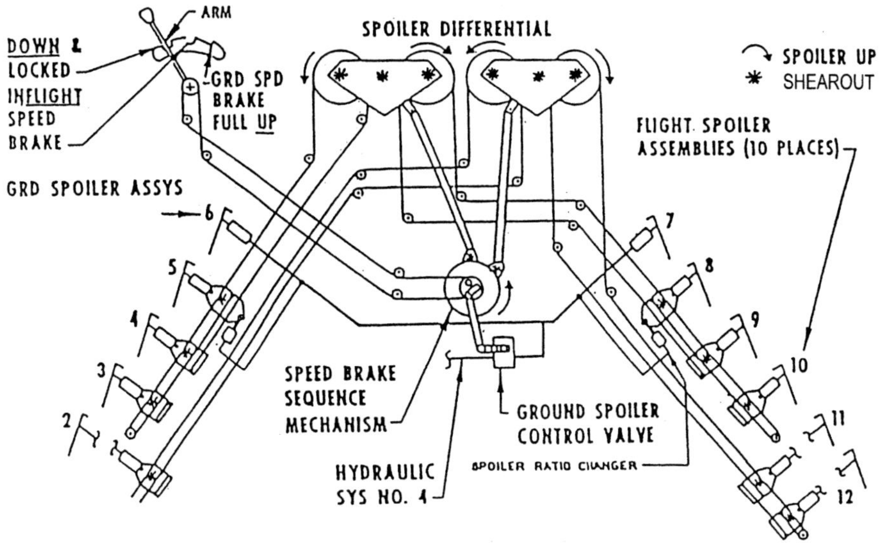

Fig. 1.7b *Spoiler and air brakes control system*

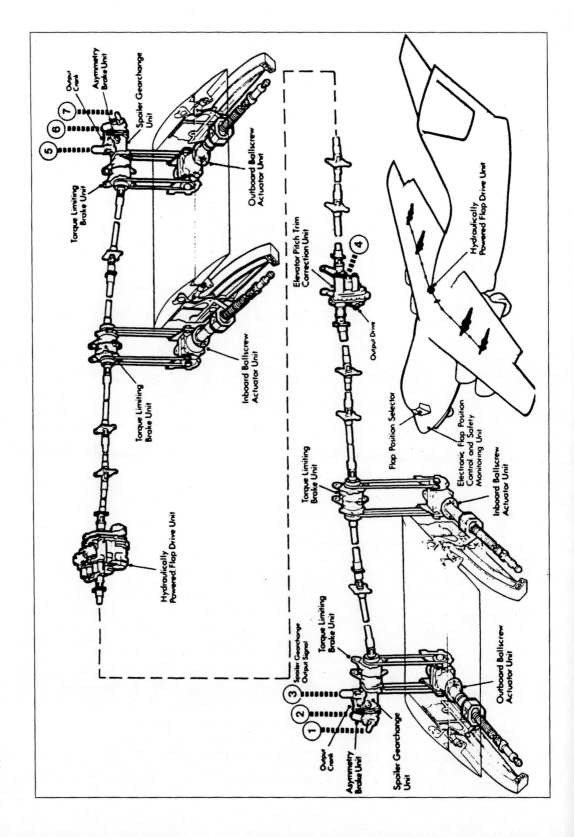

Fig. 1.8 BAE SYSTEMS 146 flap operating system (TI Group)

High-lift control systems

The example chosen to illustrate flap control is the system used on the BAE SYSTEMS 146 aircraft. This aircraft does not utilize leading-edge slats. Instead the aircraft relies upon single-section Fowler flaps which extend across 78 per cent of the inner wing trailing edge. Each flap is supported in tracks and driven by recirculating ballscrews at two locations on each wing. The ballscrews are driven by transmission shafts which run along the rear wing spar. The shafting is driven by two hydraulic motors which drive into a differential gearbox such that the failure of one motor does not inhibit the drive capability of the other. Refer to Fig. 1.8 for a diagram of the BAE SYSTEMS 146 flap operating system.

As well as the flap drive motors and flap actuation, the system includes a flap position selector switch and an electronic control unit. The electronic control unit comprises: dual identical microprocessor-based position control channels; two position control analogue safety channels; a single microprocessor-based safety channel for monitoring mechanical failures. For an excellent system description refer to the technical paper on the subject prepared by Dowty Rotol/TI Group (Reference (2)).

The slat system or leading-edge flap example chosen is that used for the Boeing 747-400. Figure 1.9 depicts the left wing leading-edge flap systems. There is a total of 28 flaps, 14 on each wing. These flaps are further divided into groups A and B. Group A flaps are those six sections outside the outboard engines; group B flaps include the five sections between inboard and outboard engines and the three sections inside the inboard engines. The inboard ones are Krueger flaps which are flat in the extended position, the remainder are of variable camber which provide an aerodynamically shaped surface when extended. The flaps are powered by Power Drive Units (PDUs); six of these drive the group A flaps and two the group B flaps. The motive power is pneumatic with electrical backup. Gearboxes reduce and transfer motion from the PDUs to rotary actuators which operate the drive linkages for each leading edge flap section. Angular position is extensively monitored throughout the system by Rotary Variable Differential Transformers (RVDTs).

Fig. 1.9 Boeing 747-400 leading-edge flap system (Boeing)

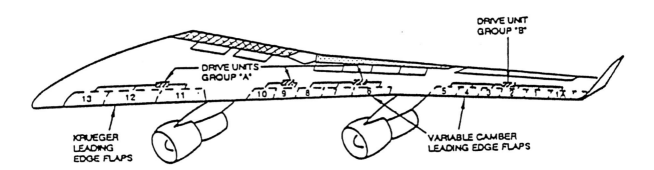

Trim and feel

The control rod example for the BAE SYSTEMS Hawk 200 aircraft shows the interconnection between the pilot's control columns and rudder bars and the hydraulically powered actuators which one would expect. However the diagram also revealed a surprising number of units associated with aircraft trim and feel. These additional units are essential in providing pleasant handling characteristics for the aircraft in all configurations throughout the flight envelope.

Trim

The need for trim actuation may be explained by recourse to a simple explanation of the aerodynamic forces which act upon the aircraft in flight. Figure 1.10 shows a simplified diagram of the pitch forces which act upon a stable aircraft trimmed for level flight. The aircraft weight, represented by the symbol W, acts downwards at the aircraft centre-of-gravity or CG. As the aircraft is stable the CG is ahead of the centre of pressure where the lift force acts and all aerodynamic perturbations should be naturally damped. The distance between the CG and the centre of pressure is a measure of how stable and also how manoeuvrable the aircraft is in pitch. The closer the CG and centre of pressure then the less stable and more manoeuvrable the aircraft. The converse is true when the CG and centre of pressure are further apart.

Examining the forces acting about the aircraft CG it can be seen that there is a counter-clockwise moment exerted by a large lift force acting quite close to the pivot point. If the aircraft is not to pitch nose-down this should be counterbalanced by a clockwise force provided by the tailplane. This will be a relatively small force acting with a large moment. If the relative positions of the aircraft CG and centre of pressure were to remain constant throughout all conditions of flight then the pilot could set up the trim and no further control inputs would be required.

In practice the CG positions may vary due to changes in the aircraft fuel load and the stores or cargo and passengers the aircraft may be carrying. Variations in the position of the aircraft CG position are allowed within carefully prescribed limits. These limits are called the forward and aft CG limits and they determine how nose-heavy or tail-heavy the aircraft may become and still be capable of safe and controllable flight. The aerodynamic centre of pressure similarly does not remain in a constant position as the aircraft flight conditions vary. If the centre of pressure moves aft then

Fig. 1.10 *Pitch forces acting in level flight*

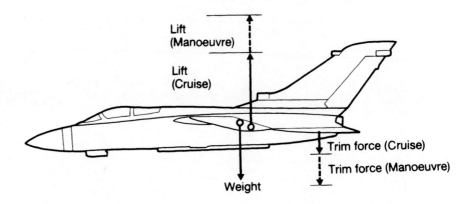

the downward trim force required of the tailplane will increase and the tailplane angle of incidence will need to be increased. This requires a movement of the pitch control run equivalent to a small nose-up pitch demand. It is inconvenient for the pilot to constantly apply the necessary backward pressure on the control column, so a pitch actuator is provided to alter the pitch control run position and effectively apply this nose-up bias. Forward movement of the centre of pressure relative to the CG would require a corresponding nose-down bias to be applied. These nose-up and nose-down biases are in fact called nose-up and nose-down trim respectively.

Pitch trim changes may occur for a variety of reasons: increase in engine power, change in airspeed, alteration of the fuel disposition, deployment of flaps or airbrakes and so on. The desired trim demands may be easily input to the flight control system by the pilot. In the case of the Hawk the pilot has a four-way trim button located on the stick top; this allows fore and aft (pitch) and lateral (roll) trim demands to be applied without moving his hand from the control column.

The example described above outlines the operation of the pitch trim system as part of overall pitch control. Roll or aileron trim is accomplished in a very similar way to pitch trim by applying trim biases to the aileron control run by means of an aileron trim actuator. Yaw or rudder trim is introduced by the separate trim actuator provided; in the Hawk this is located in the rear of the aircraft. The three trim systems therefore allow the pilot to offload variations in load forces on the aircraft controls as the conditions of flight vary.

Feel

The provision of artificial 'feel' became necessary when aircraft performance increased to the point where it was no longer physically possible for the pilot to apply the high forces needed to move the flight control surfaces. Initially with servo-boosting systems and later with powered flying controls it became necessary to provide powered assistance to attain the high control forces required. This was accentuated as the aircraft wing thickness to chord ratio became much smaller for performance reasons and the hinge moment available was correspondingly reduced. However a drawback with a pure power-assisted system is that the pilot may not be aware of the stresses being imposed on the aircraft. Furthermore, a uniform feel from the control system is not a pleasant characteristic; pilots are not alone in this regard; we are all used to handling machinery where the response and feel are sensibly related. The two types of feel commonly used in aircraft flight control systems are spring feel and 'Q' feel.

Spring feel, as the name suggests, is achieved by loading the movement of the flight control run against a spring of a predetermined stiffness. Therefore when the aircraft controls are moved the pilot encounters an increasing force proportional to the spring stiffness. According to the physical laws spring stiffness is a constant and therefore spring feel is linear unless the physical geometry of the control runs impose any non-linearities. In the Hawk 200, spring feel units are provided in the tailplane, aileron and rudder control runs. The disadvantage of spring feel units is that they only impose feel proportional to control demand and take no account of the pertaining flight conditions.

'Q' feel is a little more complicated and is more directly related to the aerodynamics and precise flight conditions that apply at the time of the control demand. As the aircraft speed increases the aerodynamic load increases in a mathematical relationship proportional to the air density and the square of velocity. The air density is relatively unimportant; the squared velocity term has a much greater effect, particularly at high speed. Therefore it is necessary to take account of this aerodynamic equation; that is the purpose of 'Q' feel. A 'Q' feel unit receives air data information from the aircraft pitot-static system. In fact the signal applied is the difference between pitot and static pressure, (known as Pt–Ps) and this signal is used to modulate the control mechanism within the 'Q' feel unit and operate a hydraulic load jack which is connected into the flight control run. In this way the pilot is given feel which is directly related to the aircraft speed and which will greatly increase with increasing air speed. It is usual to use 'Q' feel in the tailplane or rudder control runs; where this method of feel is used depends upon the aircraft aerodynamics and the desired handling or safety features. The disadvantage of 'Q' feel is that it is more complex and only becomes of real use at high speed. Figure 1.11. is a photograph of a 'Q' feel unit supplied by TI Group for the BAE SYSTEMS Harrier GR5 and Boeing AV-8B aircraft. This unit is fitted with an electrical solenoid so that the active part of the system maybe disconnected if required. This unit is designed to operate with an aircraft 20.7 MN/sq m (3,000 psi) hydraulic system pressure.

Fig. 1.11 *'Q' feel unit for GR5/AV8B (TI Group)*

The rudder control run on the Hawk 200 shown earlier in Fig. 1.6 uses both spring and 'Q' feel. It is likely that these two methods have been designed to complement each other. The spring feel will dominate at low speed and for high deflection control demands. The 'Q' feel will dominate at high speeds and low control deflections.

Power control units

The key element in the flight control system, increasingly so with the advent of fly-by-wire and active control units, is the power actuation. Actuation has always been important to the ability of the flight control system to attain its specified performance. The development of analogue and digital multiple-control lane technology has put the actuation central to performance and integrity issues. Addressing actuation in ascending order of complexity leads to the following categories:

- Simple mechanical actuation.
- Mechanical actuation with simple electromechanical features.
- Multiple-redundant electromechanical actuation with analogue control inputs and feedback.

The examination of these crudely defined categories leads more deeply into systems integration areas where boundaries between mechanical, electronic, systems and software engineering become progressively blurred.

Simple mechanical actuation

The attributes of mechanical actuation are straightforward; the system demands a control movement and the actuator satisfies that demand with a power-assisted mechanical response. The BAE SYSTEMS Hawk 200 is a good example of a system where straightforward mechanical actuation is used for most of the flight control surfaces. For most applications the mechanical actuator is able to accept hydraulic power from two identical/redundant hydraulic systems. The obvious benefit of this arrangement is that full control is retained following loss of fluid or a failure in either hydraulic system. This is important even in a simple system as the loss of one or more actuators and associated control surfaces can severely affect aircraft handling. The actuators themselves have a simple reversion mode following failure, that is to centre automatically under the influence of aerodynamic forces. This reversion mode is called aerodynamic centring and is generally preferred for obvious reasons over a control surface freezing or locking at some intermediate point in its travel. In some systems 'freezing' the flight control system may be an acceptable solution depending upon control authority and reversionary modes that the flight control system possesses. The decision to implement either of these philosophies will be a design decision based upon the system safety analysis.

Mechanical actuation may also be used for spoilers where these are mechanically rather than electrically controlled. In this case the failure mode is aerodynamic closure, that is the airflow forces the control surface to the closed position where it can subsequently have no adverse effect upon aircraft handling. Figure 1.12 illustrates the mechanical spoiler actuator supplied by Claverham/FHL for the BAE SYSTEMS 146 aircraft. This unit is simplex in operation. It produces thrust of 59.9 kN (13,460 lb) over a working stroke of 15 mm (0.6 in). It has a length of 22.4 mm (8.8 in) and weighs 8.3 kg (18.2 lb). The unit accepts hydraulic pressure at 20.7 MN/sq m (3,000 psi).

Fig. 1.12 BAE
SYSTEMS 146 spoiler
actuator
(Claverham/FHL)

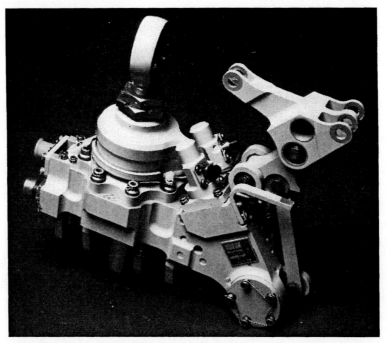

Fig. 1.12 BAE SYSTEMS 146 spoiler actuator (Claverham/FHL)

Mechanical actuation with electrical signalling

The use of mechanical actuation has already been described and is appropriate for a wide range of applications. However the majority of modern aircraft use electrical signalling and hydraulically powered (electrohydraulic) actuators for a wide range of applications with varying degrees of redundancy. The demands for electrohydraulic actuators fall into two categories: simple demand signals or autostabilization inputs.

Simple electrical demand signals are inputs from the pilots that are signalled by electrical means. For certain non-critical flight control surfaces it may be easier, cheaper and lighter to utilize an electrical link. An example of this is the airbrake actuator used on the BAE SYSTEMS 146; simplex electrical signalling is used and in the case of failure the reversion mode is aerodynamic closure.

In most cases where electrical signalling is used this will at least be duplex in implementation and for fly-by-wire systems signalling is likely to be quadruplex; these more complex actuators will be addressed later. An example of duplex electrical signalling with a simplex hydraulic supply is the spoiler actuators on the Tornado. There are four actuators fitted on the aircraft, two per wing, which are used for roll augmentation.

In general, those systems which extensively use simplex electrical signalling do so for autostabilization. In these systems the electrical demand is a stabilization signal derived within a computer unit. The simplest form of autostabilization is the yaw damper which damps out the cyclic cross-coupled oscillations which occur in roll and yaw known as 'dutch roll'. The Hawk 200 illustrated this implementation. Aircraft which require a stable platform for weapon aiming may have simplex autostabilization in pitch, roll and yaw; an example of this type of system is the Harrier/AV-8A. A similar system on the Jaguar uses simplex autostabilization in pitch and roll.

Multiple redundancy actuation

Modern flight control systems are rapidly moving towards fly-by-wire solutions as the benefits to be realized by using such a system are considerable. These benefits include a reduction in weight, improvement in handling performance and crew/passenger comfort. Concorde was the first aircraft to pioneer these techniques in the civil field using a flight control system jointly developed by GEC (now BAE SYSTEMS) and SFENA (reference (3)). The Tornado, fly-by-wire Jaguar and EAP have extended the use of these techniques; the latter two were development programmes into the regime of the totally unstable aircraft. In the civil field the Airbus A320/A330/A340 and the Boeing 777 are introducing modern state-of-the-art systems into service. For obvious reasons, a great deal of care is taken during the definition, specification, design, development and certification of these systems. Multiple-redundant architectures for the aircraft hydraulic and electrical systems must be considered as well as multiple-redundant lanes or channels of computing and actuation for control purposes. The implications of the redundancy and integrity of the other aircraft systems will be addressed elsewhere. For the present, attention will be confined to the issues affecting multiple-redundant electrohydraulic actuation.

A simplified block schematic diagram of a multiple-redundant electrohydraulic actuator is shown in Fig. 1.13. For reasons of simplicity only one lane or channel is shown; in practice the implementation is likely to be quadruplex, i.e. four identical lanes. The solenoid valve is energized to supply hydraulic power to the actuator, usually from two of the aircraft hydraulic systems. Control demands from the flight control computers are fed to the servo valves. The servo valves control the position of the first-stage valves that are mechanically summed before applying demands to the control valves. The control valves modulate the position of the control ram. Linear Variable Differential Transformers (LVDTs) measure the position of the first-stage actuator and output ram positions of each lane and these signals are fed back to the flight control computers, thereby closing the loop. Two examples of this quadruplex actuation system are given below: the Tornado quadruplex taileron and rudder actuators associated with the Control Stability Augmentation System (CSAS) and the EAP flight control system. The description given here will be confined to that part of the flight control system directly relevant to the actuator drives.

Fig. 1.13 Simplified block schematic diagram of a multiple redundant electrohydraulic actuator

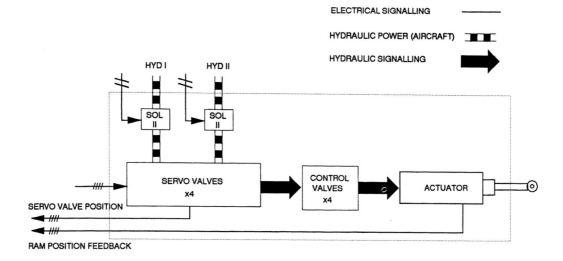

Fig. 1.14(a) Tornado
taileron/rudder CSAS
drive interface

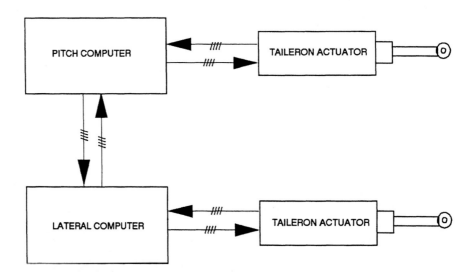

Fig. 1.14(b) Tornado
rudder actuator
(Claverham/FHL)

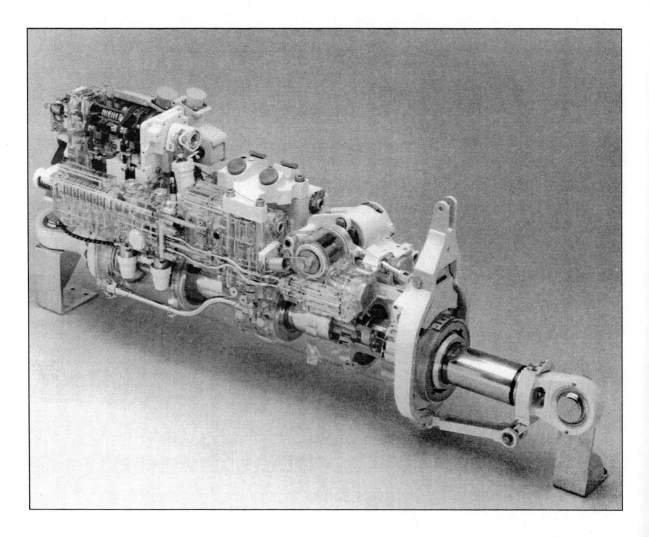

The Tornado CSAS flight control computation is provided by pitch and lateral computers supplied by GEC (now BAE SYSTEMS) and Bodenseewerk. The pitch computer predominantly handles pitch control computations and the lateral computer roll and yaw computations though there are interconnections between the two. Refer to Fig. 1.14(a). There are three computing lanes; computing is analogue in nature and there are a number of voter-monitors within the system to vote out lanes operating outside specification. The combined pitch/roll output to the taileron actuators is consolidated from three lanes to four within the pitch computer so the feed to the taileron actuators is quadruplex. The quadruplex taileron actuator is provided by Claverham/FHL and is shown in Fig. 1.14(b). This actuator provides a thrust of 339.3 kN (76,291 lb) over a working stroke of 178 mm. The actuator is 940 mm (37.0 in) long and weighs 51.0 kg and operates with the two aircraft 4,000 psi hydraulic systems. The rudder actuator similarly receives a quadruplex rudder demand from the lateral computer, also shown in Fig. 1.14(b). The rudder actuator is somewhat smaller than the taileron actuator delivering a thrust of 80.1 kN. The CSAS is designed so that following a second critical failure it is possible to revert to a mechanical link for pitch and roll. In these circumstances the rudder is locked in the central position.

The Tornado example given relates to the analogue system that comprises the CSAS. The EAP flight control system (FCS) is a quadruplex digital computing system in which all four control computations are undertaken in all four computing lanes. The system is quadruplex rather than triplex as a much higher level of integrity is required. As has been mentioned earlier the EAP is an unstable aircraft and the FCS has to be able to survive two critical failures. Figure 1.15(a) shows the relationship between the Flight Control Computers (FCCs), Actuator Drive Units (ADUs) and the actuators. The foreplane actuators are fed quadruplex analogue demands from the quadruplex digital FCCs. Demands for the left and right, inboard and outboard flaperons and the rudder are fed in quadruplex analogue form from the four ADUs. The ADUs receive the pitch, roll and yaw demands from the FCCs via dedicated serial digital links and the digital to analogue conversion is carried out within the ADUs.

The total complement of actuators supplied by TI Group for the EAP is as follows:

- Quadruplex electrohydraulic foreplane actuators – 2.
- Quadruplex electrohydraulic flaperon actuators:
  Outboard flaperons – 100 mm working stroke – 2.
  Inboard flaperons – 165 mm working stroke – 2.
- Quadruplex electrohydraulic rudder actuators – 100 mm working stroke – 1 (see Fig. 1.15(b)).

All seven actuators are fed from two independent hydraulic systems.

The EAP flight control system represents the forefront of such technology and the aircraft has continued to exceed expectations since the first flight in August 1986. Further detail regarding the EAP system and the preceding Jaguar fly-by-wire programme maybe found in a number of technical papers which have been given in recent years (references (3–8)). Most of these papers are presented from an engineering perspective. Reference (5) is a paper by Chris Yeo, Deputy Chief Test Pilot at British Aerospace at the time of the fly-by-wire programme, which includes an overview of the aircraft control laws.

*Fig. 1.15(a) EAP
actuator drive
configuration*

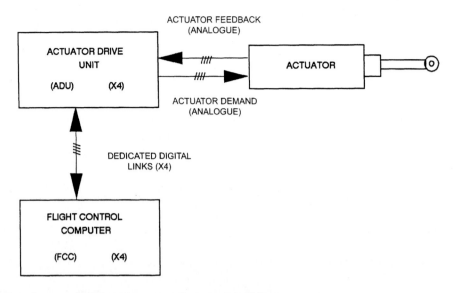

ACTUATOR FEEDBACK
(ANALOGUE)

ACTUATOR DRIVE
UNIT

(ADU)   (X4)

ACTUATOR

ACTUATOR DEMAND
(ANALOGUE)

DEDICATED DIGITAL
LINKS (X4)

FLIGHT CONTROL
COMPUTER

(FCC)   (X4)

*Fig. 1.15(b) EAP
foreplane, flaperon
and rudder actuators
(TI Group)*

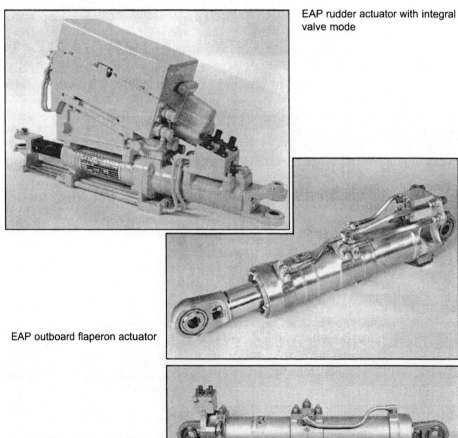

EAP rudder actuator with integral
valve mode

EAP outboard flaperon actuator

EAP foreplane actuator with
valve mode

Advanced actuation concepts

The actuation implementations described so far have all been mechanical or electrohydraulic in function using servo valves. There are a number of advanced actuation concepts under development that may supplant the existing electrohydraulic actuator. These novel types of actuation are:

- Direct drive actuation.
- The Electro Mechanical Actuator (EMA).
- The Electro Hydrostatic Actuator (EHA).

Direct drive actuation

In the electrohydraulic actuator a servo valve requires a relatively small electrical drive signal, typically in the order of 10 – 15 mA. The reason such low drive currents are possible is that the control signal is effectively amplified within the hydraulic section of the actuator. In the direct drive actuator the aim is to use an electrical drive with sufficient power to obviate the need for the servo valve/first stage valve. The main power spool is directly driven by torque motors requiring a higher signal current, hence the term 'direct drive'. Development work relating to the direct drive concept including comparison with Tornado requirements and operation with 8000 psi hydraulic systems has been investigated by Claverham/FHL (reference (6)). This paper also addresses the direct digital control of aircraft flight control actuators.

Electromechanical actuator (EMA)

The electromechanical actuator or EMA replaces the electrical signalling and power actuation of the electrohydraulic actuator with an electric motor and gearbox assembly applying the motive force to move the ram. EMAs have been used on aircraft for many years for such uses as trim and door actuation; however the power, motive force and response times have been less than that required for flight control actuation. The three main technology advancements that have improved the EMA to the point where it may be viable for flight control applications are: the use of rare earth magnetic materials in 270 VDC motors; high-power solid-state switching devices; and microprocessors for lightweight control of the actuator motor. The classical aluminium/nickel/cobalt (Alnico) magnetic materials that have been used for many years now may be replaced by samarium/cobalt magnets. Samarium/cobalt has a much higher energy product that, crudely, is a measure of the strength of the material magnetism, typical values being ten times greater than for Alnico materials. Solid-state power switching devices allow the use of pulse width modulation techniques to maintain constant motor torque over the speed range and minimize power losses. Microprocessors offer a cheap and effective means of exercising the necessary control. Microprocessors also enable easy interfacing to an aircraft digital data bus system if this feature is required. Work carried out by AiResearch using a dual 270 VDC powered actuator on a Grumman F-14 iron bird test rig has shown that this unit had a better frequency response than a conventional hydraulic actuator. At the moment EMAs may also be heavier than hydraulic actuators, however an aircraft level weight analysis may prove the installed EMA system to be lighter. One reservation which has been expressed is the reliability of the gearbox drive which is essential to the EMA. A further consideration may be the use of 270 VDC aircraft all-electrical systems which will obviously be advantageous to the EMA. This is a separate subject for consideration that is reviewed elsewhere in this book. See references (9) and (10), papers by TI Group that review the development of EMAs for aircraft systems.

Electrohydrostatic actuator (EHA)

A further option for flight control actuation which is under active development is the electrohydrostatic actuator or EHA. In the EHA an electric motor in each actuator drives a self-contained hydraulic system comprising pump and reservoir which provides the motive force to power the control surface to the demanded position. Once the control surface attains the demanded position the system 'locks up' and no further power is required while that control position is held. This has significant potential for use in high-power/sustained load applications such as a foreplane or stabilizer actuator, whereas an EMA would require power to hold the control surface position. The electric motor which drives the hydraulic pump is reversible. In EHA, the electrohydrostatic actuator matches well with the all-electric 270 VDC aircraft which has been the subject of much debate in the US for a number of years. The electronic control, loop closure, monitoring and BIT requirements also lend the EHA naturally to direct interfacing with the aircraft digital data bus.

A common feature of all three new actuator concepts outlined above is the use of microprocessors to improve control and performance. The introduction of digital control in the actuator also permits the consideration of direct digital interfacing to digital flight control computers by means of data buses (ARINC 429/ARINC 629/1553B). The direct drive developments described emphasize concentration upon the continued use of aircraft hydraulics as the power source, including the accommodation of system pressures up to 8,000 psi. The EMA and EHA developments on the other hand lend themselves to a greater use of electrical power deriving from the all-electric aircraft concept, particularly if 270 VDC power is available.

Civil system implementations

The flight control and guidance of civil transport aircraft has steadily been getting more sophisticated in recent years. Whereas Concorde was the first civil aircraft to have a fly-by-wire system, Airbus introduced a fly-by-wire system on to the A320 family and a similar system has been carried forward to the A330/340. Boeing's first fly-by-wire system on the Boeing 777 was widely believed to be a response to the Airbus technology development. The key differences between the Airbus and Boeing philosophies and implementations are described below.

Top-level comparison

The importance and integrity aspects of flight control lead to some form of monitoring function to ensure the safe operation of the control loop. Also for integrity and availability reasons, some form of redundancy is usually required. Figure 1.16 shows a top-level comparison between the Boeing and Airbus FBW implementations.

In the Boeing philosophy shown in simplified form on the right of Fig. 1.16 the system comprises three Primary Flight Computers (PFCs) each of which has three similar lanes with dissimilar hardware but the same software. Each lane has a separate role during an operating period and the roles are cycled after power-up. Voting techniques are used to detect discrepancies or disagreements between lanes and the comparison techniques used vary for different types of data. Communication with the four Actuator Control Electronics (ACE) units is by multiple A629 flight control data buses. The ACE units directly drive the flight control actuators. A separate flight control DC system is provided to power the flight control system. The schemes used on the Boeing 777 will be described in more detail later in this Chapter.

The Airbus approach is shown on the left of Fig. 1.16 Five main computers are used: three Flight Control Primary Computers (FCPCs) and two Flight Control Secondary Computers (FCSCs). Each computer comprises of command and monitor elements with different software. The primary and secondary computers have different architectures and different hardware. Command outputs from the FCSCs to ailerons, elevators and the rudder are for standby use only. Power sources and signalling lanes are segregated.

Airbus implementation

The Anglo-French Concorde apart, Airbus were the first aircraft manufacturer in recent years to introduce Fly-By-Wire (FBW) to civil transport aircraft. The original aircraft to utilize FBW was the A320 and the system has been used throughout the A319/320/321 family and more recently on the A330/340. The A320 philosophy will be described and A330/340 system briefly compared. See Reference (11).

Fig. 1.16 *Top-level Boeing and Airbus comparison*

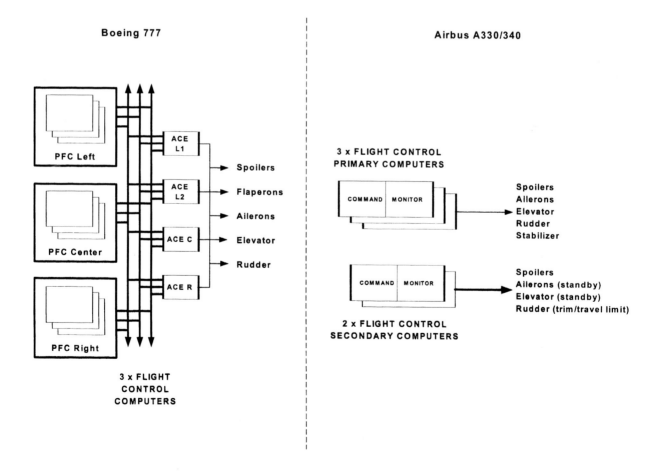

A320 FBW system

A schematic of the A320 flight control system is shown in Fig. 1.17 (see reference (11)). The flight control surfaces are all hydraulically powered and are tabulated as follows.

- Electrical control:

Elevators	2
Ailerons	2
Roll spoilers	8
Tailplane trim	1
Slats	10
Flaps	4
Speedbrakes	6
Lift Dumpers	10
Trims	

- Mechanical control:

  Rudder

  Tailplane trim (reversionary mode)

The aircraft has three independent hydraulic power systems: blue, green and yellow. Figure 1.17 shows how these systems respectively power the hydraulic flight control actuators. A total of seven computers undertake the flight control computation task as follows:

- Two Elevator/Aileron Computers (ELACs). The ELACs control the aileron and elevator actuators according to the notation in the figure.
- Three Spoiler/Elevator Computers (SECs). The SECs control all of the spoilers and in addition provide secondary control to the elevator actuators. The various spoiler sections have different functions as shown namely:

  - Ground spoiler mode: all spoilers.
  - Speed brake mode: inboard three spoiler sections.
  - Load alleviation mode: outboard two spoiler sections (plus ailerons). This function has recently been disabled and is no longer embodied in recent models.
  - Roll augmentation: outboard four spoiler sections.

- Two Flight Augmentation Computers (FACs). These provide a conventional yaw damper function, interfacing only with the yaw damper actuators.

The three aircraft hydraulic systems; green, blue and yellow provide hydraulic power to the flight control actuators according to the notation shown on the diagram.

In the very unlikely event of the failure of all computers it is still possible to fly and land the aircraft – this has been demonstrated during certification. In this case the Tailplane Horizontal Stabilizer (THS) and rudder sections are controlled directly by mechanical trim inputs – shown as M in the diagram – which allow pitch and lateral control of the aircraft to be maintained.

Another noteworthy feature of the Airbus FBW systems is that they do not use the conventional pitch and roll yoke. The pilot's pitch and roll inputs to the system are by means of a sidestick controller and this has been widely accepted by the international airline community.

In common with contemporary civil aircraft, the A320 is not an unstable aircraft like

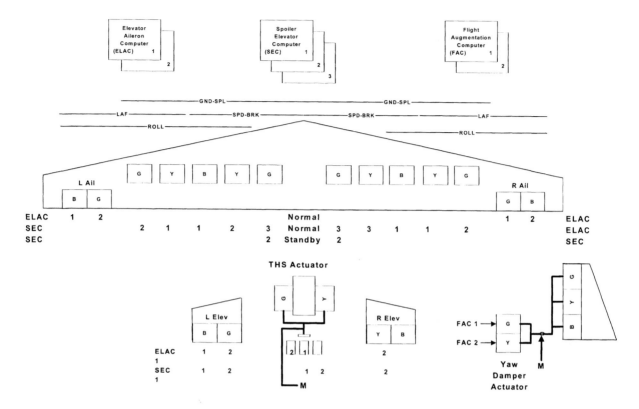

Fig. 1.17 *A320 flight control system (Interavia)*

the EAP system briefly described earlier in this chapter. Instead the aircraft operates with a longitudinal stability margin of around 5 per cent of aerodynamic mean chord or around half what would normally be expected for an aircraft of this type. This is sometimes termed relaxed stability. The A320 family can claim to be the widest application of civil FBW with over 1,000 examples delivered.

A330/340 FBW system

The A330/340 FBW system bears many similarities to the A320 heritage as might be expected.

The pilot's input to the Flight Control Primary Computers (FCPCs) and flight control secondary computers (FCSCs) are by means of the sidestick controller. The flight management guidance and envelope computers (FMGECs) provide autopilot pitch commands to the FCPC. The normal method of commanding the elevator actuators is via the FCPC although they can be controlled by the FCSC in a standby mode. Three autotrim motors may be engaged via a clutch to drive the mechanical input to the THS actuator.

For the pitch channel, the FCPCs provide primary control and the FCSCs the backup. Pilot's inputs are via the rudder pedals directly, or in the case of rudder trim, via the FCSC to the rudder trim motors.

The yaw damper function resides within the FCPCs rather than the separate Flight Augmentation Computers (FACs) used on the A320 family. Autopilot yaw demands are fed from the FMGECs to the FCPCs.

There is a variable travel limitation unit to limit the travel of the rudder input at various stages of flight. As before, the three hydraulic systems feed the rudder actuators and two yaw damper actuators as annotated on the figure.

Therefore although the implementation and notation of the flight control computers differs between the A320 and A330/340 a common philosophy can be identified between the two families.

The overall flight control system elements for the A330/340 are:

- Three Flight Control Primary Computers (FCPCs). The function of the FCPCs has been described.
- Two Flight Control Secondary Computers (FCSCs). Similarly, the function of the secondary computers has been explained.
- Two Flight Control Data Concentrators (FCDCs). The FCDCs provide data from the primary and secondary flight computers for indication, recording and maintenance purposes.
- Two Slat/Flap Control Computers (SFCCs). The SFCCs are each able to control the full-span leading-edge slats and trailing-edge flaps via the hydraulically driven slat and flap motors.

Spoiler usage on the A330/340 differs from that on the A320. There is no load alleviation function and there are six pairs of spoilers versus the five pairs on the A320. Also the functions of the various spoiler pairs differ slightly from the A320 implementation. However, overall, the philosophy is the same.

Boeing 777 implementation

Boeing ventured into the FBW field with the Boeing 777; partly it has been said, to counter the technology lead established by Airbus with the A320. Whatever the reason, Boeing have approached the job with precision and professionalism and have developed a solution quite different to the Airbus philosophy. References (12) and (13) give a detailed description of the B777 FBW system.

The Boeing 777 Primary Flight Control System (PFCS) is outlined at a system level in Fig. 1.18. The drawing shows the PFCS along the top together with the three CDUs. Most of the sensors are shown along the bottom of the diagram. The PFCS system units are interconnected by three ARINC 629 flight control data buses: left, centre and right. In total there are 76 ARINC 629 couplers on the flight control buses.

The PFCS system comprises the following control surface actuators and feel actuators:

- Four elevators. Left and right inboard and outboard.
- Elevator feel. Left and right.
- Two rudders. Upper and lower.
- Four ailerons. Left and right inboard and outboard.
- Four flaperons. Left and right inboard and outboard.
- Fourteen spoilers. Seven left and seven right.

The flight control actuators are interfaced to the three A629 flight control data buses by means of Four Actuator Control Electronics (ACE) units. These are:

- ACE Left 1
- ACE Left 2
- ACE Centre
- ACE Right

These units interface in turn with the flight control and feel actuators in accordance with the scheme shown in the centre of Fig. 1.18 – a total of 31 actuators.

The (ACE) units contain the digital-to-analogue and analogue-to-digital elements of the system. A simplified schematic for an ACE is shown in Fig. 1.19. Each ACE has a single interface with each of the A629 flight control data buses and the unit contains the signal conversion to interface the 'digital' and 'analogue' worlds.

The actuator control loop is shown in the centre-right of the diagram. The actuator demand is signalled to the Power Control Unit (PCU) which moves the actuator ram in accordance with the control demand and feeds back a ram position signal to the ACE, thereby closing the actuator control loop. The ACE also interfaces to the solenoid valve with a command to energize the solenoid valves to allow – in this example – the left hydraulic system to supply the actuator with motive power and at this point the control surface becomes 'live'.

The flight control computations are carried out in the Primary Flight Computers (PFCs) shown in Fig. 1.20. The operation of the PFCs has been briefly described earlier in the Chapter but will be recounted and amplified in this section.

Each PFC has three A629 interfaces with each of the A629 flight control buses, giving a total of nine data bus connections in all. These data bus interfaces and how they are connected and used form part of the overall Boeing 777 PFCS philosophy. The three active lanes within each PFC are embodied in dissimilar hardware. Each of the three lanes is allocated a different function as follows:

- PFC command lane. The command lane is effectively the channel in control. This lane will output the flight control commands on the appropriate A629 bus; e.g. PFC left will output commands on the left A629 bus.

Fig. 1.18 Boeing 777 primary flight control system (PFCS) (Boeing)

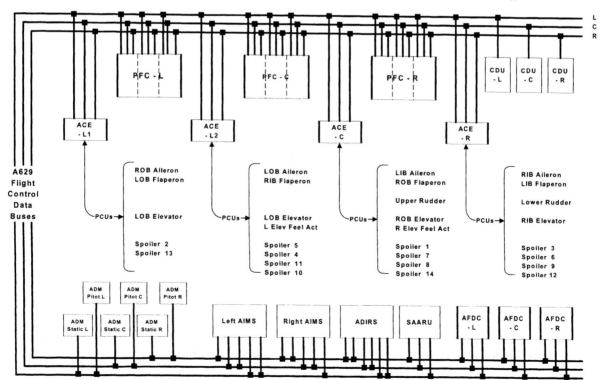

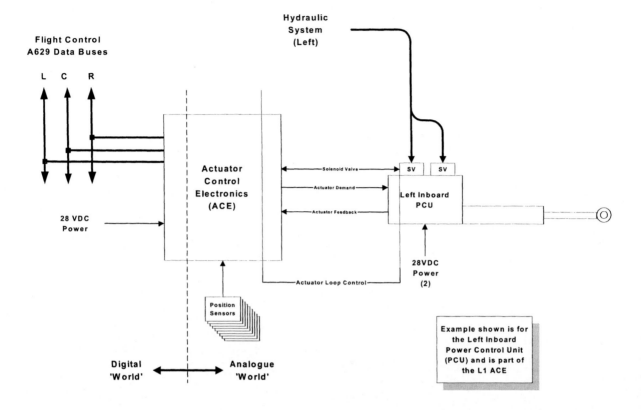

Fig. 1.19 *Actuator control electronics (ACE) unit*

- PFC stand-by lane. The stand-by lane performs the same calculations as the command lane but does not output the commands on to the A629 bus. In effect the stand-by lane is a 'hot stand-by', ready to take command in the event that the command lane fails. The stand-by lane only transmits cross-lane and cross-channel data on the A629 data bus.

- PFC Monitor Lane. The monitor lane also performs the same calculations as the command lane. The monitor lane operates in this way for both the command lane and the stand-by lane. Like the stand-by lane, it only transmits cross-lane and cross-channel data on the A629 data bus.

Figure 1.20 shows that on the data bus, each PFC will only transmit aircraft control data on the appropriate left, centre or right A629 data bus. Within each PFC the command, stand-by and monitor lane operation will be in operation as previously described and only the command channel – shown as the upper channel in the figure – will actually transmit command data.

Within this PFC and A629 architecture:

- Cross-lane comparisons are conducted via the like bus (in this case the left bus).
- Cross-channel comparisons are conducted via the unlike buses (in this case the centre and right buses).

This use of standard A629 data buses to implement the flight control integration and to host the cross-lane and cross-channel monitoring is believed to be unique in flight control. There are effectively nine lanes available to conduct the flight control function.

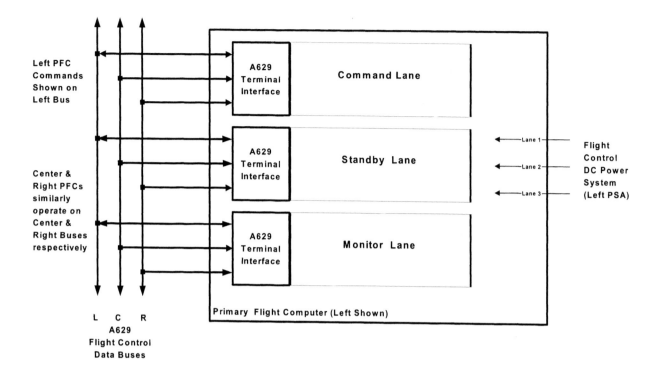

Fig. 1.20 *Boeing 777 primary flight computer (PFC)*

In the event that a single lane fails then only that lane will be shut down. Subsequent loss of a second lane within that channel will cause that channel to shut down, as simplex control is not permitted.

The aircraft may be operated indefinitely with one lane out of nine failed. The aircraft may be dispatched with two out of nine lanes failed for ten days. The aircraft may be operated for a day with one PFC channel inoperative.

The autopilot function of the Boeing 777 PFCS is undertaken by the three Autopilot Flight Director Computers (AFDCs): left, centre and right. The AFDCs have A629 interfaces on to the respective aircraft systems and flight control data buses. In other words, the left AFDC will interface on to the left A629 buses, the centre AFDC on to the centre buses and so on.

Inter-relationship of flight control, guidance and flight management

Figure 1.21 shows the generic example of the main control loops as they apply to aircraft flight control, flight guidance and flight management.

The inner loop provided by the FBW system and the pilot's controls effectively control the attitude of the aircraft.

The next outer loop is that affected by the Autopilot Flight Director System (AFDS) that controls the aircraft trajectory, that is, where the aircraft flies. Inputs to this loop are by means of the mode and datum selections on the Flight Control Unit (FCU) or equivalent control panel.

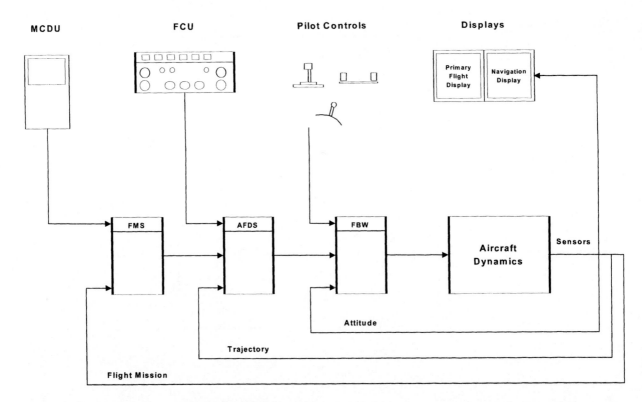

Fig. 1.21 *Definition of flight control, guidance, and management*

Finally, the Flight Management System (FMS) controls where the aircraft flies on the mission; for a civil transport aircraft this is the aircraft route. The Multipurpose Control and Display System (MCDU) controls the lateral demands of the aircraft by means of a series of waypoints within the route plan and executed by the FMS computer. Improved guidance required of 'free-flight' or CNS/ATM also requires accurate vertical or three-dimensional guidance, often with tight timing constraints upon arriving at a way-point or the entry to a terminal area.

References

(1) **BAE SYSTEMS** (1990) Hawk 200 marketing publication CO.095.0890.M5336.

(2) **Farley, B.** (1984) 'Electronic control and monitoring of aircraft secondary flying controls', *Aerospace,* March.

(3) **Howard, R.W.** (1973) Automatic flight controls in fixed wing aircraft – the first 100 years, *Aeronautical Journal*, November.

(4) **Daley, E.** and **Smith, R.B.** (1982) Flight clearance of the jaguar fly-by-wire aircraft. Royal Aeronautical Society Symposium, April.

(5) **Yeo, C.J.** (1984) 'Fly-by-wire Jaguar', *Aerospace*, March.

(6) **Kaul, H-J., Stella, F.** and **Walker, M.** (1984) The flight control system for the Experimental Aircraft Programme (EAP) Demonstrator Aircraft, 65th Flight Mechanics Panel Symposium, Toronto, October.

(7)   **Young, B.** (1987) Tornado/Jaguar/EAP experience and configuration of design, Royal Aeronautical Society Spring Convention, May.

(8)   **Snelling, K.S.** and **Corney, J.M.** (1987) The implementation of active control systems, Royal Aeronautical Society Spring Convention, May.

(9)   **Anthony, M.J.** and **Mattos, F.** (1985) Advanced flight control actuation systems and their interfaces with digital commands, A-6 Symposium, San Diego, October.

(10) **White, J.A.P.** (1978) The development of electromechanical actuation for aircraft systems, *Aerospace*, November.

(11) **Davies, C.R.** (1987) Systems aspects of applying active control technology to a civil transport aircraft, Royal Aeronautical Society Spring Convention, May.

(12) **Tucker**, **B. G. S.** Boeing 777 primary flight control computer system – philosophy and implementation, RAeS Conference – *Advanced Avionics on the A330/A340 and the Boeing 777*, November 1993.

(13) **McWha, James**, 777 – Ready for service, RAeS Conference – *The Design & Maintenance of Complex Systems on Modern Aircraft*, April 1995.

CHAPTER 2

Engine Control Systems

Introduction

In early jet aircraft the control of fuel to the combustors was performed by pneumatic and hydromechanical flow control devices. Thrust was demanded and maintained at an approximately fixed condition by the pilot adjusting the throttle lever and continuously monitoring his temperature and speed gauges. This soon proved to be totally unsatisfactory, since the wide range of ambient conditions encountered in flight meant that continual throttle adjustments were needed. Furthermore the engine had to be handled carefully to avoid flameout or surge during accelerations and decelerations.

The task of handling engines was eased by the introduction of electronic control in the form of magnetic amplifiers in early civil and military aircraft. The mag-amp allowed engines to be stabilized at any speed in the throttle range by introducing a servo-loop with engine exhaust gas temperature as a measure of engine speed and an analogue fuel valve to control fuel flow. This allowed the pilot to accelerate and decelerate the engine while the control system limited fuel flows to prevent overspeeds or excessive temperatures.

Control systems became more sophisticated with additional engine condition sensors and multiple servo-loops. Transistors, integrated circuits and high temperature semi-conductors have all played a part in the evolution of control systems from range temperature control through to full digital engine control systems. With modern FADEC systems there are no mechanical control rods or mechanical reversions, and the pilot can perform carefree handling of the engine throughout the flight envelope.

On modern aircraft the engine is supervised by a computer to allow the pilot to operate at maximum performance in a combat aircraft or at optimum fuel economy in a passenger-carrying aircraft.

Engine control evolution

The early jet engines based on a centrifugal compressor used a method of controlling fuel to the engine combustion chamber that used a fuel pump, a relief valve and a throttle valve. In series with these was a mechanical centrifugal governor. Barometric compensation of the relief valve was provided by a suitable bellows mechanism to maintain the full range of throttle movement at altitude. The design of such engines based upon Sir Frank Whittle's design was basically simple, using sound engineering practices and employing technology representing 'state of the art' of the day.

The move to the axial type of engine placed more stress on the components of the engine and, together with demands for improved performance, created a demand for more complex methods of controlling fuel flow, air flow and exhaust gas flow.

Early gas turbine control systems were initially entirely hydromechanical. As engine and materials development continued a need arose to exercise greater control of turbine speeds and temperatures to suit prevailing atmospheric conditions and to achieve surge-free operation. The latter was particularly important in military engines where handling such as changing the engine speed many times from maximum to minimum, tended to place the engine under severe conditions of operation.

To achieve the needed improvements, electronic control circuits were used to modify the basic hydromechanical fuel demands. Further developments in engine design led to the need to control more parameters and eventually led to the use of full authority analogue control systems with electrical signalling from the throttle levers.

The emergence of digital technology and serial data transmission systems, as well as higher performance electronic devices, led to the opportunity to integrate the control systems with the aircraft avionics and flight control systems, and to consider the mounting of complex electronic control units on the engine itself.

The control problem

The basic control action is to control a flow of fuel and air to the engine to allow it to operate at its optimum efficiency over a wide range of forward speeds, altitudes and temperatures whilst allowing the pilot to handle the engine without fear of malfunction. The degree of control required depends to a large extent upon the type of engine and the type of aircraft in which it is installed.

The military aircraft is usually specified to operate in world-wide conditions, and is expected to experience a wide range of operating temperatures. To be successful in combat the aircraft must be manoeuvrable. The pilot, therefore, expects to be able to demand minimum or maximum power with optimum acceleration rates, as well as to make small adjustments with equal ease, without fear of surge, stall, flame-out, overspeed or over-temperature. The pilot also needs a fairly linear relationship between throttle lever position and thrust.

The civil operator requires reliable, economical and long-term operation under clearly defined predictable conditions with minimum risk to passengers and schedules To obtain these objectives, control can be exercised over the following aspects of engine control:

- Fuel flow – to allow varying engine speeds to be demanded and to allow the engine to be handled without damage by limiting rotating assembly speeds, rates of acceleration and temperatures.

- Air flow – to allow the engine to be operated efficiently throughout the aircraft flight envelope and with adequate safety margins.
- Exhaust gas flow – by burning the exhaust gases and varying the nozzle area to provide additional thrust.

Electronic control has been applied in all these cases with varying degrees of complexity and control authority. Such control can take the form of simple limiter functions through to sophisticated multi-variable, full authority control systems closely integrated with other aircraft systems.

Fuel flow control

The control of power or thrust of the gas turbine engine is obtained by regulating the quantity of fuel injected into the combustion system. When a higher thrust is required the throttle is opened and the fuel pressure to the burners increases due to the higher fuel flow. This has the effect of increasing the gas temperature which, in turn, increases the acceleration of the gases through the turbine to give a higher engine speed and correspondingly greater air flow, resulting in an increase in thrust.

The relationship between the airflow induced through the engine and the fuel supplied is, however, complicated by changes in altitude, air temperature and aircraft speed. These variables change the density of the air at the engine intake and consequently the mass of air flowing through the engine.

To meet this change in air flow a similar change in fuel flow must occur, otherwise the ratio of air to fuel will alter and the engine speed will increase or decrease from that originally selected by the pilot in setting the throttle lever position. Fuel flow must, therefore, be controlled to maintain the conditions demanded by the pilot whatever the changes in the outside world. Failure to do so would mean that the pilot would constantly need to make minor adjustments to the throttle lever position, increasing his work load and distracting his attention from other aspects of aircraft operation.

The usual method of providing such control is by means of a Fuel Control Unit (FCU). The FCU is a hydromechanical device mounted on the engine. It is a complex engineering mechanism containing valves to direct fuel and to restrict fuel flow, pneumatic capsules to modify flows according to prevailing atmospheric conditions, and dashpot/spring/damper combinations to control acceleration and deceleration rates. An excellent description of the principles of operation of the jet engine and turbo-prop can be found in the Rolls-Royce Book of the Jet Engine (reference (1)).

The engine speed must be controlled from idle to maximum rating. Overspeed must be avoided to reduce stresses in the rotating assemblies, and over-temperature must be avoided to prevent blade damage and to reduce thermal creep. The engine must be allowed to accelerate and decelerate smoothly with no risk of surge.

Such control influences are difficult to achieve manually. Therefore the FCU has, over the generations of jet engines, been designed to accommodate control inputs from external electronic devices. Electrical valves in the FCU can be connected to electronic control units to allow more precise and continuous automatic control of fuel flows in response to throttle demands, using measurements derived from the engine, to achieve steady state and transient control of the engine without fear of malfunction.

Air flow control

It is sometimes necessary to control the flow of air through to the engine to ensure efficient operation over a wide range of environmental and usage conditions to maintain a safe margin from the engine surge line. To do this some engines, mainly those in aircraft expected to perform manoeuvres at high speeds, have control systems and moving surfaces in the engine intakes to provide optimum flows under all conditions.

Control systems

The number of variables that affect engine performance is high and the nature of the variables is dynamic, so that the pilot cannot be expected constantly to adjust the throttle lever to compensate for changes, particularly in multi-engined aircraft. In the first gas turbine engined aircraft, however, the pilot was expected to do just that.

A throttle movement causes a change in the fuel flow to the combustion chamber spray nozzles. This, in turn, causes a change in engine speed and in exhaust gas temperature. Both of these parameters are measured; engine speed by means of a gearbox mounted speed probe and exhaust gas temperature (or Turbine Gas Temperature – TGT) by means of thermocouples, and presented to the pilot as analogue readings on cockpit mounted indicators. The pilot can monitor the readings and move the throttle to adjust the conditions to suit his own requirements or to meet the maximum settings recommended by the engine manufacturer. The FCU with its internal capsules is able to adjust fuel flow to compensate for changes in atmospheric conditions and to maintain a constant engine speed.

In the dynamic conditions of an aircraft in flight at different altitudes, temperatures and speeds, continual adjustment by the pilot soon becomes impractical. He cannot be expected to continuously monitor the engine conditions safely for a flight of any significant duration. For this reason some form of automatic control is essential.

Control system parameters

To perform any of the control functions automatically requires devices to sense engine operating conditions and to perform a controlling function. These can usually be conveniently subdivided into input and output devices producing input and output signals to the control system.

To put the control problem into perspective the control system can be regarded as a box on a block diagram receiving input signals from the aircraft and the engine and providing outputs to the engine and the aircraft systems. This system is shown diagramatically in Fig. 2.1.

The input signals provide information from the aircraft and the engine to be used in control algorithms, while the output signals provide the ability to perform a control function. Further signals derived from output devices provide feedback to allow loop closure and stable control. Typical inputs and outputs are described below.

Input signals

- Throttle position – A transducer connected to the pilot's throttle lever allows thrust demand to be determined. The transducer may be connected directly to the throttle lever with electrical signalling to the control unit, or connected to the end of control rods to maintain mechanical operation as far as possible. The transducer

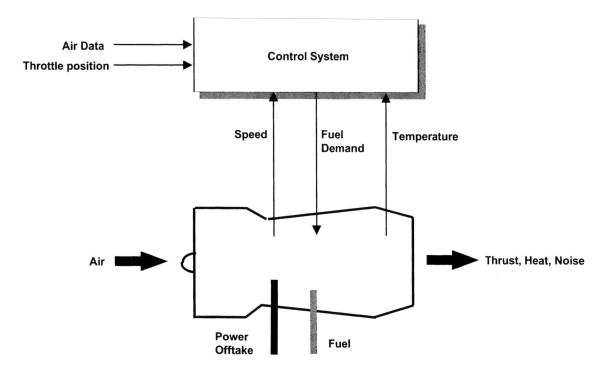

Fig. 2.1 *Engine control system basic inputs and outputs*

may be a potentiometer providing a DC signal or a variable transformer to provide an AC signal. To provide suitable integrity of the signal a number of transducers will be used to ensure that a single failure does not lead to an uncommanded change in engine demand.

- Air data – Air speed and altitude can be obtained as electrical signals representing the pressure signals derived from airframe mounted capsule units. These can be obtained from the aircraft systems such as an Air Data Computer (ADC) or from the flight control system air data sensors. The latter have the advantage that they are likely to be multiple-redundant and safety monitored.

- Total temperature – A total temperature probe mounted at the engine face provides the ideal signal. Temperature probes mounted on the airframe are usually provided, either in the intakes or on the aircraft structure.

- Engine speed – The speed of rotation of the shafts of the engine is usually sensed by pulse probes located in such a way as to have their magnetic field interrupted by moving metallic parts of the engine or gearbox. The blades of the turbine or compressor, or gearbox teeth, passing in front of a magnetic pole piece induce pulses into a coil or a number of coils wound around a magnet. The resulting pulses are detected and used in the control system as a measure of engine speed.

- Engine temperature – The operating temperature of the engine cannot be measured directly since the conditions are too severe for any measuring device. The temperature can, however, be inferred from measurements taken elsewhere in the engine. The traditional method is to measure the temperature of the engine exhaust gas using thermocouples protruding into the gas stream. The thermocouples are usually arranged as a ring of parallel connected thermocouples to obtain a measurement of mean gas temperature and are usually of chromel-alumel junctions. A cold junction is provided to obtain a reference voltage. An alternative method is to measure the temperature of the turbine blades with an optical pyrometer. This takes the form of a fibre optic with a lens mounted on the engine casing and a semiconductor sensor mounted in a remote and cooler environment. Both of these temperatures can be used to determine an approximation of turbine entry temperature, which is the parameter on which the temperature control loop should ideally be closed.

- Nozzle position – For those aircraft fitted with reheat (or afterburning) the position of the reheat nozzle may be measured using position sensors connected to the nozzle actuation mechanism or to the nozzle itself. An inductive pick-off is usually used since such types are relatively insensitive to temperature variations, an important point because of the harsh environment of the reheat exhaust.
- Fuel flow – Fuel flow is measured by means of a turbine type flowmeter installed in the fuel pipework to obtain a measure of fuel inlet flow as close to the engine as possible.
- Pressure ratio – The ratio of selected pressures between different stages of the engine can be measured by feeding pressure to both sides of a diaphragm-operated device.

Output signals

- Fuel flow control – The fuel supply to the engine can be varied in a number of ways depending on the type of fuel control unit used. Solenoid-operated devices, torque motor or stepper motor devices have all been employed on different engine types. Each device has its own particular failure modes and its own adherents.
- Air flow control – The control of air flow at different stages of the engine can be applied by the use of guide vanes at the engine inlet, or by the use of bleed valves between engine stages. These can be operated manually or automatically to attempt to preserve a controlled flow of air.

Example systems

Using various combinations of input and output devices to obtain information from the engine and the airframe environment, a control system can be designed to maintain the engine conditions stable throughout a range of operating conditions. The input signals and output servo demands can be combined in varying degrees of complexity to suit the type of engine, the type of aircraft, and the manner in which the aircraft is to be operated. Thus the systems of civil airliners, military trainers and high-speed combat aircraft will differ significantly.

In a simple control system, such as may be used in a single-engine trainer aircraft the primary pilot demand for thrust is made by movements of a throttle lever. Rods and levers connect the throttle lever to a FCU so that its position corresponds to a particular engine condition, say rpm or thrust. Under varying conditions of temperature and altitude this condition will not normally stay constant, but will increase or decrease according to air density, fuel temperature or demands for take-off power. To obtain a constant engine condition, the pilot would have continually to adjust the throttle lever, as was the case in the early days of jet engines. Such a system with the pilot in the loop is shown in Fig. 2.2.

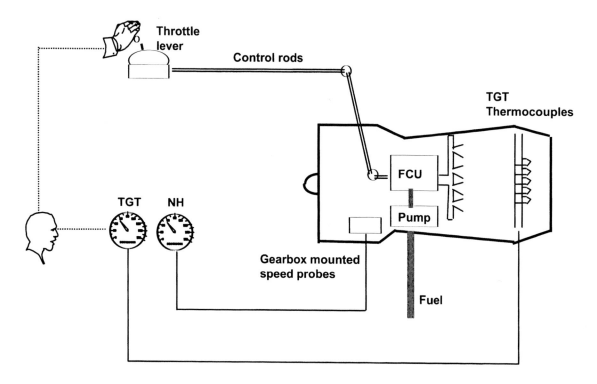

The flow of fuel to the combustion chambers can be modified by an electrical valve in the FCU that has either an infinitely variable characteristic, or moves in a large number of discrete steps to adjust fuel flow. This valve is situated in the engine fuel feed line so that flow is constricted, or is by-passed and returned to the fuel tanks, so that the amount of fuel entering the engine is different from that selected.

Fig. 2.2 Simple control system with the pilot in the loop

This valve forms part of a servo loop in the control system so that continuous small variations of fuel flow stabilize the engine condition around that demanded by the pilot. This will allow the system to compensate for varying atmospheric and barometric conditions, to ensure predictable acceleration and deceleration rates; and to prevent over-temperature or overspeed conditions occurring over the available range – acting as a range speed governor; Fig. 2.3 illustrates such a control system. It can be seen that the pilot shown in Fig. 2.2, now acts in a supervisory role, relying on the control system to maintain basic control conditions whilst he monitors the indicators for signs of overspeed or over-temperature.

Even this task can be reduced considerably by incorporating an automatic means of signalling an overspeed or over-temperature. This can be performed in the control unit by setting a datum related to a particular engine type, or by setting a variable 'bug' on the cockpit indicator. If either pre-set datum is exceeded a signal is sent to the aircraft warning system to warn the pilot by means of a red light and signal tone (see Chapter 9). This principle is illustrated in Fig. 2.4 which shows warning systems for both over-temperature and overspeed conditions.

In this diagram the overspeed warning is provided by a mechanism in the Turbine Gas Temperature (TGT) indicator. A knob on the indicator allows the pilot to set a 'bug' to a particular temperature. When the indicator pointer exceeds that setting a pair of contacts in the indicator close and provide a signal to the aircraft central warning

Fig. 2.3 A simple
limited authority engine
control system

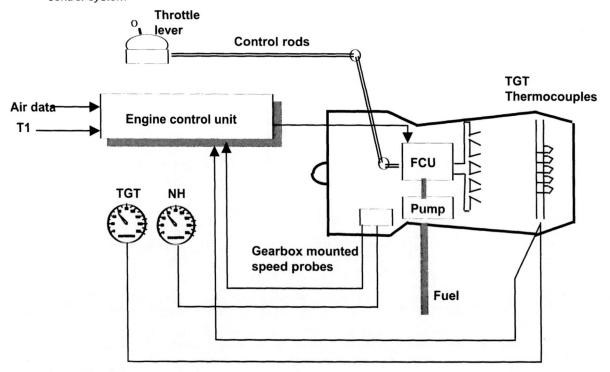

Fig. 2.4 Engine
control system with
NH and TGT
exceedance warnings

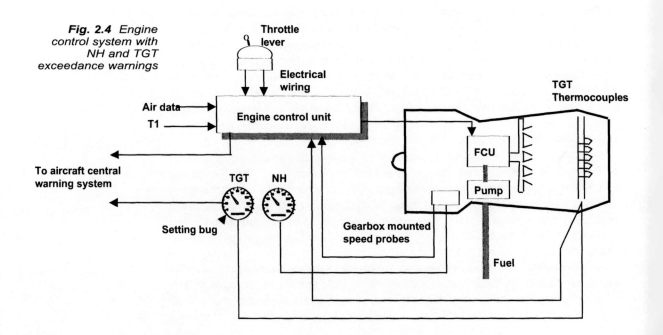

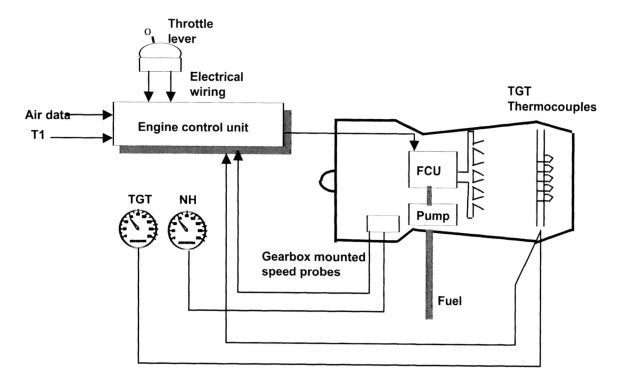

system. The overspeed warning is provided by a pair of contacts in the engine control
unit. In practice either one method or the other is used in one aircraft type. In today's
digital systems such signals will be generated electronically and passed to the cockpit
as digital data over one of the standard aerospace data buses; ARINC 429, MIL-STD-
1553B – refer to Chapter 12 for a description of these data buses.

In many modern aircraft the simple throttle signalling system is retained, but with
the replacement of rods and levers by electrical signalling from the throttle levers. This
reduces friction and eliminates the possibility of jamming in the control rod circuit. An
example of a system with electrical throttle signalling is illustrated in Fig. 2.5. The
removal of any mechanical links between the pilot and the engine means that the control
unit has full authority control. There is nothing the pilot can do to correct an engine
malfunction other than to shut down the engine. Because of this the throttle signalling
circuit (like the rest of the control system) is designed with great care to ensure that all
failures are detected and taken care of by the control system. For example, additional
windings on the Tornado throttle position transducer enable the control system to detect
open circuits and short circuits and to take corrective action.

For multiple engine types of similar complexity, the system is duplicated with no
cross-connection between the systems to reduce the risk of common mode failures.

More functions can be added to the system to enable the engine to operate in more
demanding situations. For example, air bleed valves between engine stages can be
opened or closed to stabilize the engine as a function of speed or acceleration. The
ignition system can be switched on during periods of heavy rain or icing; and all
conditions can be signalled to the crew by cockpit instruments or warning lights.

The system illustrated in Fig. 2.3 is typical of many systems engineered in the 1950s
and 1960s. The Canberra and Lightning aircraft contained engine control systems based

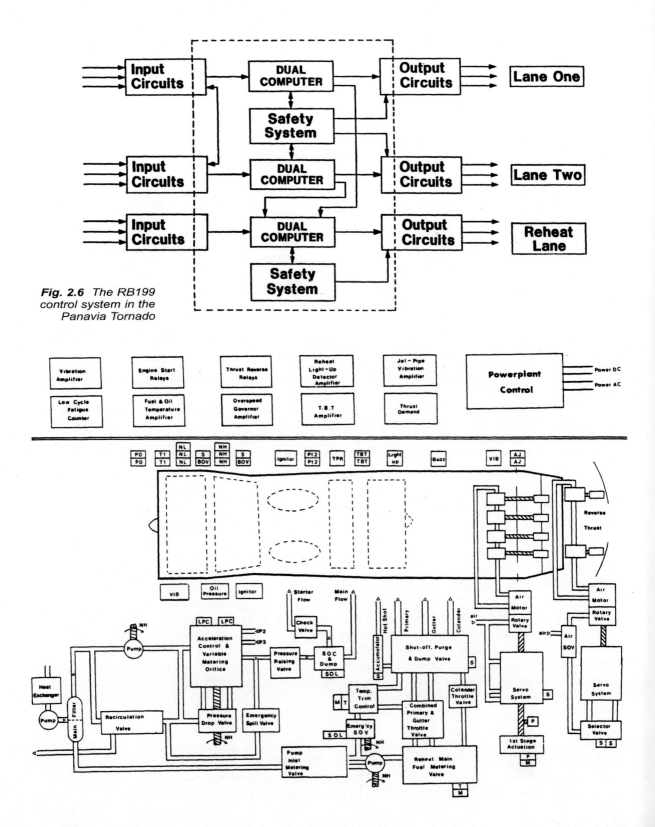

Fig. 2.6 *The RB199 control system in the Panavia Tornado*

on magnetic amplifiers used as an analogue control system. Developments in semiconductor technology led to the introduction of transistorized analogue amplifiers such as that used in the control unit for the Adour engine installed in the Sepecat Jaguar.

Jaguar was an early venture into European collaboration between BAE SYSTEMS (then British Aircraft Corporation) and Dassault (then Avions Louis Breguet). The engine control unit was manufactured by Elecma in France to control the Rolls-Royce/Turbomeca Adour twin engine combination. Each engine had its own control unit mounted on the airframe in a ventral bay between the two engines. Provision was made for the connection of test equipment and for adjustments to the unit to allow the engine to be set up correctly on engine ground runs.

Concorde made full use of electronic technology for the control of its four reheated Olympus 593 engines. The control system for each engine was designed as a full authority self-monitoring system, completely independent of the others. The control units were mounted on the airframe and provided control for the main engine and reheat functions. This analogue system went into each of the production Concorde aircraft. A separate system provided control of the intake ramps to provide a suitable air flow to the engines under all flight conditions.

The Turbo Union RB199 engines in the Panavia Tornado made full use of the experience gained on Concorde. Each engine was controlled by a single Main Engine Control Unit (MECU). Each MECU contained two independent lanes of dry engine control and a single reheat control lane. A single engine system is shown in Fig. 2.6.

The RB199 is a complex engine, and a number of separate input conditioning units were required to provide the completed control and indication package. Instead of TGT, engine temperature was measured using an optical pyrometer monitoring the infra-red radiation of the turbine blades. This required a Turbine Blade Temperature (TBT) amplifier which not only converted the pyrometer signal into a form suitable for connection to the MECU, but also provided a signal to the TBT indicator in the cockpit. The TBT indicator provided a signal to the aircraft central warning panel in the event of an over-temperature. This system is shown in Fig. 2.7.

Other individual electronic units were provided for monitoring vibration using piezoelectric transducers, for detecting the light-up of the reheat system using ultra-violet detectors, for providing an independent overspeed governor circuit for both HP and LP turbines, and for controlling reverse thrust. Throttle position was signalled using dual winding AC pick-offs.

All electronic units were airframe mounted in the aircraft front fuselage avionics bays. This required long lengths of multiple cable harnesses to run almost the full length of the aircraft. The harnesses had to be designed to allow physical separation, not only of each engine harness, but also each control lane, and for electromagnetic health reasons. This resulted in a large weight of wiring in the aircraft and required a large number of connectors to allow the wiring to cross between the engine and the airframe. This was a heavy and costly arrangement, but one which was necessary because semiconductor technology was insufficiently advanced at that time to allow electronic control units to be mounted in the high temperature and vibration environment of the engine bay. There was an absolute limit on some devices that would be destroyed by high internal temperatures; the environment would lead to unacceptable low reliability for complex units.

TRW Lucas Aerospace made considerable advances in technology in the development of integrated circuits mounted on ceramic multi-layer boards to provide a

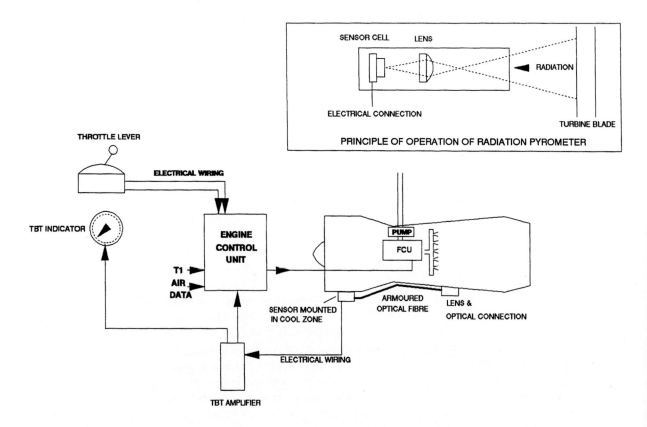

Fig. 2.7 *The RB199 turbine blade temperature system*

highly reliable engine control system. Roll-Royce, MTU and FIAT formed a joint engine company – Turbo Union, which designed and manufactured the engine, and acted as prime contractor for the engine control system.

In the early 1960s Rolls-Royce began to experiment with the use of digital control systems which led to a demonstration of such a system on a test rig. However, by the 1970s sufficient work had been done to enable them and TRW Lucas Aerospace to design and build an experimental full-authority control system for use with multiple spool engines. Such a system was flown connected to a single engine of Concorde 002 in July 1976; (reference (2)). This advance in technology went through several stages of design and approval before it became accepted as a suitable system for use in the Tornado, and the MECU was replaced by the Digital Electronic Control Unit (DECU).

The concept of full-authority digital control went a stage further in the BAE SYSTEMS Experimental Aircraft Programme (EAP) in which the DECU became integrated with the aircraft avionics. A system had been installed in EAP to provide digital control and monitoring of all the aircraft utility systems. This system was known as utility systems management (USM) and was essentially a multi-computer system interconnected with a MIL-STD-1553B (Def Stan 00-18 Part 2) serial data transmission system. A simple and economic method of incorporating the RB199 and its control system into the aircraft structure of data buses and multi-function cockpit displays was to provide a means of interconnection through USM; (reference (3)). This system is illustrated in Fig. 2.8.

This first step at integration allowed the pilot to handle the engines in a high-speed, highly manoeuvrable aircraft without any engine displays – the engine parameters were displayed on demand only and not permanently presented on individual indicators as in previous combat aircraft. The resulting displays can be seen in Fig. 2.9.

This was an important step on the path to full integration of flight and engine control systems which is the subject of current research studies.

Full-Authority Digital Engine Control (FADEC) is now common on many engines, and semiconductor and equipment-cooling technology has advanced so that control units can now be mounted on the engine and still provide highly reliable operation for long periods. An example of a typical modern installation is the FADEC mounted on the Pratt and Whitney PW305 engine in a small business jet aircraft. This installation is shown in Fig. 2.10.

Fig. 2.8 The RB199 control system in the BAE SYSTEMS EAP

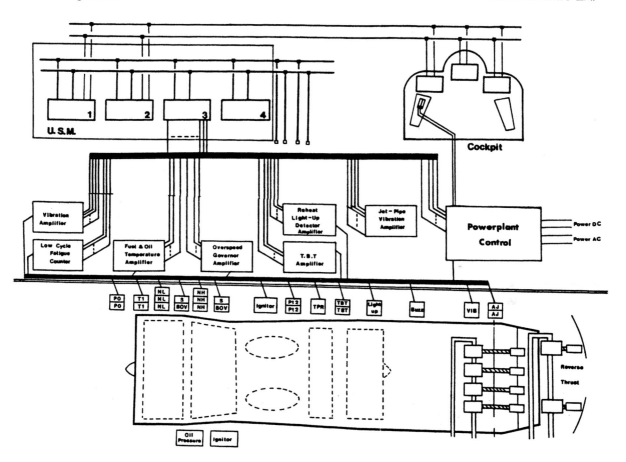

Design criteria

The engine and its control system are considered to be safety critical. That is to say that a failure may hazard the aircraft and the lives of the crew, passengers and people on the ground. For this reason the system is generally designed to eliminate common mode failures, to reduce the risk of single failures leading to engine failure and to contain the risk of failure within levels considered to be acceptable by engineering and certification authorities. As an example the Civil Aviation Authority set the integrity requirements for the Concorde engine control system (reference (2)). These were:

Fig. 2.9 Engine displays
in the EAP cockpit (BAE
SYSTEMS)(see also
colour plate section)

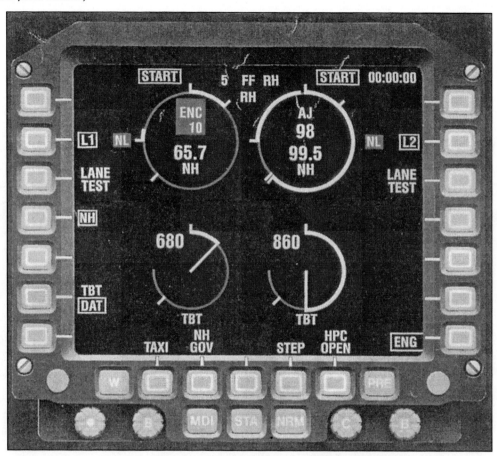

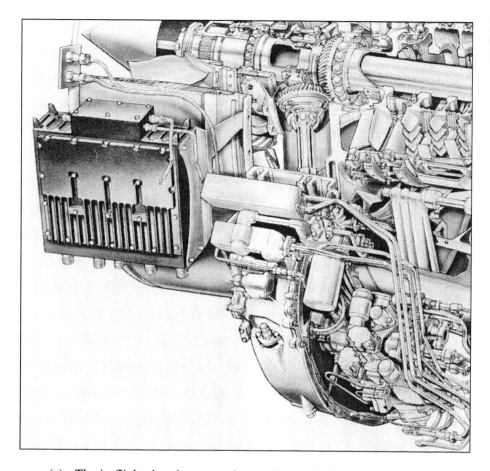

Fig. 2.10 *Installation of FADEC on the PW 305 engine (United Technologies)*

(a)   The in-flight shut-down rate due to electronics failure must not exceed 2.3 x 10⁻⁶ per engine hour.

(b)   The upward runaway rate due to electronics failure must not exceed 1 x 10⁻⁶ per engine hour.

(c)   The downwards runaway rate due to electronics failure must not exceed 2.7 x 10⁻⁶ per engine hour.

Similar design targets are set for every project and they are based upon what the certification authorities consider to be an acceptable failure rate. They are used by the engineer as targets that should never be exceeded, and are used as a budget from which individual control system components and modules can be allocated individual targets. The sum of all individual modules must never exceed the budget. A wise engineer will ensure that an adequate safety margin exists at the beginning of the design.

The design failure rate targets are based upon the well-known random failure properties of hardware. Every item of electronic hardware has a failure rate that can be obtained from a design handbook or from the component manufacturer's literature. This rate is based upon statistical evidence gathered from long-term tests under varying conditions, and may be factored by practical results from the use of components in service. The designer selects the correct components, ensures that they are not overstressed in use and observes scrupulous quality control in design and manufacture.

Refer to Chapter 11 for further amplification of reliability assessment methods and redundant system architectures.

On the airframe side similar care is taken in the provision of cooling, freedom from vibration and by providing high quality power supplies.

Nevertheless, failures will occur, albeit rarely. Techniques have been established to ensure that the effects of failure on system operation are minimised. A common method of reducing the effect of failures is to introduce redundancy into a system. Concorde, for example had four engines, therefore a failure of at least one could be tolerated, even at take-off. Each engine had a separate control unit with no physical interconnections, each control unit has two independent lanes of control, and duplicated input signals were obtained from separate sources. The wiring harnesses were widely separated in the airframe to reduce the risk of mechanical damage or electromagnetic interference affecting more than one system. In addition a separate overspeed governor was provided to ensure that the HP turbine was never allowed to over-speed and suffer catastrophic failure.

It is important that the entire system is designed to be fail-safe. For an aircraft, fail-safe means that the system must be able to detect failures and to react to them by either failing to a condition of existing demand (fail-frozen) or to a condition of maximum demand. This is to ensure that a failure in a critical regime of flight, such as take-off, will enable the pilot to continue with the take-off safely. For this reason fuel valves generally fail to the open position.

These techniques are used on many multiple engine combinations with electronic control systems. The techniques are well established, well understood and can be analysed numerically to provide evidence of sound design.

Difficulties began to occur when digital control was introduced. Software does not have a numerical failure rate. If failures are present they will be caused by inadequate design, and not discovered during testing. The design process for software used in safety critical applications, such as engine control, should ensure that there are no incipient design faults.

In a multiple-engine aircraft, as explained above, each engine will have its own independent control unit. For an analogue system the random failure characteristics of electronic components means that failures will generally be detectable and will be contained within one engine control system The possibility of two failures occurring in the same flight is extremely unlikely to occur on a second engine. For example – taking the case of a twin-engined aircraft – if the failure rate of a component leading to an engine failure or shut-down is predicted as 1×10^{-6} per flying hour, then the probability of a similar failure occurring on the remaining engine is 1×10^{-12} per flying hour. The design of systems and the efforts taken to prevent common mode failures and fault propagation has resulted in failure rates which are considered acceptable to allow transatlantic flights on twin-engine aircraft, provided that the time at risk after the first failure is sufficient to allow the aircraft to land at a diversionary airfield.

It is argued that, since software in independent digital control units is identical (hence a common mode failure potential), then it is possible for undetected design faults to manifest themselves with particular combinations of data and instruction. As the software and control systems are identical, then in theory the same set of conditions could occur on the same flight and may result in multiple engine shut-down.

To counteract this effect a number of techniques were used in some systems. Dissimilar redundancy was one such technique in which different teams of engineers designed and coded the software in different control lanes or control units. This was an extremely costly method, requiring two design teams, two test programmes, and two certification programmes. An alternative was to provide a mechanical reversionary mode that allowed the pilot to effect rudimentary control over the flow of fuel to the engine by means of a switch and solenoid valve.

However, the best method of producing sound software is to establish sound design principles. For this reason modern techniques of software design include structured methods of requirements analysis, software design, modular coding and thorough testing, as well as such techniques as static code analysis. Modern engine control systems are now well-established and trusted and have achieved many trouble-free flying hours.

Engine starting

To start the engines a sequence of events is required to allow fuel flow, to rotate the engine and to provide ignition energy. For a particular type of aircraft this sequence is unvarying, and can be performed manually with the pilot referring to a manual to ensure correct operation, or automatically by the engine control unit. Before describing a typical sequence of events, an explanation of some of the controls will be given.

Fuel control

Fuel from the tanks to the engine feed line is interrupted by two shut-off cocks. The first is in the low pressure feed lines, at which fuel pressure is determined by the fuel boost pumps (See Chapter 3, Fuel Systems). The valve, known as the LP cock or fire wall shut-off cock, is situated close to the engine fire wall. Its primary purpose is to isolate the engine in the event of a fire. It is usually a motor-driven valve controlled by a switch in the cockpit and, once opened, cannot be shut except by means of the switch. The switch is usually covered by a guard so that two actions are needed to select the switch to either open or close the cock. This helps to prevent inadvertent actions that may lead to accidental engine shut down.

The second valve, called the HP cock, is in the high-pressure fuel line, in which the fuel pressure is determined by an engine-driven pump. The function of this valve is to open and close the fuel feed close to the engine inlet at the fuel control unit. It is opened manually by the pilot, or automatically by the engine control unit at an appropriate stage in the engine start cycle. The location of these valves is shown in Fig. 2.11.

Ignition control

The ignition system consists of a high-energy ignitor which is switched on for a period during the start cycle. The ignitors initiate combustion of the fuel vapour in the combustion chamber. An ignitor plug is supplied with electrical energy by an ignition exciter that produces stored energy from 1 to 6 joules depending on the type required. High-energy systems are used for starting, and low-energy systems can be provided to maintain engine ignition during aircraft operations in heavy rain, slushy runways or icing conditions. Some examples of typical ignition equipment are shown in Fig. 2.12.

Engine rotation

During the starting cycle the engine needs to be rotated until the fuel has ignited and the temperature of combustion is sufficient for the engine to rotate without assistance. At

Fig. 2.11 *Typical
location of LP and HP
fuel cocks*

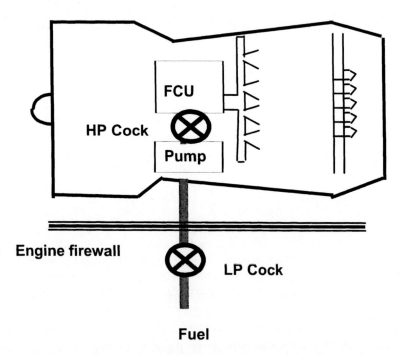

Fuel

this point the engine is said to be self-sustaining. A number of methods are in current use for providing assistance by means of air, electrical energy or chemical energy. A number of these methods are illustrated in Fig. 2.13.

Air at high pressure can be provided by an external air compressor trolley connected to the engine by ground crew, or by air supplied by an on-board Auxiliary Power Unit (APU). This is a small gas turbine that is started prior to engine start. It has the advantage of making the aircraft independent of ground support and is useful at remote airfields. It is also used to provide electrical and hydraulic energy for other aircraft services.

A DC motor mounted on the engine can be supplied with energy from an external battery truck or from the aircraft internal battery.

Chemical energy can be provided by the use of cartridges or a mono-fuel such as Iso-Propyl-Nitrate (IPN) to rotate a small turbine connected to the engine.

Throttle levers

The throttle lever assembly is often designed to incorporate HP cock switches so that the pilot has instinctive control of the fuel supply to the engine. Microswitches are located in the throttle box so that the throttle levers actuate the switches to shut the valves when the levers are at their aft end of travel. Pushing the levers forward automatically operates the switches to open the fuel cocks, which remain open during the normal operating range of the levers. Two distinct actions are required to actuate the switches again. The throttle lever must be pulled back to its aft position and a mechanical latch operated, or a detent (hard point) overcome, to allow the lever to travel further and shut off the fuel valve. The throttle lever for the BAE SYSTEMS 146 is shown in Fig. 2.14 showing four levers and four latches.

Fig. 2.12 *Some examples of high-energy ignition equipment (TRW Lucas Aerospace)*

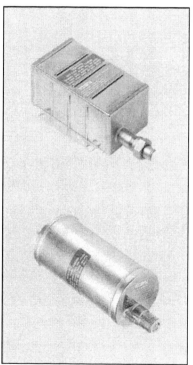

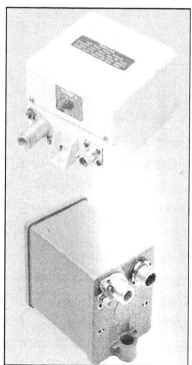

Starting sequence

A typical start sequence is:

- Open LP cocks
- Rotate engine
- Supply ignition energy
- Set throttle levers to idle – open HP cocks
- When self-sustaining – switch off ignition
- Switch off or disconnect rotation power source

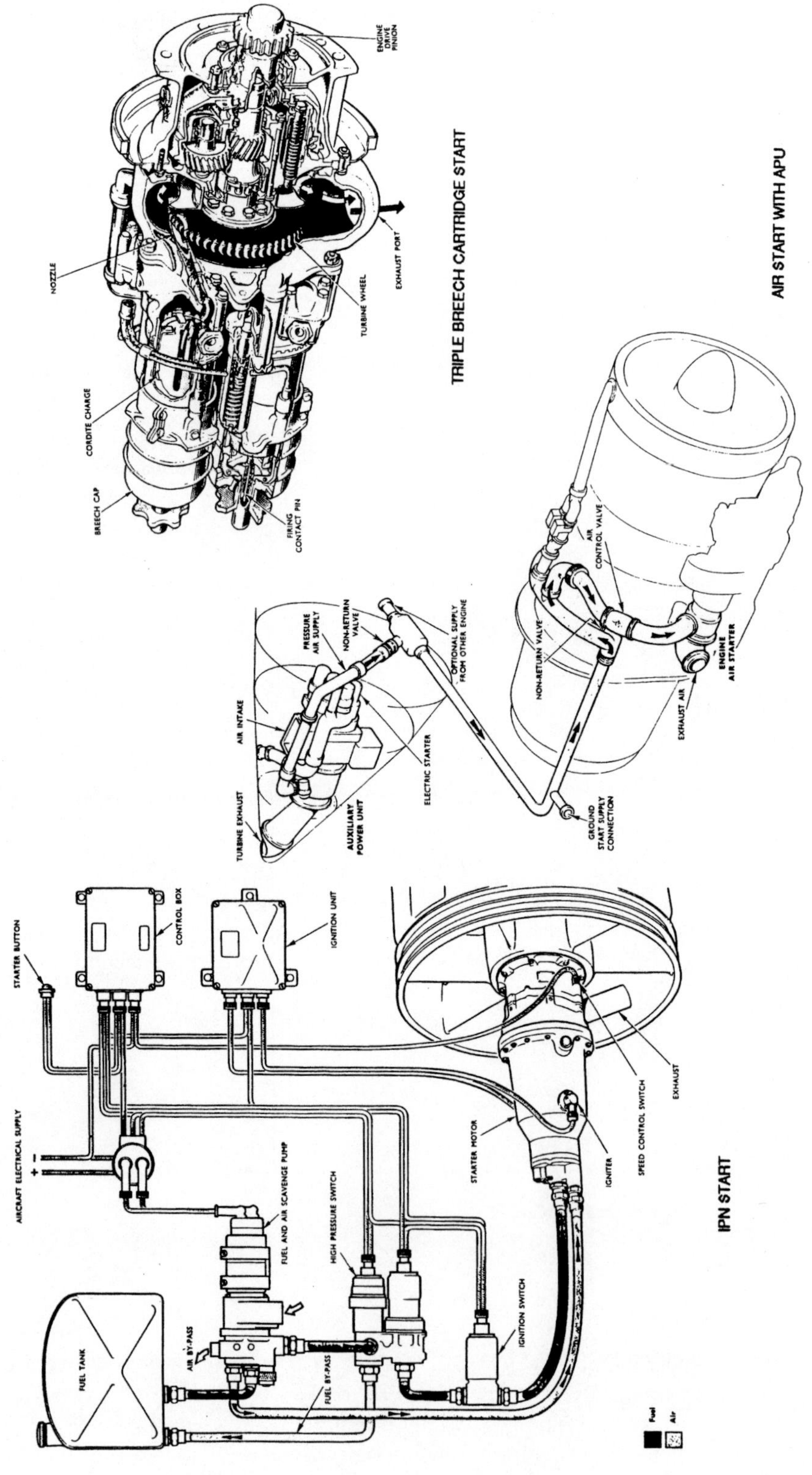

Fig. 2.13 Methods of rotating a gas turbine engine (Rolls-Royce)

TRIPLE BREECH CARTRIDGE START

ENGINE DRIVE PINION

NOZZLE

CORDITE CHARGE

BREECH CAP

FIRING CONTACT PIN

TURBINE WHEEL

EXHAUST PORT

AIR START WITH APU

PRESSURE AIR SUPPLY

NON-RETURN VALVE

OPTIONAL SUPPLY FROM OTHER ENGINE

AIR INTAKE

ELECTRIC STARTER

TURBINE EXHAUST

AUXILIARY POWER UNIT

GROUND START SUPPLY CONNECTION

AIR CONTROL VALVE

NON-RETURN VALVE

EXHAUST AIR

ENGINE AIR STARTER

IPN START

STARTER BUTTON

CONTROL BOX

IGNITION UNIT

AIRCRAFT ELECTRICAL SUPPLY

FUEL AND AIR SCAVENGE PUMP

HIGH PRESSURE SWITCH

AIR BY-PASS

FUEL TANK

FUEL BY-PASS

IGNITION SWITCH

STARTER MOTOR

IGNITER

SPEED CONTROL SWITCH

EXHAUST

Fuel

Air

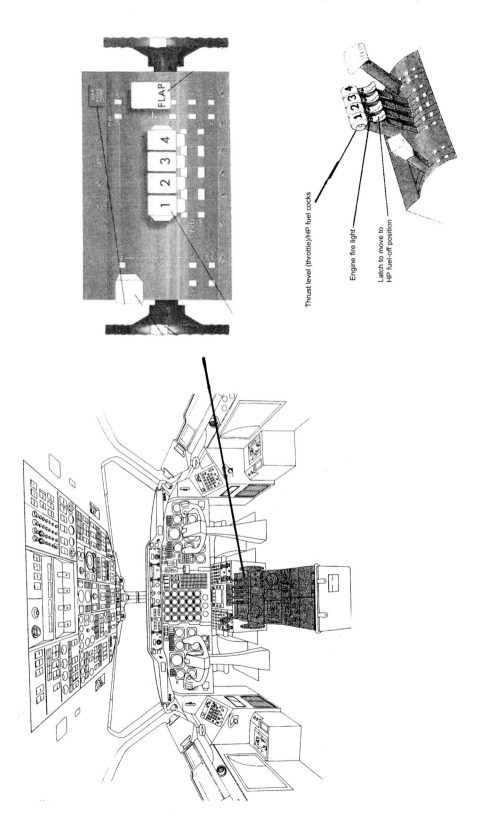

Thrust level (throttle)/HP fuel cocks

Engine fire light

Latch to move to
HP fuel-off position

FLAP

Fig. 2.14 *The 146
throttle levers and HP
cock latches (BAE
SYSTEMS)*

Together with status and warning lights to indicate 'start in progress', 'failed start' and 'engine fire' the pilot is provided with information on indicators of engine speeds, temperatures and pressures that he can use to monitor the engine start cycle.

In many modern aircraft the start cycle is automated so that the pilot has only to select START for the complete sequence to be conducted with no further intervention.

Engine indications

Despite the fact that engine control systems have become very comprehensive in maintaining operating conditions at the most economic or highest performance, depending on the application, there is still a need to provide the pilot with an indication of certain engine parameters.

Under normal conditions the pilot is interested in engine condition only at the start and when something goes wrong. The engine control system, with its monitoring and warning capability, should inform the pilot when something untoward does happen. However, there may be circumstances when human intuition wins the day.

During engine start the pilot monitors (and checks with his co-pilot in a multi-crew aircraft) that start progresses satisfactorily with no observed sluggish accelerations, no low oil pressures or over-temperatures. Much of this monitoring involves pilot familiarity with the aircraft type and engine type, incurred over many starts. The crew may accept certain criteria that an automatic system would not.

During normal operation the control system should provide sufficient high integrity observation by self-monitoring and by checking certain parameters against pre-set values. In this way the system can monitor accelerations, rates of change, value exceedance and changes of state and issue the necessary warning.

Control systems are good at detecting sudden changes of level or state. However, slow, gradual but persistent drift and transient or intermittent changes of state are a designer's nightmare. The first may be due to degradation in performance of a component, e.g. a component becoming temperature sensitive, a gradually blocking filter or the partial occlusion of a pipe or duct. The second may be due to a loose connection somewhere in the system.

The pilot can observe the effects of these circumstances. In a four-engine aircraft, for example, one indicator reading differently to three others can be easily seen because the indicators are grouped with just such a purpose in mind.

Until recently all aircraft had at least one panel dedicated to engine instruments. These were in view at all times and took the form of circular pointer instruments, or occasionally vertical strip scales, reading such parameters as:

- Engine speed – NH and NL
- Engine temperature
- Pressure ratio
- Engine vibration
- Thrust (or torque)

In modern aircraft cockpits the individual indicator has largely given way to the visual display unit (VDU). With a VDU any information can be shown in any format, in full colour, at any time. This facility is often exploited to ensure that the pilot is only given the information that is essential for a particular phase of flight. This means that engine displays may occur on a single screen or page that is automatically presented to the pilot at certain times, say starting, take-off and landing, but is hidden at all other times. Provision is made for the pilot to select any page so that he can check from time to time, and an engine warning may automatically trigger the engine page to appear.

Engine indications are obtained from the same type of sensors and transducers that provide the inputs to the control system, as described earlier. However, for integrity reasons at least two sources of signal are required – one (or more) for control, another for the indicator. For example the engine rpm signal will be obtained from two separate coils of a speed sensor. This guards against a common mode failure that would otherwise affect both the control system and the indication system.

Engine control on a modern civil aircraft

A typical civil engine is shown in Fig. 2.15. Most are twin-shaft engines with LP and HP shafts. Some Rolls-Royce engines such as the RB211 and Trent family are triple-shaft engines with LP, IP and HP shafts. A high proportion of air by-passes the engine core on a modern gas turbine engine; the ratio of by-pass air to engine core air is called the by-pass ratio. The by-pass ratio for most civil engines is in the ratio of 4:1 to 5:1.

Most modern civil engines use a Full-Authority Digital Engine Control System (FADEC), mounted on the fan casing to perform all the functions of powerplant management and control. A highly simplified diagram showing all the functions to be performed on the aircraft's large, high by-pass engines is illustrated.

The key areas of monitoring and control are:

- Various speed probes (N1, N2); temperature and pressure sensors (P2/T2, P2.5/T2.5, and T3); Exhaust Gas Temperature (EGT) and oil temperature and pressure sensors are shown.
- The turbine case cooling loops – high pressure (HP) and low pressure (LP).
- Engine start.
- Fuel control for control of engine speed and, therefore, thrust.

Fig. 2.15 *Typical civil engine*

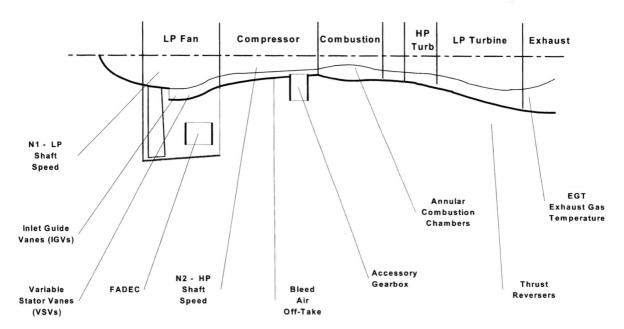

- The engine Permanent Magnet Alternators (PMAs) are small dedicated generators that supply primary power on the engine for critical control functions.
- Various turbine blade cooling, Inlet Guide Vanes (IGVs), Variable Stator Vanes (VSVs) and bleed air controls.

The engine supplies bleed air for a variety of functions as described in Chapter 6, Pneumatic Systems. Bleed air provides the actuator motive power for some of the controls on the engine as well as supplying medium-pressure air to the airframe for a variety of functions such as anti-icing, cabin pressurization, and cabin temperature control among other functions.

An idea of the complexity of other engine off-takes may be gathered from Fig. 2.16 which shows a typical engine accessory gearbox.

It can be seen that many of the drives off the accessory gearbox are for the use of the engine:

- LP and HP fuel pumps
- Oil scavenge pumps; oil is used to cool the electrical generator as well as lubricate the engine
- PMAs to supply 28 VDC power for the dual channel FADEC
- Oil breather

Interfaces with the aircraft include:

- Supply of 3-phase 115 VAC, 400 Hz electrical power – rated in the range from 40 to 90 kVA per channel on most civil transport aircraft; 120 kVA per channel on B777 and B767-400
- Supply of 3,000 psi hydraulic power
- Engine tachometer and other engine indications
- Input of bleed air from a suitable air source to start the engine. This can be a ground power cart, the APU or air from the other engine if that has already been started

An important feature of commercial aircraft operations is the increasing use of two-engine aircraft flying Extended Range Twin Operations (ETOPS) routes across trans-oceanic or wilderness terrain. The majority of trans-Atlantic flights today are ETOPS operations. The integrity of the engines and related systems is clearly vital for these operations and the engine In-Flight Shut-Down (IFSD) rate is central to determining whether 120 min or 180 min ETOPS approval may be granted by the certification authorities. Reference (5) is consulted for ETOPS clearance. It mandates that the engine IFSD needed for ETOPS approval is < 50 per million flight hours and < 20 per million flight hours for 120 min and 180 min respectively and the actual rate achieved in service is well below these minima.

Recently efforts have been made by Boeing to extend this to 208 min to take full account of the extended range of later versions of the Boeing 777.

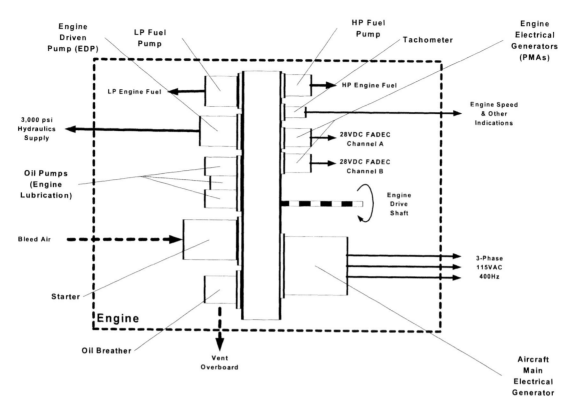

Fig. 2.16 Typical engine accessory gearbox

References

(1)   The Rolls-Royce Book of the Jet Engine, Rolls-Royce Ltd.

(2)   **McNamara, J., Legge, C.J.** and **Roberts, E.** (1979) Experimental full-authority digital engine control on Concorde, AGARDConference on *Advanced Control Systems for Aircraft Powerplants*, (CP 274) October.

(3)   **McNamara, J.** and **Seabridge, A.G.** (1982) Integrated aircraft avionics and powerplant control and management systems. ASME International Gas Turbine Conference, April.

(4)   FAA Advisory Circular AC 120-42A.

CHAPTER 3

Fuel Systems

Introduction

At the onset of aviation aircraft fuel systems were remarkably simple affairs. Fuel was gravity fed to the engine in most cases though higher performance engines would have an engine-mounted fuel pump. Tank configurations were extremely simple and fuel contents were by means of float driven indications.

Higher performance gave rise to more complexity within the fuel system. The need for transfer and booster pumps accompanied the arrival of high-performance aircraft. More complex tank configurations introduced the need for multi-valve systems such that the flight crew could move fuel around the fuel tanks according to the needs at the time.

The arrival of jet turbine powered aircraft brought a range of engines that were much thirstier than their piston-engined predecessors: the early jet aircraft in general had a very short sortie length. More accurate fuel gauging systems were required to give the pilot advanced and accurate information regarding the aircraft fuel state in order that recovery to an airfield could be accomplished before running out of fuel. The higher performance jet engine also required considerably greater fuel delivery pressures to avoid cavitation and flame-out.

A further effect of the high fuel consumption was the use of under-wing or under-fuselage ventral tanks to enhance the range of the aircraft. These additional tanks further complicated the fuel system and tank pressurization systems were developed to transfer the external fuel to the aircraft internal tanks. These systems brought the requirement for further valves to control tank pressurization and ensure that the tanks could not be damaged by excessive pressure.

Fuel gauging systems became more complex as greater gauging accuracies were sought and achieved. Most systems are based upon capacitance measurement of the fuel level within the aircraft, using fuel probes placed at various locations within the fuel tanks. A large system may require some 30 or 40 probes or more to measure the contents accurately. Typical figures for the airliners of today are in the region of 1–2 per cent accuracy, depending upon the sophistication of the systems, some of which can compensate for fuel temperature and density, aircraft attitude, fuel height and a variety of other variables.

Although not a new concept, the development of in-flight refuelling techniques has further extended the range of military aircraft and enhanced the flexibility of air power leading to a 'force-multiplier' effect. Military actions in the Falklands in the early 1980s and in the Persian Gulf in 1991 have underlined the vital nature of in-flight refuelling, (see Figs 3.1 and 3.2) and not just for fighter aircraft. In-flight refuelling has also been used to speed the pace of development programmes, especially in the US where the B-2, YF-22A and YF-23A flight test programmes all used the technique to extend sortie length soon after first flight.

Fig. 3.1 Handley Page W 10 tanker refuelling Sir Alan Cobham's Airspeed Courier in October 1934. S/Ldr W. Helmore had the draughty task of handling the fuel hose (Flight Refuelling Ltd)

Modern aircraft fuel management and gauging systems are based upon a plethora of valves, pumps, probes, level sensors, switches etc. controlled by microprocessor based systems. This has led to more capable and more reliable systems needed for the aircraft to meet the exacting demands placed upon them.

Characteristics of aircraft fuel systems

Fig. 3.2 Tornado GR1s refuelling from a Vickers VC 10 tanker during the 1991 Gulf War (BAE SYSTEMS)

The purpose of an aircraft fuel system is primarily to provide a reliable supply of fuel to the engines. Without the motive power provided by them the aircraft is unable to sustain flight. Therefore the fuel system is an essential element in the overall suite of systems required to assure safe flight. Modern aircraft fuels are hydrocarbon fuels similar to those used in the automobile. Piston-engined aircraft use a higher octane fuel called AVGAS in aviation parlance. Jet engines use a cruder fuel with a wider distillation cut and with a lower flashpoint. AVTAG and AVTUR are typical jet engine fuels. The specific gravity of aviation fuels is around 0.8, that is about eight-tenths of the density of water. Therefore fuel may be quantified by reference to either volume (gallons or litres) or weight (pounds or kilograms). As the density of fuel varies according to temperature both may be used. The volume of an aircraft fuel tankage is fixed and therefore it will not be able to accommodate the same weight of fuel at high temperature when the fuel density is lower. For most practical purposes a gallon of fuel may be assumed to weigh around 8 lb (as opposed to 10 lb for a gallon of water).

The essential characteristics of a modern aircraft fuel management system may embrace some or all of the following modes of operation:

- Fuel pressurization
- Engine feed
- Fuel transfer
- Refuel/defuel

- Vent systems
- Use of fuel as heat sink
- Fuel jettison
- In-flight refuelling

Before describing the operation of these typical modes of operation it is worth examining one and outlining the primary components that comprise such a system.

Descriptions of fuel system components

Fuel transfer pumps

Fuel transfer pumps perform the task of transferring fuel between the aircraft fuel tanks to ensure that the engine fuel feed requirement is satisfied. On a fighter this will require the supply of fuel to collector tanks which carry out the obvious task of collecting or consolidating fuel before engine feed. Transfer pumps may also be required to transfer fuel around the aircraft to maintain pitch or lateral trim. In the case of pitch trim this requirement is becoming more critical for unstable control configured aircraft where the task of active CG control may be placed upon the fuel management system. Similarly on civil aircraft there is a requirement to transfer fuel from wing tanks to the fuselage centre tank where fuel may typically be consolidated before engine feed. There are FAA/JAA regulations which require independent engine feed systems. On more recent civil aircraft such as the Airbus A340 the horizontal stabilizer may contain fuel which has to be transferred to maintain the aircraft CG within acceptable limits. However older aircraft such as the Vickers VC10 also contain fuel in the empennage, in this case

Fig. 3.3 *Jaguar fuel transfer pump (BAE SYSTEMS)*

the fin, to increase fuel capacity. In these cases gravity feed or pumps are also required to transfer fuel forward to a centre tank for consolidation. A typical aircraft system will have a number of transfer pumps for the purposes of redundancy, as will be seen in the examples given later in this chapter.

An example of a fuel transfer pump is shown in Fig. 3.3, this particular example being used on the Anglo-French Jaguar fighter. This is a fuel-lubricated pump; a feature shared by most aircraft fuel pumps. The pump has the capability of safely running dry in the event that no fuel should remain in the tank for any reason. Thermal protection is also incorporated to prevent overheating. This particular pump is designed to supply in the region of 400 lb/min at a pressure of 10 psi.

Fuel booster pumps

Fuel booster pumps, sometimes called engine feed pumps, are used to boost the fuel flow from the aircraft fuel system to the engine. One of the reasons for this is to prevent aeration (i.e. air in the fuel lines that could cause an engine 'flame-out' with consequent loss of power). Another reason in the case of military aircraft is to prevent 'cavitation' at high altitudes. Cavitation is a process in which the combination of high altitude, relatively high fuel temperature and high engine demand produce a set of circumstances where the fuel is inclined to vaporize. Vaporization is a result of the combination of low fuel vapour pressure and high temperature. The effect is drastically to reduce the flow of fuel to the engine which can cause a flame-out in the same way as aeration (as may be caused by air in the fuel).

Booster pumps are usually electrically driven; for smaller aircraft such as the Jet Provost and the Harrier the pump is driven from the aircraft 28 VDC system with delivery pressures in the range 10–15 psi and flow rates up to 2.5 kg/sec of fuel. The booster pumps of larger, high-performance aircraft with higher fuel consumption are powered by three-phase AC motors; in the case of Tornado delivering 5 kg/sec. Booster pumps are cooled and lubricated by the fuel in which they are located in a similar way

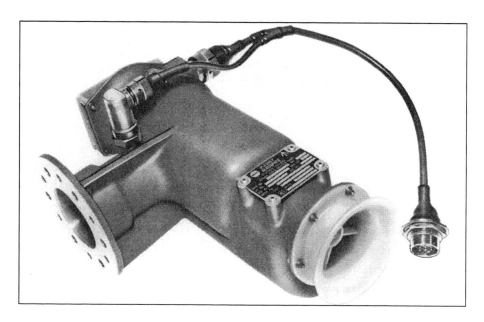

Fig. 3.4 *Tornado double-ended booster pump (BAE SYSTEMS)*

to transfer pumps, and may be specified to run for several hours in a 'dry' environment. Fuel pumps can also be hydraulically driven or, in certain cases, ram air turbine driven, such as the VC10 tanker in-flight re-fuelling pump. The example of a booster pump shown in Fig. 3.4 is the double-ended pump used in the Tornado to provide uninterrupted fuel supply during normal and inverted flight/negative-g manoeuvres.

Fuel transfer valves

A variety of fuel valves will typically be utilized in an aircraft fuel system. Shut-off valves perform the obvious function of shutting off fuel flow when required. This might involve stemming the flow of fuel to an engine, or it may involve the prevention of fuel transfer from one tank to another. Refuel/defuel valves are used during aircraft fuel replenishment to allow flow from the refuelling gallery to the fuel tanks. These valves will be controlled so that they shut off once the desired fuel load has been taken on board. Similarly, during defuelling the valves will be used so that the load may be reduced to the desired level. Cross-feed valves are used when the fuel is required to be fed from one side of the aircraft to the other.

Fuel dump valves perform the critical function of dumping excess fuel from the aircraft tanks in an emergency. These valves are critical in operation in the sense that they are required to operate and dump fuel to reduce the fuel contents to the required

Fig. 3.5 Transfer valve driven by a rotary actuator (High Temp Engineers)

levels during an in-flight emergency. Conversely, the valves are not required to operate and inadvertently dump fuel during normal flight.

The majority of the functions described are performed by motorized valves that are driven from position to position by small electric motors. Other valves with a discrete on/off function may be switched by electrically operated solenoids. Figure 3.5 shows an example of a transfer valve driven by a DC powered rotary actuator. An actuator of this type may be two-position (90 degrees) or three-position (270 degrees) or continually modulating over 90 degrees.

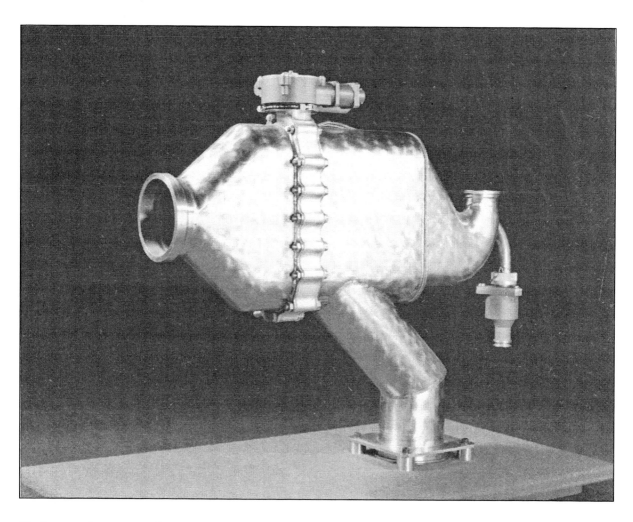

Fuel vent valves are used to vent the aircraft fuel tanks of air during the refuelling process; they may also be used to vent excess fuel from the tanks in flight. An example of such a valve is shown in Fig. 3.6. This valve permits inward or outward venting of around 20–25 lb of air per minute during flight/pressure refuelling as appropriate. The valve also permits venting of fuel (in the event of a refuelling valve failing to shut off) of about 800 lb/min or 100 gal/min.

Fig. 3.6 *Typical fuel vent valve (High Temp Engineers)*

Non-return valves (NRVs)

A variety of non-return valves or check valves are required in an aircraft fuel system to preserve the fluid logic of the system. Non-return valves as the name suggests prevent the flow of fuel in the reverse sense. The use of non-return valves together with the various transfer and shut-off valves utilized around the system ensure correct system operation in the system modes listed above and which will be described in more detail later in the chapter.

Fuel quantity measurement

Level sensors

Level sensors measure the fuel level in a particular tank and thereby influence fuel management system decisions. Level sensors are used to prevent fuel tank overfill during refuelling. Level sensors are also used for the critical low-level sensing and display function to ensure that fuel levels do not drop below flight critical levels where the aircraft has insufficient fuel to return to a suitable airfield. Level sensors may be one of a number of types: Float-operated; optical; ultrasonic; or Zener diode – two of which are described below.

Float level sensors

Float level sensors act in a similar way to a domestic toilet cistern connected to the water supply shut-off valve that is closed as the float rises. The refuelling valve, operating in the same way, is a simple but effective way of measuring the fuel level but it has the disadvantage that, having moving parts, the float arm may stick or jam.

Zener diode level sensors

By using simple solid state techniques it is possible to determine fluid levels accurately. The principle is based upon a positive temperature coefficient directly heated Zener diode. The response time when sensing from air to liquid is less than 2 sec (refuelling valve) and from liquid to air less than 7 sec (low level warning). Fluid level may be sensed to an accuracy of about plus/minus 2 mm and the power required is around 27 mA per channel at 28 VDC. The sensor operates in conjunction with an amplifier within a control unit and can accommodate multi-channel requirements. A typical fluid sensor of this type is shown in Fig. 3.7. The advantage of this method of level sensing is accuracy and the fact that there are no moving parts.

Fuel gauging probes

Many of the aircraft functions relating to fuel are concerned with the measurement of fuel quantity on board the aircraft. For example, the attainment of a particular fuel level could result in a number of differing actions depending upon the circumstances: opening or closing fuel valves or turning on/off fuel pumps in order to achieve the desired system state. Quantity measurement is usually accomplished by a number of probes based upon the principle of fuel capacitance measurement at various locations throughout the tanks. Air and fuel have different dielectric values and by measuring the capacitance of a probe the fuel level may be inferred. These locations of the fuel probes are carefully chosen such that the effects of aircraft pitch and roll attitude changes are minimized as far as

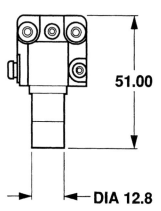

51.00

DIA 12.8

Fig. 3.7 *Solid-state level sensor (Smiths Industries)*

quantity measurement is concerned. Additional probes may cater for differences in fuel density and permittivity when uplifting fuel at differing airfields around the world as well as for fuel at different temperatures. Fuel gauging, or fuel quantity indication systems (FQIS) as they are sometimes known, are therefore an essential element in providing the flight and ground crews with adequate information relating to the amount of fuel contained within the aircraft tanks.

Fuel gauging probes are concentric cylindrical tubes with a diameter of about 1 in. Despite experiments with glass-fibre probes, metal ones have been found to be the most reliable for minimum weight. Plastic, non-conducting cross-pins maintain the concentricity of the tubes while providing the necessary electrical insulation. Tank units may be either internally or externally mounted on straight or angled flanges, for both rigid and flexible tanks.

A number of factors may affect fuel measurement accuracy.

- Tank geometry. The optimum number of probes for a given tank is established by means of computerized techniques to model the tank and probe geometry. Each probe may then be 'characterized' to achieve a linear characteristic of the gauging system. This may be done by mechanical profiling to account for tank shape and provide a linear output. This is an expensive and repetitive manufacturing process which may be more effectively achieved by using 'linear' probes with the correction being derived in computer software for some of the more advanced microprocessor driven fuel gauging systems.
- Permittivity variations. Variations in the permittivity of the fuel may adversely affect gauging accuracy. Reference units may be used to compensate for the varying temperature within the fuel. These may be separate stand-alone units or may be incorporated into the probe itself.

Examples of particular tank probes are shown in Fig. 3.8.

Fig. 3.8 *Examples of fuel probe units (Smiths Industries)*

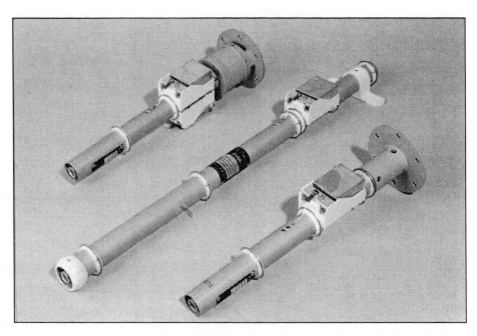

Fuel quantity measurement systems

Fuel quantity measurement systems using capacitance probes of the type already described may be implemented in one of two ways. These relate to the signalling techniques used to convey the fuel tank capacitance (and therefore tank contents) to the fuel indicator or computer:

(1) AC system.
(2) DC system.

AC systems

In an AC system the tank unit information is conveyed by means of an AC voltage modulated by the measured tank capacitance and therefore fuel quantity. The problem with the AC signalling technique is that there is a greater risk of electromagnetic interference (EMI) so that coaxial cables and connectors are required making the installation more complex, expensive and difficult to maintain. Therefore although individual AC tank units may be lighter, cheaper and more reliable (being simpler in construction) than the DC tank unit equivalent, the overall system penalties in terms of weight and cost may be greater.

DC systems

In the DC system the probes are fed by a constant voltage/frequency probe drive and utilize automatic fuel probe diode temperature compensation. Fuel probe signals are rectified by the diodes and the resulting signal proportional to fuel contents returned to the processor as a DC analogue signal. The more complex coaxial cables and connectors of the AC system are not required. The overall system weight and cost of the DC system is therefore usually less than an AC system, overall system reliability is usually better than for the DC system. There is an increasing tendency for modern systems to adopt the DC system due to the inherent benefits. A disadvantage of a DC system is the need for additional components within the fuel tank.

In reality the choice between AC and DC systems will be heavily biased by the experience accrued by a specific airframe manufacturer.

DC fuel gauging system examples – Fokker F50/F100 and Airbus

Two examples of DC systems which have recently entered service are the systems used on board the Fokker F50/F100 and the Airbus A320.

Fokker F50/F100

The diagrammatic layout of this system and the system architecture are shown in Figs 3.9(a) and 3.9(b) respectively.

Data from the DC fuel probes in the wing and fuselage tanks are summed and conditioned in the Combined Processor Totaliser (CPT) and fed to the fuel indicator portion of the unit. Dual 8-bit microprocessors process the information into serial digital form for transmission on ARINC 429 data buses to the Total Contents Display (TCD) in the cockpit and the Fuel Control Panel (FCP) in the right wing root. The system displays individual tank contents to the crew. The FCP enables the aircraft to be automatically refuelled to preset fuel quantities without operator intervention. The

Fig. 3.9(a) Fokker
100 diagrammatic
layout (Smiths
Industries)

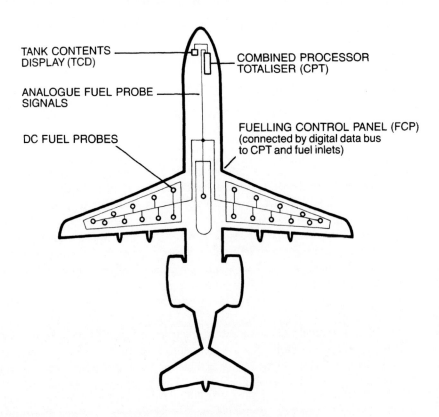

TANK CONTENTS
DISPLAY (TCD)

ANALOGUE FUEL PROBE
SIGNALS

DC FUEL PROBES

COMBINED PROCESSOR
TOTALISER (CPT)

FUELLING CONTROL PANEL (FCP)
(connected by digital data bus
to CPT and fuel inlets)

Fig. 3.9(b) Fokker
F50/F100 system
architecture (Smiths
Industries)

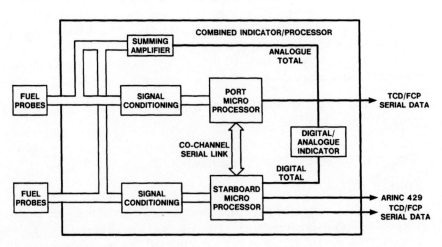

COMBINED INDICATOR/PROCESSOR

SUMMING
AMPLIFIER

ANALOGUE
TOTAL

FUEL
PROBES

SIGNAL
CONDITIONING

PORT
MICRO
PROCESSOR

TCD/FCP
SERIAL DATA

DIGITAL/
ANALOGUE
INDICATOR

CO-CHANNEL
SERIAL LINK

DIGITAL
TOTAL

FUEL
PROBES

SIGNAL
CONDITIONING

STARBOARD
MICRO
PROCESSOR

ARINC 429
TCD/FCP
SERIAL DATA

accuracy of this type of system is of the order of 2 per cent. The system is designed so that no single failure will cause total loss of all fuel gauging information.

Airbus A320

The DC fuel system used on the Airbus A320 is shown in Figs 3.10(a) and 3.10(b).

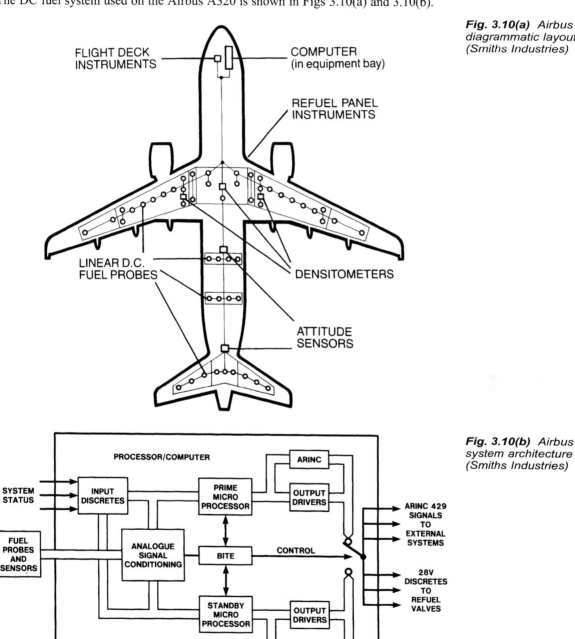

Fig. 3.10(a) *Airbus diagrammatic layout (Smiths Industries)*

Fig. 3.10(b) *Airbus system architecture (Smiths Industries)*

The A320 example is more complex than the Fokker F50/F100 system. Linear DC probes are located in the two wing tanks, and three fuselage tanks; later models such as the A340 will also have a tank located in the rear fuselage or tailplane. Densitometers are fitted in the wing and centre fuselage tanks. The system also uses attitude data supplied by the aircraft systems. The system is based upon a dual redundant computer architecture using Motorola 68,000 microprocessors: each processor handles identical data and in the event of one processor failing the other automatically takes over the computation tasks without any loss of continuity. The system is designed to fail with 'graceful degradation', that is to degrade gently in accuracy while informing the crew.

In this system data relating to the tank geometry is stored in memory together with the computed fuel density, permittivity, fuel temperatures, aircraft attitude and other relevant aircraft information. The computers then use various algorithms to calculate the true mass of fuel. Multiple ARINC 429 serial data buses provide data to the flight management computer and the various displays. In this system discrete signal outputs are used to control the operation of refuelling valves or transfer valves. The overall accuracy of this system is in the order of 1 per cent.

Further information regarding these systems is given in references (1) and (2).

'Smart' probes

A further variation on the theme of capacitance probes is the 'smart' probe used on the Eurofighter and Nimrod aircraft. The probes are active or 'smart' in that each probe has dedicated electronics associated with the probe. Each is supplied with a regulated and protected DC voltage supply to power the local electronics. The local electronics process the capacitance value to produce a pulse train, the period of which is proportional to the capacitance sensed and therefore the fuel level measured by the probe. The benefit of this type of system is to provide a means of reducing the EMI susceptibility of the fuel probe transmission system. Twisted, screened three-wire signal lines are used which are simpler than coaxial cables but nonetheless expensive in wiring terms. A disadvantage is the need to provide electronics for each individual probe in a relatively hostile environment within the airframe.

Ultrasonic probes

All of the above systems use capacitive measurement techniques to sense fuel level. Ultrasonic techniques are now being developed which utilize ultrasonic transducers to measure fuel level instead of the conventional capacitive means. The sensor is located at the bottom of the waveguide. The waveguide arrangement at the base of the tank directs the ultrasonic transmission back to the transducer. To measure height with ultrasonics the speed of sound in the fuel medium is required. This is generally measured using a fixed reference in the waveguide. A portion of the ultrasonic wave is reflected directly back to the transducer and serves as a reference signal. The time taken for the signal to be reflected back from the fuel surface is measured and by using a simple ratiometric calculation the fuel height may be determined. Fuel level may be measured by comparing the time of propagation for the reference signal with that for the fuel level reflected signal. This type of quantity measuring system was introduced on the Boeing 777 airliner which first flew on revenue service in June 1995 and of which there are over two hundred examples in service today.

Fuel system operating modes

The modes of operation described in the following paragraphs are typical of many aircraft fuel systems. Each is described as an example in a particular fuel system. Any system may exhibit many but probably not all of these modes. In an aircraft the fuel tanks and components have to compete with other systems, notably structure and engines for the useful volume contained within the aircraft profile. Therefore fuel tanks are irregular shapes and the layman would be surprised by how many tanks there are, particularly within the fuselage where competition for usable volume is more fierce. The proliferation of tanks increases the complexity of the interconnecting pipes and certainly does not ease the task of accurate fuel measurement. As an example of a typical fighter aircraft fuel tank configuration see Fig. 3.11 which shows the internal fuel tank configuration for EAP.

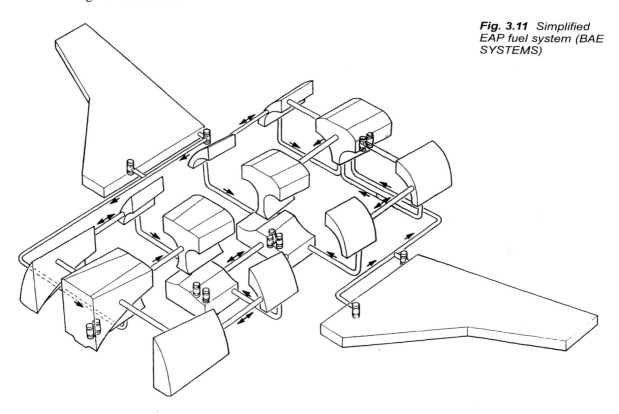

Fig. 3.11 *Simplified EAP fuel system (BAE SYSTEMS)*

This is a simplified diagram showing only the main fuel transfer lines; refuelling and vent lines have been omitted for clarity. Whereas the wing fuel tanks are fairly straightforward in shape, the fuselage tanks are more numerous and of more complex geometry than might be supposed. The segregation of fuel tanks into smaller tanks longitudinally (fore and aft) is due to the need to avoid aircraft structural members. The shape of most of the fuselage tanks also shows clearly the impositions caused by the engine intakes. Furthermore as an experimental aircraft EAP was not equipped for in-flight refuelling nor were any external under-wing or ventral tanks fitted. It can be seen that a fully operational fighter would have a correspondingly more complicated fuel system than the one shown.

Fuel pressurization

Fuel pressurization is sometimes required to assist in forcing the fuel under relatively low pressure from certain tanks to others that are more strategically placed within the system. On some aircraft there may be no need for a pressurization system at all; it may be sufficient to gravity feed the fuel or rely on transfer pumps to move it around the system. On other aircraft ram air pressure may be utilized to give a low but positive pressure differential. Some fighter aircraft have a dedicated pressurization system using high-pressure air derived from the engine bleed system.

The engine bleed air pressure in this case would be reduced by means of pressure-reducing valves (PRVs) to a more acceptable level. For a combat aircraft which may have a number of external fuel tanks fitted the relative regulating pressure settings of the PRVs may be used to effectively sequence the transfer of fuel from the external and internal tanks in the desired manner. For example, on an aircraft fitted with under-wing and under-fuselage (ventral) tanks it may be required to feed from under-wing, then the ventral and finally the internal wing/fuselage tanks. The PRVs may be set to ensure that this sequence is preserved, by applying a higher differential pressure to those tanks required to transfer fuel first.

In some aircraft such as the F-22, inert gas is used to pressurize the fuel tanks. Inert gas for this purpose can be obtained from an on-board inert gas generating system (OBIGGS).

Engine feed

The supply of fuel to the engines is by far the most critical element of the fuel system. Fuel is usually collected or consolidated before being fed into the engine feed lines. The example in Fig. 3.12 shows a typical combat aircraft, the fuel is consolidated in two collector tanks; one for each engine. This schematic diagram may be reconciled with the Experimental Aircraft Programme (EAP) example depicted in Fig. 3.11. The fuel transfer from the aircraft fuel tanks into the collector tanks is fully described in the fuel transfer section.

The collector tanks may hold sufficient fuel for several minutes of flying, depending upon the engine throttle settings at the time. The contents of these tanks will be gauged as part of the overall fuel contents measuring system. However, due to the criticality to the engine of the engine fuel feed function additional measurement sensors are added. It is usual to provide low-level sensors (shown as ▼) that measure and indicate when the collector tanks are almost empty. These low-level sensors generate critical warnings to inform the pilot that he is about to run out of fuel and that the engine will subsequently flame-out. The low-level warnings are a last ditch indication that the pilot should be preparing to evacuate the aircraft if he is not already doing so.

The collector tanks contain the booster pumps (shown as Ⓑ) that are pressurizing the flow of fuel to the engines. It is usual for two booster pumps to be provided so that one is always available in the event that the other should fail. Booster pumps are immersed in the fuel and for a combat aircraft the scavenge pipes feeding fuel to the pump inlets will have a provision such that a feed is maintained during inverted or negative-g flight. Note that the booster pump example shown in Fig. 3.4 had such a facility. Booster pumps are usually powered by 115 VAC three-phase motors of the type described in Chapter 5, Electrical Systems. However the motor itself is controlled by a three-phase relay, the relay coil being energized by a 28 VDC supply. An auxiliary

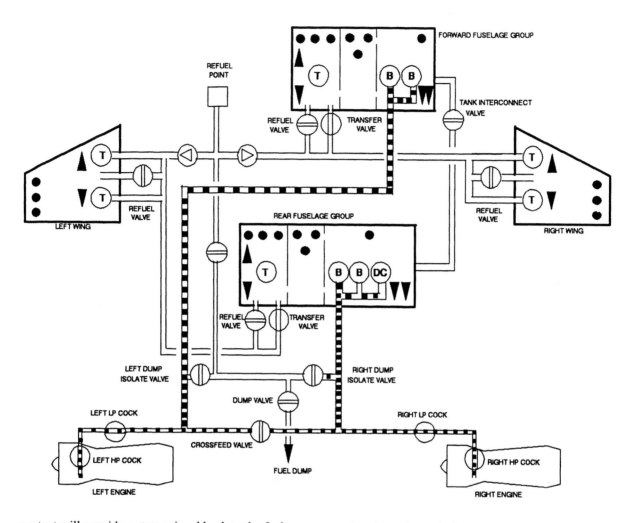

Fig. 3.12 *Typical fighter aircraft engine feed*

contact will provide a status signal back to the fuel management system, alternatively a pressure switch or measuring sensor may be located in the delivery outlet of the pump which can indicate that the pump is supplying normal delivery pressure. Booster pumps are fuel lubricated and also have the capability of running dry should that be necessary.

Downstream of the booster pump is the engine high-pressure (HP) pump which is driven by the engine accessory gearbox. Engine HP pumps are two-stage pumps; the first stage provides pressure to pass the fuel through heat exchangers and filters and to provide a positive inlet pressure to the second stage. The second stage supplies high pressure fuel (around 1,500 to 2,000 psi) to the engine fuel control system.

A number of shut-off valves are associated with the control of fuel to the engine. A pilot-operated low-pressure (LP) cock provides the means of isolating the fuel supply between the booster pump and the HP engine-driven pump. This valve may also be associated with a firewall shut-off function which isolates the supply of fuel to the engine compartment in the event of an engine fire. A cross-feed valve located upstream of the LP cocks provides the capability of feeding both engines from one collector tank if necessary; in most cases the cross-feed valve would be closed as shown in Fig. 3.12. The pilot may also operate a high-pressure (HP) cock that has the ability to isolate the fuel supply on the engine itself. In normal operation both the LP and HP cocks are open allowing an unimpeded supply of fuel to the engine. The cocks are only closed in the case of normal engine shut-down or in flight following an engine fire.

Fuel transfer

The task of fuel transfer is to move fuel from the main wing and fuselage tanks to the collector tanks. In commercial transport there tend to be fewer tanks of more regular shape and transfer pumps may merely be used for redistributing fuel around the tanks. In the example given in Fig. 3.13 the fuselage and wing tanks for the EAP are shown. The main tankage comprises left and right wing tanks and forward and rear fuselage tanks.

Two transfer pumps (shown as ⓣ) are provided in each wing tank and two in each of the fuselage groups. Transfer pumps are usually activated by the level of fuel in the tank that they supply. Once the fuel has reached a certain level measured by the fuel gauging system, or possibly by the use of level sensors, the pumps will run and transfer fuel until the tank level is restored to the desired level. In the EAP this means that the

Fig. 3.13 EAP fuel transfer operation

forward and rear groups are replenished from the left and right wing tanks respectively in normal operation. The fuselage groups in turn top up the collector tanks with the aid

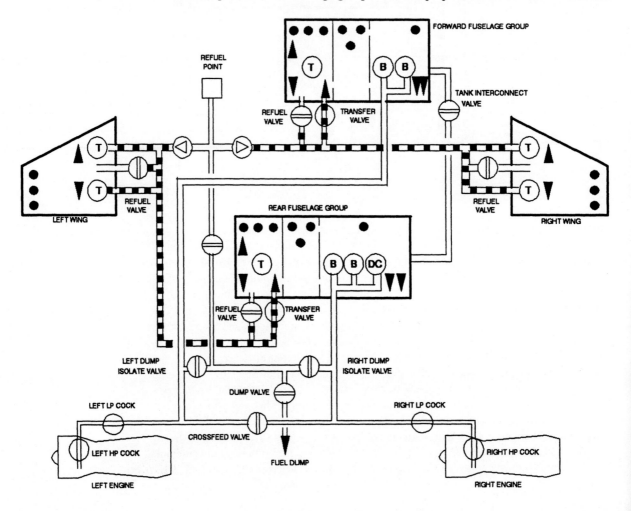

of further transfer pumps. The tank interconnect valve also provides for fuel crossfeed from one fuel system (left/forward) to the other (right/rear) which allows fuel to be balanced between left and right or permits one system to feed both engines if the need arises. Transfer pumps operate in a similar fashion to booster pumps; they are also electrically operated by 115 VAC 3-phase electrical power driving an induction motor. The duty cycle of the transfer pumps is not continuous like the booster pumps, rather their operation is a periodic on-off cycle as they are required to top up the relevant aircraft tanks subject to fuel demand.

Refuel/Defuel

Aircraft refuelling and defuelling is controlled by a separate subsystem within the overall fuel system. Refer to Fig. 3.14. The aircraft is fuelled by means of a refuelling receptacle that connects to the refuelling tanker. From the receptacle it enters a refuelling gallery which distributes the incoming fuel to the various aircraft tanks. The control of fuel entry into each tank is undertaken by valves that are under the control of the fuel management system. In the crudest sense fuel will enter the tanks until they are full, whereupon the refuelling valve will be shut off preventing the entry of any more.

Fig. 3.14 *Refuelling operation schematic*

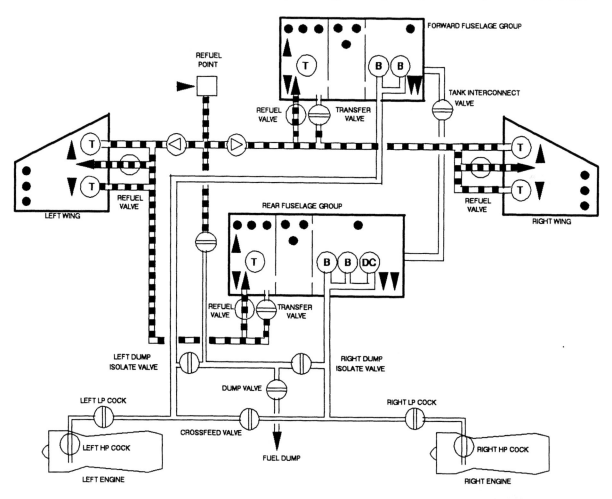

In a very simple system this shut-off may be accomplished by means of a simple float-operated mechanical valve. In more sophisticated systems the fuel management system has control over the operation of the refuelling valve, usually by electrical means such as a solenoid operated or motorized valve. A typical system may comprise a mixture of both types. In most cases the aircraft is not filled to capacity, rather the maintenance crew select a fuel load and set the appropriate levels at the refuel/defuel panel adjacent to the refuelling receptacle – often located under the aircraft wing in an accessible position.

The defuelling process is almost the reverse of that for refuelling. It may be necessary to defuel the aircraft for maintenance reasons. In general defuelling is carried out relatively infrequently compared to refuelling.

In some simpler aircraft it is possible to carry out over-wing refuelling. This is undertaken at remote airstrips where there may not be any dedicated refuelling machinery such as a fuel bowser and the fuel is provided in drums. In this situation an over-wing panel is removed and fuel is poured manually into the wing tanks.

Certain aircraft, usually commuter and commercial types, have devices called magnetic level indicators (MLIs) which are equivalent to a fluid level dipstick. The MLIs are mounted under the wing and when a simple catch is released the indicator drops until the upper portion is level with the fuel surface. The extended portion of the MLI is graduated so that the amount by which the device extends can be measured. And hence the level of fuel in the tanks can be deduced and cross-checked with the level indicated by the aircraft fuel gauges. For an example the BAE SYSTEMS ATP has a total of eight MLIs fitted, four for each wing tank

Vent systems

When an aircraft is being refuelled, or during fuel transfer, large quantities of air in the tanks can be displaced by fuel very quickly, particularly during pressure refuelling. Pressure refuelling involves a relatively high positive pressure being applied to speed the refuelling process. Typical pressures are of the order of 50 psi. With pressures of this magnitude it is possible to damage the aircraft tanks as 50 psi acting over a large area can exert a considerable force and the excess air needs to be dumped or vented overboard. The vent system may also be required to allow air into the tanks as the fuel is used though this is not true of pressurized fuel systems. The excess air is vented by means of valves fitted at the top of the fuel tanks. Vent valves separate fuel and air so that only air is vented overboard and not fuel. On a civil aircraft this function is undertaken by the vent system.

Use of fuel as a heat sink

In certain aircraft such as high performance jet fighters and Concorde the aircraft fuel performs the very important function of acting as a heat sink for heat generated within the aircraft during flight. For Concorde the kinetic heat is generated by air friction during prolonged flight at very high speeds (Mach 2) in the cruise. In the case of fighter aircraft prolonged operation at high speeds is not likely because of the punitive fuel consumption. The aircraft will generate a lot of heat, particularly from the hydraulic and environmental control system, which needs to be 'sunk' in the fuel.

External fuel tanks

Combat aircraft increase range by the use of external fuel tanks. These are usually mounted underwing but have also been belly mounted (ventral tanks) and over-wing mounted. The Lightning Mk 6 had a ventral tank fitted for normal operation and over-wing long-range ferry tanks as shown in Fig. 3.15. The ventral tank had a capacity of 609 gallons/4,872 lb while the over-wing ferry tanks had a capacity of 540 gal/4,320 lb each. This compares to the aircraft internal fuel capacity of 716 gal/5,728 lb.

Fig. 3.15 *Lightning F6 with over-wing tanks (BAE SYSTEMS)*

The Boeing F-15 Eagle fighter usually carries under-wing tanks but can also carry close-fitting ventral tanks called conformal tanks to further extend range. In this case the under-wing tanks add a capacity of 1,484 gal/11,869 lb and the conformal tanks add 1,216 gal/9,728 lb. The internal fuel capacity of the F-15 is 1,637 gal/13,094 lb. Figure 3.16 shows a F-15 with a centreline and conformal tanks fitted.

External fuel tanks have a disadvantage in that they cause significant additional drag, thereby reducing range and the benefits of the extra fuel they provide. Some fuel tanks are not stressed for supersonic flight and an aircraft operating with external tanks may be subject to a 'q' or airspeed limitation as well as a 'g' limit due to the higher weight and accompanying higher structural loading. It is common for an aircraft to jettison under-wing tanks before combat though this is clearly expensive and may cause logistic difficulties during a prolonged conflict.

Fig. 3.16 *F-15E Eagle (Boeing)*

Fuel jettison

Fuel constitutes a large portion of overall aircraft weight, particularly at the beginning of a flight. Therefore if an aircraft suffers an emergency or malfunction shortly after take-off it may prove necessary to jettison a large proportion of the fuel in order to reduce weight rapidly. This may be to reduce the aircraft weight from close to maximum All-Up Weight (AUW) to a level that is acceptable for landing; many aircraft are not stressed to land with a full fuel load. Alternatively if an engine has failed the fuel may need to be jettisoned merely to remain airborne. On an aircraft such as EAP the fuel jettison valves are tapped off from the engine feed lines with left and right jettison valves feeding fuel from the left and right engine feed lines respectively. Refer to Fig. 3.17.

A fuel jettison master valve is provided downstream to prevent inadvertent fuel jettison which could itself present a flight safety hazard. Only when both left and right and master valves are opened will fuel be jettisoned overboard.

On a civil transport fuel dumping is likely to be achieved by different means with the fuel being ejected from jettison masts situated at the rear of each wingtip. On an aircraft such as EAP the jettison valves are electrically-operated motorized valves as are many of the valves in the fuel system.

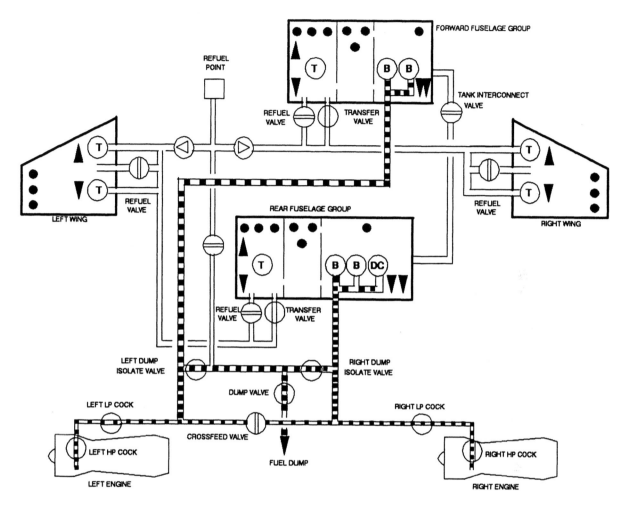

In-flight refuelling

Fig. 3.17 *EAP fuel jettison system*

For many years the principle of in-flight refuelling has been known. In fact the first demonstration of in-flight refuelling occurred in April 1934 (Fig. 3.1). Today it is an important and inherent method of operating military aircraft. The use of the principle was first widely applied to fighter aircraft because of their high rates of fuel consumption and short sortie length. However more recently, and particularly during the Falklands campaign the use of in-flight refuelling was extended to transports (Hercules and VC10), maritime patrol aircraft (Nimrod), and tankers (Tristar and VC10). The ability to refuel an aircraft in the air greatly adds to the flexibility of air power giving what is termed in military parlance a 'force multiplier' effect. In the Falklands campaign it was the sheer distance between Ascension Island and the Falklands themselves with virtually no diversions in between that required extensive use of in-flight refuelling. For fighter aircraft maintaining a combat air patrol over a specific objective the operational advantage is gained by keeping armed aircraft in the air, around the clock if necessary.

There are two methods of in-flight refuelling widely in use today. One – the probe and drogue method – is that generally favoured by the Royal Air Force. The other – the boom and receptacle – is used almost exclusively by the US military. In the former the tanker aircraft trails a refuelling hose with a large drogue attached, behind the aircraft. The recipient is fitted with a fuel probe that may be either fixed or retractable when not in use. The pilot of the receiving aircraft has the responsibility of inserting the refuelling probe into the tanker drogue. When positive pressure is exerted on the drogue by the refuelling probe fuel is able to pass to the receiving aircraft. The transfer of fuel is monitored by the tanker and by the gauging system of the recipient. Contact is broken when the receiving aircraft drops back and the positive pressure between probe and drogue is lost. At this point the refuelling operation is complete. Royal Air Force tankers usually operate with one drogue from the aircraft centreline and one from under-wing refuelling pods, so a total of three stations is available. It is possible to refuel more than one aircraft at a time using this method. See Fig. 3.18 for an example of probe and drogue in-flight refuelling.

Fig. 3.18 *Probe and drogue in-flight refuelling (BAE SYSTEMS)*

In the boom method, sometimes called the flying boom, the technique is different. The responsibility for making contact is that of the boom operator in the tanker – typically a KC-135 or a KC-10 – who flies the boom so that the recipient makes contact in a similar manner to the drogue method. The receiving aircraft has a receptacle on its upper surface into which the refuelling boom is inserted. A tanker has one boom mounted on the centreline from the rear of the aircraft and therefore the number of aircraft refuelling using this method maybe limited. See Fig. 3.19.

Air-to-air refuelling is now extensively used during aircraft flight test programmes in the UK where it is possible to extend the duration of flight tests and effectively accelerate programme completion. The Northrop B-2 stealth bomber, the competing Lockheed/Boeing/General Dynamics YF-22A and the Northrop/McDonnell Douglas YF-23A Advanced Tactical Fighter prototypes used this technique during their respective development programmes. This is a graphic illustration of how commonplace this activity has now become.

Fig. 3.19 Boom in-flight refuelling (US Air Force)

In terms of interfacing with the normal refuelling system, the air-to-air refuelling probe feeds into the refuelling lines via a Non-Return Valve (NRV) which only permits flow from the probe into the system and not vice versa. Therefore once probe contact has been made and is maintained, air-to-air refuelling continues in an identical fashion to the normal refuelling operation except that the aircrew determine when to halt the process.

Integrated civil aircraft systems

The integration of aircraft civil fuel systems has become more prevalent over the last decade or so using digital data buses and the supply of hardware from one or more manufacturers. Most civil aircraft have a fuel tank configuration as shown in Figure 3.20.

This configuration comprises left and right wing tanks and a centre tank. However it is also possible for aircraft to have an aft or trim tank. The major transfer modes are:

- Engine and APU feed
- Fuel transfer
- Refuel/defuel
- Fuel jettison

Depending upon the aircraft configuration and the degree of control, the aft or trim tank

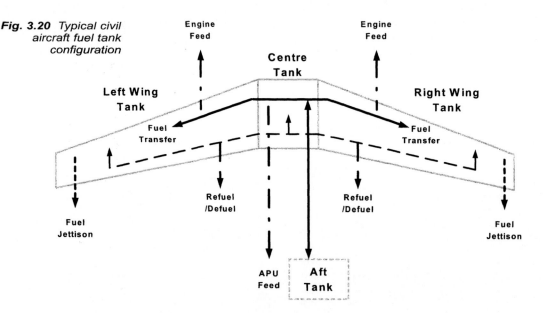

Fig. 3.20 *Typical civil aircraft fuel tank configuration*

may be used as a means of controlling the aircraft centre of gravity (CG). Altering the contents of a trim tank can reduce trim drag and improve aircraft range; it is also possible to reduce the structural weight of the tailplane. Most aircraft have variations on this basic topology although the number of wing tanks may also be dictated by the wing structure, the number of engines, or the need to partition fuel to cater for engine turbine disc burst zones.

This section addresses two examples:

- Bombardier Global Express
- Boeing 777

Bombardier Global Express

The Fuel Management & Quantity Gauging System (FMQGS) developed by Parker Aerospace for the Bombardier Global Express is typical of a family of systems which may be found fitted to regional aircraft and business jets. The Global Express has a true intercontinental range capability approaching 6,000 mi and is cleared to 51,000 ft. The system has interfaces to:

- Engine Indication and Crew Alerting System (EICAS) and ground crew via A429 data buses.
- Cockpit control panel for APU and engine selector switches and fire handles.
- Cockpit fuel panel for fuel system mode selections.
- Electrical load management system for supplying power to the electrically powered pumps and valves. The system receives status discretes from fuel pumps and valves.
- Cockpit and wing Refuel/Defuel Control Panels (RDCPs).

Refer to Fig. 3.21.

The heart of the system is the dual channel FMQCG which embraces the following functions.

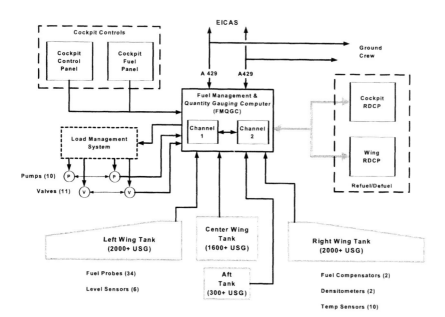

Fig. 3.21 *Simplified Global Express fuel system (Parker Aerospace)*

Fuel management

The fuel management function provides the following:

- Control, status and built-in test (BIT) of all system pumps, valves and pressure sensors.
- Fuel transfer – burn sequence and lateral balance.
- Flight crew and ground crew interface.
- Automatic/manual refuel/defuel operation.
- BIT fault detection and annunciation.

Optional thermal management

The operation of the aircraft for long periods at altitude provides extreme cold soak conditions. The system provides control of the return of warm fuel from the engine oil coolers to the wing tanks when extreme low temperature operation might be encountered.

Fuel quantity gauging

Fuel quantity gauging using the following sensors:

- Linear AC capacitance fuel probes (34).
- Level sensors – software adjustable (6).
- Fuel compensators (2).
- Self-calibrating densitometers (2).
- Temperature sensors (10).

The FMQGS is an ARINC 600 LRU designed to meet the DO160C environment. The unit contains a dual-channel microprocessor architecture hosting software to DO178B Level B. On this system Parker Aerospace performed the role of systems integrator, taking responsibility for design and development, controlling configuration and certifying the system (Reference (3)).

Boeing 777

Fig. 3.22 Simplified portrayal of Boeing 777 fuel gauging/fuel management (Smiths Industries)

The Boeing 777 in contrast uses an integrated architecture based upon A429 and A629 data buses as shown in Fig. 3.22. This diagram emphasizes the refuel function which is controlled via the Electrical Load Management System (ELMS) P310 stand-by power management panel in association with the integrated refuel panel and the Fuel Quantity Processor Unit (FQPU).

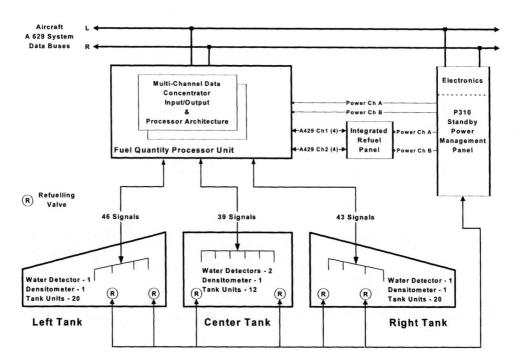

There are six refuelling valves, marked as R on the diagram, two in each of the left wing, centre and right wing tanks. The P310 panel provides power to the FQPU, integrated refuelling panel and controls the operation of the refuelling valves. The FQPU and refuelling panel communicate by means of dual A429 data links. The top level integration of the FQPU and ELMS P310 panel is via the aircraft system's left and right A629 data buses. This system permits the automatic refuelling of the aircraft to a preset value, as the FQPU senses the fuel tank quantities reaching their assigned value, messages are sent to the ELMS to shut off the refuelling valves until all three tanks have attained the correct level.

The function of the Boeing 777 ELMS is described in Chapter 5, Electrical Systems. In this mode of operation the ELMS is able to power up the necessary components of the fuel system to accomplish refuelling during ground maintenance operations without the need to power the entire aircraft.

The FQPU is a multi-channel multi-processor controller which processes the fuel quantity information provided by a total of 52 tank units (probes), four water detectors and three densitometers located in the three fuel tanks. The Boeing 777 uses ultrasonic fuel probes, the first civil airliner to do so.

The ELMS, FQPU and integrated refuelling panel are supplied by Smiths Industries.

References

(1) Smiths Industries Marketing Publication SAV 247X Issue 2.
(2) Smiths Industries Marketing Publication SIA 663.
(3) Parker Aerospace Marketing Publication GPDS9709-FMCG.

CHAPTER 4

Hydraulic Systems

Introduction

Hydraulic systems made their appearance on aircraft in the early 1930s when the retractable undercarriage was introduced. Since that time an increasing number of tasks have been performed by the application of hydraulic power and the power demand has consequently increased greatly. Hydraulic power was seen as an efficient means of transferring power from small low energy movements in the cockpit to high energy demands in the aircraft. Hydraulic systems now have an important role to play in all modern aircraft, both military and civil.

The introduction of powered flying controls was an obvious application for hydraulic power by which the pilot was able to move the control surfaces with ever-increasing speeds and demands for manoeuvrability. This application brought hydraulics in the area of safety critical systems in which single failures could not be allowed to hazard the aircraft. The system developed to take account of this using multiple pumps, accumulators to store energy and methods of isolating leaks.

The hydraulic system today remains a most effective source of power for both primary and secondary flying controls, and for undercarriage, braking and anti-skid systems. However it will become apparent later in the book that more-electric systems are being considered to replace hydraulically powered systems in some areas.

From the beginning the use of hydraulics as a means of transmitting power has not gone unchallenged. Of the various alternatives considered the chief contender has been the use of the electrical systems. The lure of the all-electric aeroplane has been a tempting prize, and numerous technical papers have evaluated the relative merits over at least the last thirty years. Hydraulics power has nevertheless maintained its position due to a unique combination of desirable features, not least of which is low weight per unit power. Even with the advent of rare earth magnetic materials, the electric motor cannot yet match the power-to-weight ratio of a hydraulic actuator, particularly above 3 kW.

In choosing any type of system certain general characteristics, often conflicting, are sought. The principal requirements are low weight, low volume, low initial cost, high reliability and low maintenance. The latter two are the crucial constituents of low cost of ownership. Hydraulic systems meet all these requirements reasonably well, and have additional attractions. The small pipe diameters lend themselves to flexibility of installation, the use of oil as the working fluid provides a degree of lubrication, and the system overloads can be withstood without damage. Within the limits of their structural strength, actuators can stall and in some cases actually reverse direction. They will return to working condition perfectly normally on removal of the overload. Many mechanical engineers consider that these attractions make the hydraulic system more flexible and more robust than an electrical actuation system with the same power demand.

The last decades have seen the ever-accelerating introduction of digital processing systems, both for monitoring system performance and to perform control functions. This has proved to be a major step forward, permitting some previous shortcomings to be overcome and opening the way to so-called 'smart' pumps and valves.

Circuit design

The majority of aircraft in use today need hydraulic power for a number of tasks. Many of the functions to be performed affect the safe operation of the aircraft and must not operate incorrectly, i.e. must operate when commanded, must not operate when not commanded and must not fail totally under single failure conditions.

These requirements together with the type of aircraft, determine the design of a hydraulic system. When starting the design of any new hydraulic system the engineer must first determine the functions to be performed, and secondly he must assess their importance to flight safety. Thus a list of functions may appear as:

Primary flight controls:	Elevators
	Rudders
	Ailerons
	Canards
Secondary flight controls:	Flaps
	Slats
	Spoilers
	Airbrakes
Utility systems:	Undercarriage
	Wheelbrakes
	Nosewheel steering
	In-flight re-fuelling probe
	Cargo doors
	Loading ramp
	Passenger stairs

Many other functions are carried out on various aircraft by hydraulics, but those listed above may be used as a typical example of modern aircraft systems. The wise designer will always allow for the addition of further functions during the development of an aircraft.

From the above list the designer may conclude that all primary flight controls are critical to flight safety and consequently no single failures must be allowed to prevent, or even momentarily interrupt their operation. This does not necessarily mean that their performance cannot be allowed to degrade to some pre-determined level, but that the degradation must always be controlled systematically and the pilot must be made aware of the state of the system. The same reasoning may apply to some secondary flight controls, for example flaps and slats.

Other functions, commonly known as 'services' or 'utilities', may be considered expendable after a failure, or may be needed to operate in just one direction after a positive emergency selection by the pilot. In this case the designer must provide for the emergency movement to take place in the correct direction, for example, undercarriages must go down when selected and flight refuelling probes must go out when selected. It is not essential for them to return to their previous position in an emergency, since the aircraft can land and take on fuel – both safe conditions.

Wheelbrakes tend to be a special case where power is frequently provided automatically or on selection, from three sources. One of these is a stored energy source which also allows a parking brake function to be provided.

In its simplest form a hydraulic system is shown in Fig. 4.1. The primary source of power on an aircraft is the engine, and the hydraulic pump is connected to the engine gearbox. The pump causes a flow of fluid at a certain pressure, through stainless steel pipes to various actuating devices. A reservoir ensures that sufficient fluid is available under all conditions of demand.

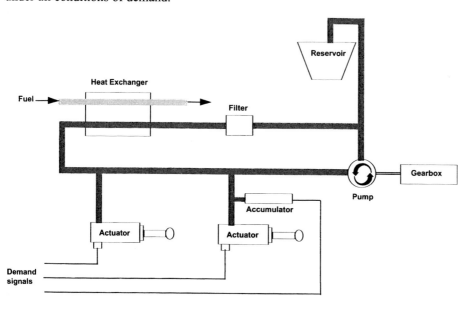

Fig. 4.1 *A simple hydraulic system*

This simple system is unlikely to satisfy the condition stated above, and in practice most aircraft contain multiple pumps and connections of pipes to ensure that single failures and leaks do not deplete the whole system of power. A more complex system, although still not adequate in practice is shown in Fig. 4.2 as a simple example to describe the various components of a hydraulic system before going on to show some real-life examples.

To achieve the levels of safety described above requires at least two hydraulic circuits as shown in Fig. 4.2. The degree of redundancy necessary is very largely controlled by specifications and mandatory regulations issued by the national and international bodies charged with air safety. The requirements differ considerably between military and civil aircraft. Military aircraft frequently have two independent circuits, large civil transports and passenger aircraft invariably have three or more. In both types additional auxiliary power units and means of transferring power from one system to another are usually provided.

Fig. 4.2 *A typical hydraulic system*

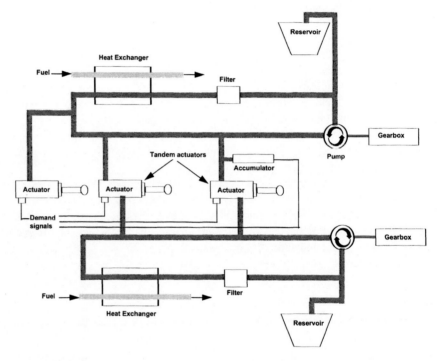

Actuation

On military aircraft the primary flight control actuator normally consists of two pistons in tandem on a common ram as illustrated in Fig. 4.3.

Each piston acts within its own cylinder and is connected to a different hydraulic system. The ram is connected at a single point to a control. The philosophy is different on civil aircraft where each control surface is split into two or more independent parts. Each part has its own control actuator, each of which is connected to a different hydraulic system as shown in Fig. 4.4.

The majority of actuators remain in a quiescent state, either fully extended or fully closed. They control devices which have two discrete positions, for example airbrakes or in-flight refuelling probes that are either IN or OUT, or undercarriages which are either UP or DOWN. Although there is obviously a finite time during which these devices are travelling, it is usually undesirable that they should stop whilst in transition. They are essentially two-state devices.

The actuator can be commanded to one or other of its states by a mechanical or electrical demand. This demand moves a valve that allows the hydraulic fluid at pressure to enter the actuator and move the ram in either direction.

A mechanical system can be commanded by direct rod, lever or cable connection

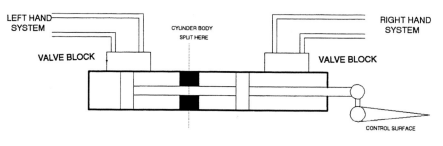

LEFT HAND SYSTEM

CYLINDER BODY
SPLIT HERE

RIGHT HAND SYSTEM

VALVE BLOCK

VALVE BLOCK

CONTROL SURFACE

ACTUATOR SCHEMATIC

Fig. 4.3 *A flight control actuator (Claverham/FHL)*

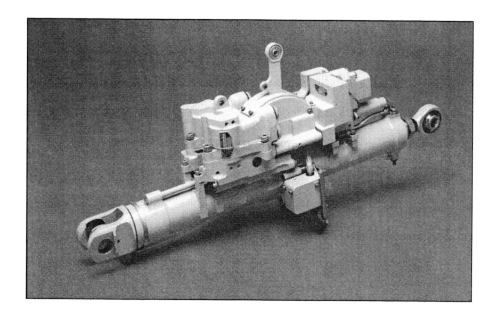

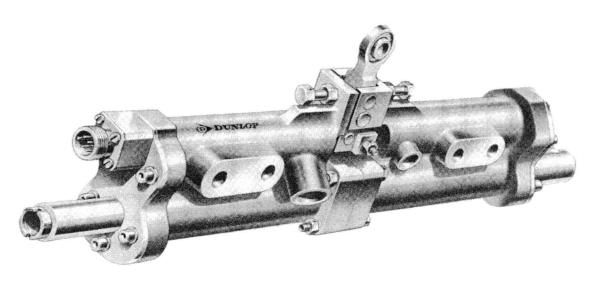

Fig. 4.4 *Civil aircraft
control surface
actuation*

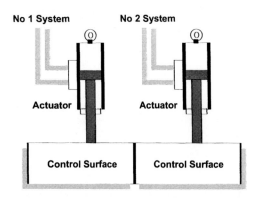

Fig. 4.4 *Civil aircraft control surface actuation*

from a pilot control lever to the actuator. An electrical system can be connected by means of a solenoid or motor that is operated by a pilot or by a computer output.

In some instances it is necessary to signal the position of the actuator, and hence the device it moves, back to the pilot. This can be achieved by connecting a continuous position sensor such as a potentiometer, or by using microswitches at each end of travel to power a lamp or magnetic indicator. Some devices, however, are not simply two-state, but are continuously variable. Examples are active primary flight control surfaces or engine reheat nozzles. These devices need to be variable and are usually controlled electrically by computers, which drive torque motors or stepper motors connected to a variable valve on the actuator. This allows the actuator to be driven to any point in its range, stopped, advanced or reversed as often and as rapidly as required.

A continuous position sensor connected into the computer servo loop allows the computer to drive the actuator accurately in accordance with the demands of the control system. Like the computers driving the actuator, the motors and position sensors must be multiple redundant. In the case of a quadruplex flight control system, an actuator will be equipped with four torque motors and four position sensors, each connected to a different computer (refer to Chapter 1, Flight Control Systems).

Hydraulic fluid

The working fluid will be considered as a physical medium for transmitting power, and the conditions under which it is expected to work, for example maximum temperature and maximum flow rate are described.

Safety regulations bring about some differences between military and civil aircraft fluids. With very few exceptions modern military aircraft have, until recently, operated exclusively on a mineral-based fluid known variously as:

DTD 585	in the UK
MIL-H-5606	in the USA
AIR 320	in France
H 515	NATO

This fluid has many advantages. It is freely available throughout the world, reasonably priced, and has a low rate of change of viscosity with respect to temperature compared to other fluids. Unfortunately, being a petroleum based fluid, it is flammable and is limited to a working temperature of about 130 °C. One of the rare departures from DTD 585 was made to overcome this upper temperature limit. This led to the use of DP 47, known also as Silcodyne, in the ill-fated TSR2.

Since the Vietnam War much industry research has been directed to the task of finding a fluid with reduced flammability, hence improving aircraft safety following accident or damage, particularly battle damage in combat aircraft. This work has resulted in the introduction of MIL-H-83282, an entirely synthetic fluid, now adopted for all US Navy aircraft. It is miscible with DTD 585 and, although slightly more viscous below 20 °C, it compares well enough.

In real terms the designer of military aircraft hydraulic systems has little or no choice of fluid since Defence Ministries of the purchasing nations will specify the fluid to be used for their particular project. Most specifications now ask for systems to be compatible with both DTD 585 and MIL-H-5606.

Fluid pressure

Similarly little choice is available with respect to working pressure. Systems have become standardized at 3,000 psi or 4,000 psi. These have been chosen to keep weight to a minimum, while staying within the body of experience built up for pumping and containing the fluid. Many studies have been undertaken by industry to raise the standard working pressure. Pressure targets have varied from 5,000 psi to 8,000 psi, and all resulting systems studies claim to show reduced system component mass and volume. Interestingly DTD 585 cannot be used above 5,000 psi because of shear breakdown within the fluid.

A detailed study would show that the optimum pressure will differ for every aircraft design. This is obviously impractical and would preclude the common use of well proven components and test equipment.

Fluid temperature

With fast jet aircraft capable of sustained operation above Mach 1, there are advantages in operating the system at high temperatures, but this is limited by the fluid used. For many years the use of DTD 585 has limited temperatures to about 130 °C, and components and seals have been qualified accordingly. The use of MIL-H-83282 has raised this limit to 200 °C and many other fluids have been used from time to time, for example on Concorde and TSR2, to allow high temperature systems to be used.

Fluid flow rate

Determination of the flow rate is a more difficult problem. When the nominal system pressure is chosen it must be remembered that this is, in effect, a stall pressure. That is to say, that apart from some very low quiescent leakage, no flow will be present in the circuit. The designer must allocate some realistic pressure drop that can be achieved in full flow conditions from pump outlet to reservoir. This is usually about 20–25 percent of nominal pressure.

Having established this, the pressure drop across each actuator will be known. The aerodynamic loads and flight control laws will determine the piston area and rate of movement. The designer must then decide which actuators will be required to act simultaneously and at what speed they will move. The sum of these will give the maximum flow rate demanded of the system. It is important also to know at what part of the flight this demand takes place.

It is normal to represent the flow demands at various phases of the flight – take-off, cruise etc. – graphically. The maximum flow rate does not necessarily size the pump to be used. It is frequently found that the flow required on approach provides the design case, when the engine rpm, and hence pump rpm, are low.

It may be found that the absolute maximum flow demand is of very short duration, involving very small volumes of oil at very high velocities. In this case sizing a pump to meet this demand may not be justified. An accumulator can be used to augment the flow available, but care must be taken. An accumulator contains a compressed gas cylinder, and the gas is used to provide energy to augment system pressure. Therefore, the fluid volume and pressure available will depend on the gas temperature. In a situation where the flow demanded will exceed the pump capabilities the system pressure is controlled by the accumulator, not the pump. This case will influence the circuit pressure drop calculations if the necessary pressure across the actuator piston is to be maintained.

The frequency of maximum demand must also be known, and time must be available for the pump to recharge the accumulator if it is not eventually to empty by repeated use.

Hydraulic piping

When the system architecture is defined for all aircraft systems using hydraulic power, then it is possible to design the pipe layout in the aircraft. This layout will take into account the need to separate pipes to avoid common mode failures as a result of accidental damage or the effect of battle damage in a military aircraft. Once this layout has been obtained it is possible to measure the lengths of pipe and to calculate the flow rate in each section and branch of pipe. It is likely that the first attempts to define a layout will result in straight lines only, but this is adequate for a reasonably accurate initial calculation.

If an allowable pressure drop of 25 per cent has been selected throughout the system, this may now be further divided between pressure pipes, return pipes and components. The designer will eventually control the specifications for the components, and in this sense he can allocate any value he chooses for pressure drop across each component. It must be appreciated, however, that these values must eventually be achieved without excessive penalties, being incurred by over-large porting or body sizes.

Once pipe lengths, flow rates and permissible pressure drops are known, pipe diameters can be calculated using the normal expression governing friction flow in pipes. It is normal to assume a fluid temperature of 0 °C for calculations, and in most cases flow in aircraft hydraulic systems is turbulent. Pressure losses in the system piping can be significant and care should be taken to determine accurately pipe diameters. Theoretical sizes will be modified by the need to use standard pipe ranges, and this must be taken into account.

Hydraulic pump

A system will contain one or more hydraulic pumps depending on the type of aircraft and the conclusions reached after a thorough safety analysis and the consequent need for redundancy.

The pump is normally mounted on the engine gearbox. The pump speed is therefore directly related to engine speed, and must therefore be capable of working over a wide speed range. The degree of gearing between the pump and the engine varies between engine types, and is chosen from a specified range of preferred values. A typical maximum continuous speed for a modern military aircraft is 6,000 rpm, but this is largely influenced by pump size, the smallest pumps running fastest.

The universally used pump type is known as variable delivery, constant pressure. Demand on the pump tends to be continuous throughout a flight, but frequently varying in magnitude. This type of pump makes it possible to meet this sort of demand pattern without too much wastage of power. Within the flow capabilities of these pumps the pressure can be maintained within 5 per cent of nominal except during the short transitional stages from low flow to high flow. This also helps to optimize the overall efficiency of the system. A characteristic curve for a nominally constant pressure pump is shown in Fig. 4.5.

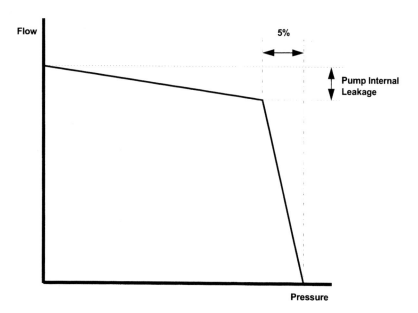

Fig. 4.5
Characteristic curve for a 'constant pressure' pump

The pumps are designed to sense outlet pressure and feed back this signal to a plate carrying the reciprocating pistons. The plate is free to move at an angle to the longitudinal axis of the rotating drive shaft. There are normally nine pistons arranged diametrically around the plate. The position of the plate therefore varies the amount of reciprocating movement of each piston.

A diagrammatic representation of a pump showing the working principle is shown in Fig. 4.6 and some commercial examples of hydraulic pumps can be seen in Fig. 4.7. When the plate is at 90 degrees to the linear axis, there is no linear displacement of the pistons. Up to its maximum limit the plate will move to displace the volume needed to maintain nominal system pressure. When flow demands beyond maximum displacement are made the system pressure drops rapidly to zero. For short periods pressure can be maintained by means of an accumulator as described above. Examples of accumulators can be seen in Fig. 4.8.

Fig. 4.6 *Working principle of hydraulic pump*

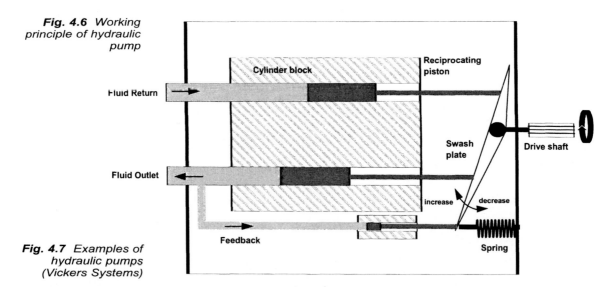

Fig. 4.7 *Examples of hydraulic pumps (Vickers Systems)*

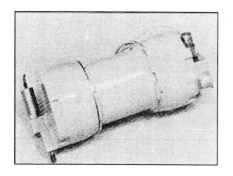
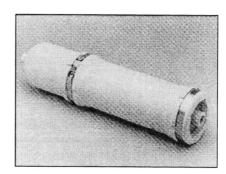

Fig. 4.8 *Examples of hydraulic accumulators (Claverham/FHL)*

Fluid conditioning

Under normal working conditions hydraulic fluid needs cooling and cleaning. Occasionally it is necessary to de-aerate by the connection of ground equipment, although increasingly modern systems are being produced with devices to bleed off any air accumulating in the reservoir.

For cooling purposes the fuel/hydraulic heat exchanger is used. This ensures that cooling on the ground is available. Further air/fluid cooling may be provided once the aircraft is in flight. Since heat exchangers are low-pressure devices they are normally situated in the return line to the actuator/service.

When a pump is running off-load, all the heat generated by its inefficiencies is carried away by the pump case drain line. The heat exchanger should therefore be positioned to cool this flow before its entry into the reservoir. Care must be taken to determine the maximum pressure experienced by the heat exchanger and to ensure that, not only is adequate strength present to prevent external burst, but in addition no failure occurs across the matrix between fuel and hydraulic fluid.

The introduction of servo-valves with very fine clearances emphasized the need for very clean fluids. The filter manufacturers responded to this by developing filter elements made of resin bonded paper supported by arrangements of metal tubes and wire mesh. This produces filter elements of high strength capable of withstanding differential pressures of one and a half times the system pressure.

These filters are capable, under carefully designed test conditions, of stopping all particles of contaminant above five microns in size, and a high percentage of particles below this size. This characteristic has led to filter elements becoming known by an absolute rating, the two examples above being five micron absolute.

More recent work is based on the ratio of particles upstream and downstream of the filter unit. This is referred to as the 'beta' rating. When specifying and choosing filter elements it is most important to specify the test method to be used.

Several standards exist defining the cleanliness of the fluid and these are based on a number of particles in the series of size ranges. Typically these are: 5–15 microns, 15–25 microns, 25–50 microns, 50–100 microns and above 100 microns, to be found in 100 ml of liquid. Unfortunately there is no way of calculating the relationship between the element's absolute rating and the desired cleanliness level. The choice of elements rests entirely on past experience and test results. In most cases it has been found that an adequate level of cleanliness can be achieved and maintained by the use of a 5 micron absolute return line filter in combination with a 15 micron pressure line filter. This combination also gives acceptable element life. Filters are not used in the

FILTER, RETURN LINE

FILTER, CASE DRAIN

FILTER, LOW PRESSURE

Fig. 4.9 Some typical filter units (Claverham/FHL)

pump inlet line. Figure 4.9 shows various filter units.

A further consequence of the demand for clean fluid has been a need for a means of measuring the cleanliness levels achieved. Electronic automatic counters are now available that are capable of providing rapid counts with a repeatability to within 5 per cent in a form suitable for rapid interrogation by ground servicing crews.

The reservoir

The requirements for this component vary depending on the type of aircraft involved. For most military aircraft the reservoir must be fully aerobatic. This means that the fluid must be fully contained, with no air/fluid interfaces, and a supply of fluid must be maintained in all aircraft attitudes and g conditions. In order to achieve a good volumetric efficiency from the pump, reservoir pressure must be sufficient to accelerate a full charge of fluid into each cylinder while it is open to the inlet port. The need to meet pump response times may double the pressure required for stabilized flow conditions.

The volume of the reservoir is controlled by national specifications and includes all differential volumes in the system, allowance for thermal expansion and a generous emergency margin.

It is common practice to isolate certain parts of the system when the reservoir level falls below a predetermined point. This is an attempt to isolate leaks within the system and to provide further protection for flight safety critical subsystems. The cut-off point must ensure sufficient volume for the remaining systems under all conditions. The reservoir will be protected by a pressure relief valve which can dump fluid overboard.

Warnings and status

Several instruments are normally situated in the hydraulic power generation system to monitor continuously its performance. Pressure transducers monitor system pressure and transmit this signal to gauges in the cockpit. Pressure switches are also incorporated to provide a warning of low pressure in the system on the central warning panel. Filter blockage indicators show the condition of the filter elements to ground servicing personnel, and a fluid temperature warning may be given to the aircrew. With increasing use of microprocessor-based system management units, more in-depth health monitoring of all major components is possible with data displayed to ground crews on a maintenance data panel.

Emergency power sources

All hydraulic systems have some form of emergency power source. In its simplest form this will be an accumulator. It is mandatory for wheel-brake systems to have a stand-by accumulator capable of supplying power for a predetermined number of brake applications when all other sources of power are inoperative. Cockpit canopies are frequently opened and closed hydraulically and emergency opening can be achieved by the use of accumulator stored energy.

Accumulators may also be used to provide sufficient flight control actuator movement to recover the aircraft to straight and level flight so that the crew can eject safely in the event of total systems failure.

To supply emergency power for longer periods an electric motor driven pump may be provided. Battery size and weight are the main limitations in this case, and to minimize these factors, the flow available is usually kept as low as possible to operate only those devices considered indispensable. Frequently it is also possible to operate at some pressure below nominal system pressure, even so it is unlikely that an acceptable installation can be achieved which will provide power for more than five or six minutes.

Weight may be kept to a minimum by the use of a one-shot battery. This allows the latest battery technology to be exploited without any concessions being made to obtain recharge capabilities. Selection will be automatic from a pressure switch with additional cockpit selection also being available.

For continuous emergency supply a Ram Air Turbine (RAT) may be used. This carries with it several disadvantages. Space must be found to stow the turbine and carriage assembly, a small accumulator is needed to deploy the turbine in emergency, and because speed governing and blade feathering are employed the assembly is complicated. Hydraulic pumps and/or emergency electrical generators can be mounted immediately behind the turbine on the same shaft. It is, however, more common to mount them at the bottom of the carriage arm close to the deployment hinge axis. This involves the use of drive shafts and gears. To keep the turbine blade swept diameter at a reasonable figure, the power developed must be kept low and it may be difficult to mount the assembly on the airframe so that the air flow is not impeded by the fuselage at peculiar aircraft attitudes. Deployment of the RAT is as for the electric motor driven pump.

In spite of these drawbacks, ram air turbines have several times proved their worth, particularly on civil aircraft, providing the only means of hydraulic power until an emergency has been dealt with and the aircraft has been recovered to a safe attitude.

Proof of design

All the effort put into designing an hydraulic system culminates in the testing to prove that the design works in the required way. All the systems in an aircraft must be qualified before the aircraft is approved for flight. The qualification is built up through a series of steps starting with demonstrations that each individual component meets its specification. This will include proof and burst pressure tests, fatigue, vibration, acceleration and functional tests. These may be complemented by accelerated life tests.

Satisfactory completion of the tests is formalized in a Declaration of Design and Performance Certificate signed by the specialist company responsible for design and manufacture of the component, and by the company designing the aircraft.

The entire hydraulic system is then built up into a test rig. The rig consists of a steel structure representing the aircraft into which the hydraulic piping and all components are mounted in their correct relationship to each other. The pipes will be the correct diameter, shape and length. Flight standard pumps will provide the correct flows and pressures. The rig will incorporate loading devices to simulate aerodynamic and other loads on the undercarriage and other surface actuators. Strain gauging and other load techniques are used to measure forces and stresses as required. It is normal to 'fly' the rig for several hundred hours in advance of actual flight hours on the prototype aircraft.

Ultimately, before a customer accepts the aircraft into service, the hydraulic system can be declared fully qualified on the basis of the evidence obtained from the rig plus flight testing.

The cost and effort involved is considerable, but a well designed and operated hydraulic test rig is crucial to the process of formal qualification and certification of the aircraft. A typical test rig is shown in Fig. 4.10.

Fig. 4.10 *Hydraulic system rig (BAE SYSTEMS)*

Aircraft applications

Since the range of hydraulic system design is dependent on the type of aircraft, it would not be sensible to give a single example. The following applications cover a range of single- and multiple-engine aircraft of both civil and military types.

The BAE SYSTEM 146 family hydraulic system

The BAE SYSTEMS family consists of aircraft seating from 70 to 128 (and later the RJ Avro RJ70, RJ85 and RJ100) passengers. The 146 is a four-engine regional jet airliner

Fig. 4.11 BAE SYSTEMS 146 regional jet hydraulic system schematic (BAE SYSTEMS)

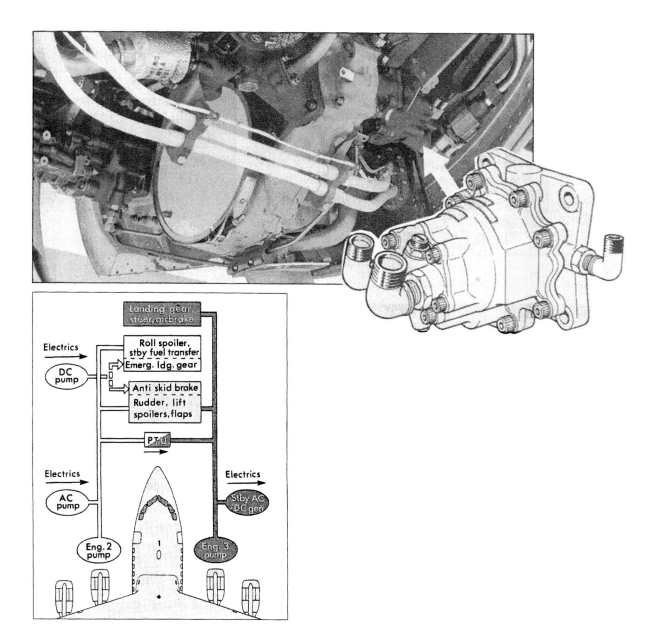

designed for world-wide operations. Its hydraulic system has been designed to combine the lightness and simplicity of a two-engine design with the back-up levels associated with a four-engine system.

Two independent systems each operate at a nominal 3,000 psi. Hydraulic system controls and annunciations are located on the pilot's overhead panel. An amber caption on the master warning panel, plus a single audio chime draws attention to fault warnings on the overhead panel. Figure 4.11 shows the 146 family hydraulic system schematic.

The systems are designated Yellow and Green and are normally pressurized by a self-regulating engine-driven pump on the inboard engines. Each system has an independent hydraulic reservoir, pressurized by regulated airbleed from its respective engine. Flareless pipe couplings with swaged fittings are used throughout for reliability and ease of repair.

Yellow and Green systems are geographically segregated as far as possible. The Yellow system is on the left of the aircraft and the Green on the right. Back-up power for the Yellow system is provided by an AC electric pump, and back-up for power for the Green system is provided by a Power Transfer Unit (PTU) driven by the Yellow system.

An electrically operated DC pump, fed from a segregated hydraulic supply, provides emergency lowering of the landing gear and operation of the brakes in the event of failures in both the Yellow and Green systems.

The AC pump, PTU, hydraulic reservoirs etc. are housed in a pressurized and vented hydraulic equipment bay and are fully protected from foreign object damage. The primary power generation components of the Yellow system are as follows.

- Engine-Driven Pump (EDP) on No. 2 engine.
- Stand-by AC powered hydraulic pump.
- Emergency DC powered hydraulic pump.
- Accumulator.
- Reservoir.

All these components, except for the EDP, are located in the hydraulics equipment bay. The components are shown in Fig. 4.12.

Yellow system

The Yellow system powers the following services.

- 1 flap motor
- Flap asymmetry brakes
- Roll spoilers
- 2 lift spoilers (inner spoilers on the left and right wing)
- 1 rudder servo control
- Stand-by fuel pumps (left and right)
- Landing-gear emergency lock-down
- Wheel brakes including park brake
- Airstairs through the AC pump
- Power Transfer Unit (PTU)

Yellow system stand-by AC pump

In the event of an EDP failure, the Yellow system is supported by a stand-by AC pump. The pump is continuously rated and is capable of maintaining the system pressure at

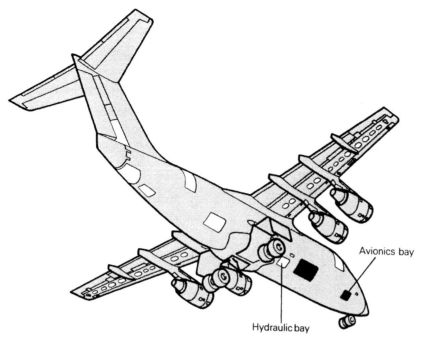

Avionics bay

Hydraulic bay

Fig. 4.12 *Yellow system components in the hydraulic bay (BAE SYSTEMS)*

3,000 psi. The AC pump is controlled by a three-position switch on the hydraulics overhead panel on the flight deck. This panel also includes the amber pump high temperature and failure annunciators.

The pump may be selected ON or OFF manually, but normally operates in automatic mode. In this mode a pressure switch in the Yellow and Green systems switches and latches the pump ON if the delivery pressure of either EDP falls below 1,500 psi. The stand-by pump therefore supports the Yellow system directly and the Green system indirectly via the PTU.

Yellow system emergency

Back-up DC pump

In the event of a failure of both Yellow and Green systems the DC back-up pump provides emergency lock-down of the main landing-gear and operation of the Yellow system wheel brakes. On the ground it can provide brake pressure in the Yellow system for parking, starting or towing.

The system has its own DC powered hydraulic pump, fluid supply and an accumulator. The DC pump is controlled from the hydraulics overhead panel on the flight-deck and is supplied from the emergency DC busbar. Hydraulic fluid is supplied from a segregated reservoir in the Yellow tank system.

The Yellow system accumulator is connected to the Yellow system wheel brakes and is protected from all other services by non-return valves. The accumulator stabilizes the system and assists the DC pump. The accumulator is pressurized by the Yellow EDP, AC pump or DC pump.

Yellow system reservoir

A 15.5 l reservoir is provided for the Yellow system. It is pressurized by bleed air regulated to 50 psi from the engine HP compressor. The reservoir incorporates the following:

- A pressure gauge
- A sight-glass
- An air low-pressure switch
- Inward and outward relief valves
- A bursting disc to protect against manual failure of the outward relief valve
- A ground charge connection and manual pressure release lever
- A contents transmitter

Indications of tank contents are provided on the flight-deck overhead panel that also includes amber low quantity and high temperature annunciators.

Engine-driven pump

The Yellow system Engine-Driven Pump (EDP) is mounted on the left inner engine auxiliary gearbox at the bottom of the engine to ensure easy maintenance access. The EDP has an associated motorized isolation valve. When the valve is closed it isolates the pump from the tank and provides an idling circuit to off-load the pump. If the engine fire handle is pulled to its fullest extent the valve closes automatically, preventing more fluid reaching the pump.

A two-position switch on the overhead hydraulic panel controls the position of the EDP isolation valve. An amber annunciator on the overhead panel illuminates when the valve is travelling and remains ON until it reaches the selected position. The EDP also has an associated relief valve which opens to allow excess pressure back to the tank at 3,500 psi.

Green system

The primary power generation components of the Green system are:

- Engine-Driven Pump (EDP) on No. 3 engine
- Power Transfer Unit (PTU)
- Hydraulic reservoir
- Accumulator

All components, except for the EDP, are located in the hydraulic equipment bay. The Green system powers the following:

- 1 flap motor
- 4 lift spoilers (centre and outer spoilers on the left and right wing)
- Air brakes
- Landing-gear – normal
- Nose-gear steering
- Wheel brakes excluding park brake

Green system stand-by PTU

The PTU is an alternative power source for the Green system. The PTU is a back-to-back hydraulic motor and pump. It can support all Green system services except for the stand-by AC/DC generator. The motor is powered by the Yellow system pressure and is connected by a drive shaft to a pump in the Green system. The PTU is controlled from the hydraulics overhead panel by a two-position switch. When the switch is in the ON position it is automatically activated if Green system pressure falls below 2,600 psi.

With the switch in the OFF position, the motor is isolated from the Yellow system by a motorized valve. Movement of the valve is indicated by an amber PTU VALVE annunciator on the flight-deck hydraulics panel. The PTU may also be used during ground servicing to pressurize the Green hydraulic system, provided the hydraulic reservoir is fully charged with air.

Green system stand-by AC/DC generator

The Green hydraulic system can support the electrical system in the event of low electrical power. A stand-by AC/DC generator, driven by a hydraulic motor is powered by the Green system and is controlled by a three-position switch on the flight-deck overhead electrical panel. The generator can be selected ON or OFF manually but is usually in automatic stand-by or ARM mode. The generator is normally isolated from the system pressure by a solenoid-operated selector valve.

When the stand-by AC/DC pump is operating its selector valve is opened, and at the same time Green system services are isolated by their shut-off valve. Green system services are therefore not available while the generator is operating and the Green system LO PRESS annunciator is indicated by a white light on the overhead electrical panel.

Green system reservoir

The Green system reservoir has the same capacity as the Yellow system and is charged with bleed air from No. 3 engine. Its features are exactly the same as the Yellow system reservoir.

Accumulator

The Green system accumulator is identical to the Yellow system accumulator. It maintains stability in the Green systems during operation of the PTU and also assists the EDP for initial run-up of the stand-by AC/DC generator.

The BAE SYSTEMS Hawk 200 hydraulic system

The BAE SYSTEMS Hawk 200 is a single-engine, single-seat multi-role attack aircraft in which the hydraulic power is provided by two independent systems. Both power the flying controls by means of tandem actuators at the ailerons and tailplane. The number 1 system provides power to the rudder, which can also be manually operated.

The number 1 system also provides power for utility services such as flaps, air brakes, landing-gear and wheel brakes. The number 2 system is dedicated to the operation of the flying control surfaces. In the event of engine or hydraulic pump failure, a ram air turbine driven pump automatically extends from the top rear fuselage into the airstream. This powers the flying control system down to landing speed.

Fig. 4.14 The BAE SYSTEMS Hawk 200 ram air turbine extended (BAE SYSTEMS)

A pressurized nitrogen accumulator is provided to operate the flaps and landing-gear in an emergency, and wheel brake pressure is maintained by a separate accumulator. The Hawk 200 hydraulic system is shown in Fig. 4.13 (see colour plate section) and the ram air turbine is shown in Fig. 4.14.

The Panavia Tornado hydraulic system

The Panavia Tornado is a twin-engine, two-seat, high-performance aircraft designed for ground attack as the IDS version, or for air defence as the ADV version. Its hydraulic system is a 4,000 psi fully duplicated system shown diagramatically in Fig. 4.15. The high operating pressure allows the use of small-diameter piping, and the system is low-weight despite the duplicated pipe routings required for battle damage tolerance. The two pumps are mounted on the engine gearboxes and incorporate depressurizing valves. During engine start the hydraulic system is depressurized to reduce engine power off-take to allow rapid engine starting. A cross-drive is provided between the two RB 199 engines, which allows either engine to power both hydraulic pumps should one engine fail.

The pumps are driven by two independent accessory drive gearboxes, one connected by a power off-take shaft to the right-hand engine, and the other similarly connected to the left-hand engine. This allows the hydraulic pumps, together with the fuel pumps and independent drive generators, to be mounted on the airframe and separated from the engine by a fire-wall. This means that the Tornado hydraulic system is completely contained within the airframe. Not only is this a safety improvement, but it also improves engine change time, since the engine can be removed without the need to disconnect hydraulic pipe couplings.

The engine intake ramp, taileron, wing-sweep, flap and slat actuators are all fed from both systems. Should any part of the utility system become damaged, isolating valves operate to give priority to the primary control actuators.

The undercarriage is powered by the number 2 system and in the event of a failure the gear can be lowered by means of an emergency nitrogen bottle. A hand pump is provided to charge the brake and canopy actuators.

Skin mounted pressure and contents gauges are provided adjacent to the charging points and all filters are hand-tightened.

Civil transport comparison

The use of 3,000 psi hydraulics systems in civil transports is widespread and the BAE SYSTEMS Avro RJ systems have been described in depth. However as a way of examining different philosophies a comparison is made between an Airbus narrow body – the A320 family, and a Boeing wide body – the Boeing 767. It is usual for three independent hydraulic systems to be employed, since the hydraulic power is needed for flight control system actuation. Hydraulic power is produced by pumps driven by one of the following methods of motive power.

- Engine driven
- Electrically driven
- Air turbine/bleed air driven
- Ram air turbine driven

Airbus A320

The aircraft is equipped with three continuously operating hydraulic systems called Blue, Green and Yellow. Each system has its own hydraulic reservoir as a source of hydraulic fluid

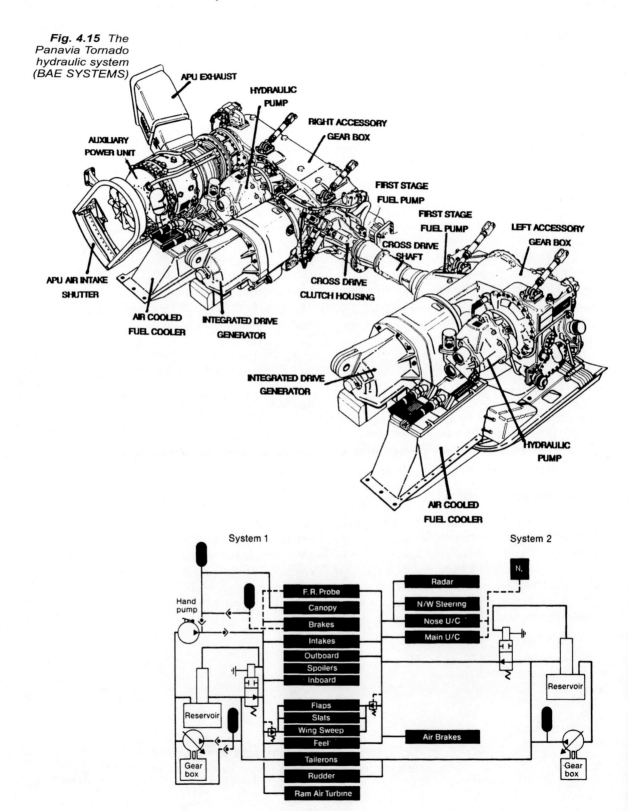

Fig. 4.15 *The Panavia Tornado hydraulic system (BAE SYSTEMS)*

- The Green system (System 1) is pressurized by an Engine-Driven Pump (EDP) located on No 1 engine which may deliver 37 gal/min (US gpm) or 140 l/min.
- The Blue system (System 2) is pressurized by an electric motor driven pump capable of delivering 6.1 gpm or 23 l/min. A Ram Air Turbine (RAT) can provide up to 20.6 gpm or 78 l/min at 2175 psi in emergency conditions.
- The Yellow system (System 3) is pressurized by an EDP driven by No 2 engine. An electric motor driven pump is provided which is capable of delivering 6.1 gpm or 23 l/min for ground servicing operations. This system also has a hand pump to pressurize the system for cargo door operation when the aircraft is on the ground with electrical power unavailable.

Each channel has the provision for the supply of ground-based hydraulic pressure during maintenance operations. Each main system has a hydraulic accumulator to maintain system pressure in the event of transients. Refer to Fig. 4.16.

Fig. 4.16 *Simplified A320 family hydraulic system*

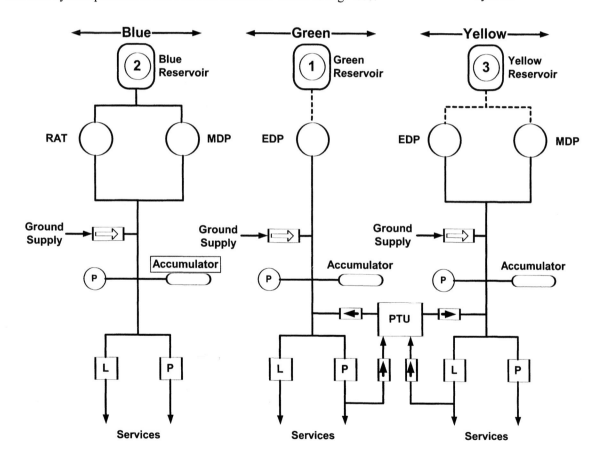

Each system includes a leak measurement valve (shown as L in a square on the diagram), and a priority valve (shown as P in a square).

- The leak measurement valve is positioned upstream of the primary flight controls and is used for the measurement of leakage in each flight control system circuit. They are operated from the ground maintenance panel.

- In the event of a low hydraulic pressure, the priority valve maintains pressure supply to essential systems by cutting of the supply to heavy load users.

The bi-directional Power Transfer Unit (PTU) enables the Green or the Yellow systems to power each other without the transfer of fluid. In flight, in the event that only one engine is running, the PTU will automatically operate when the differential pressure between the systems is greater than 500 psi. On the ground, while operating the Yellow system using the electric motor driven pump, the PTU will also allow the Green system to be pressurized.

The RAT extends automatically in flight in the event of failure of both engines and the APU. In the event of an engine fire, a fire valve in the suction line between the EDP and the appropriate hydraulic reservoir may be closed, isolating the supply of hydraulic fluid to the engine.

Pressure and status readings are taken at various points around the systems which allows the composition of a hydraulic system display to be shown on the Electronic Centralized Aircraft Monitor (ECAM).

Boeing 767

The Boeing 767 also has three full-time independent hydraulic systems to assure the supply of hydraulic pressure to the flight controls and other users. These are the left, right and centre systems serviced by a total of eight hydraulic pumps.

- The left system (Red System) is pressurised by an EDP capable of delivering 37.5 gpm or 142 l/min. A secondary or demand electric motor driven pump capable of delivering 7 gpm or 26.5 l/min is turned on automatically in the event that the primary pump cannot maintain pressure.
- The right system (Green System) has a similar configuration to the left system.
- The centre system (Blue System) uses two electric driven motor pumps, each with the capability of delivering 7 gpm or 26.5 l/min as the primary supply. An air-driven pump (ADP) with a capacity of 37 gpm or 140.2 l/min is used as a secondary or demand pump for the centre system. The centre system also has an emergency RAT rated at 11.3 gpm or 42.8 l/min at 2,140 psi.

Refer to Fig. 4.17 for a simplified diagram of the Boeing 767 hydraulic system. Primary flight control actuators, autopilot servo-valves and spoilers receive hydraulic power from each of the three independent hydraulic systems. The stabilizer, yaw dampers, elevator feel units and the brakes are operated from two systems. A Power Transfer Unit (PTU) between the left and right systems provides a third source of power to the horizontal stabilizer.

A motorized valve (M) located between the delivery of ACMP No 1 and ACMP No 2 may be closed to act as an isolation valve between the ACMP No 1 and ACMP No 2/ADP delivery outputs.

Hydraulic systems status and a synoptic display may be portrayed on the Engine Indication and Crew Alerting System (EICAS) displays situated between the Captain and First Officer on the instrument console. A number of maintenance pages may also be displayed.

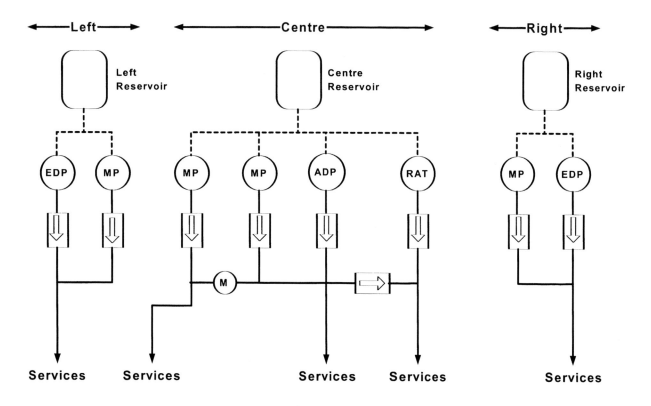

The supply schedule for the different pumps is given in Table 4.1:

Fig. 4.17 Simplified B767 hydraulic system

Table 4.1 *Boeing 767 simplified hydraulic schedule*

	Hydraulic power summary		
System	**Pump continuous**	**Pump demand**	**Operating conditions**
Left or Right	EDP		Basic system pressure
		ACMP	Supplements EDP
	ACMP No 1		Basic system pressure – maintains isolated system pressure
Centre			
	ACMP No 2		Basic system pressure – does not operate when one engine is out or left and right ACMPs on
Centre		ADP	Supplements ACMPs No 1 & No 2
Centre (Emergency)		RAT	Operates when deployed

The RAT supplies emergency power in flight once the engine speed (N2) has fallen below 50 per cent on both engines and the air speed is in excess of 80 kts. The RAT may only be restowed on the ground.

While this description outlines the Boeing 767 system at a top level, the systems on the Boeing 747-400 and Boeing 777 also use a combination of engine driven (EDP), air driven (ADP) and electric motor driven pumps and a RAT albeit in different architectures with a different pump configuration. The Boeing philosophy appears to favour fewer accumulators but use more pumps with a more diverse selection of prime pump energy.

A very useful Reference (1) summarizes the key hydraulic system characteristics of virtually all wide-body, narrow-body and turboprop/commuter aircraft flying today.

Landing-gear systems

The Raytheon 1000 is representative of many modern aircraft; its landing-gear is shown in Figs 4.18 and 4.19. It consists of the undercarriage legs and doors, steering and wheels and brakes and anti-skid system. All of these functions can be operated hydraulically in response to pilot demands at cockpit-mounted controls.

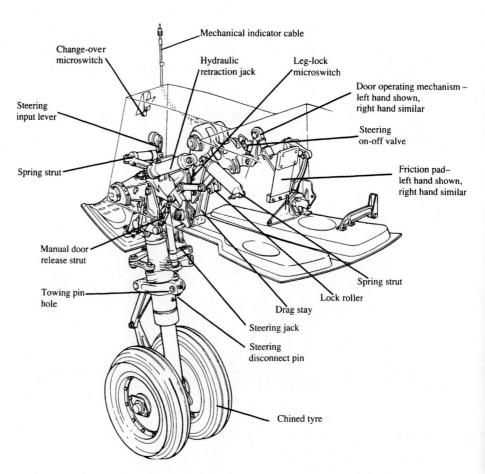

Fig. 4.18 *The Raytheon 1000 nose landing-gear (Raytheon)*

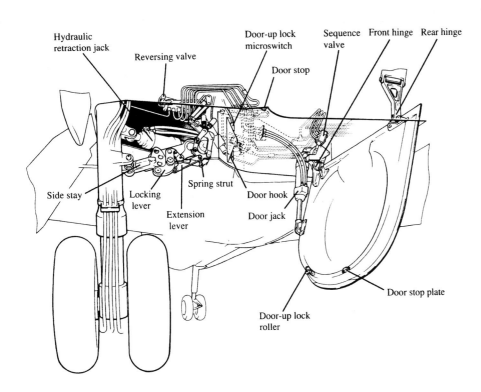

Hydraulic retraction jack

Reversing valve

Door-up lock microswitch

Door stop

Sequence valve

Front hinge

Rear hinge

Side stay

Locking lever

Spring strut

Extension lever

Door hook

Door jack

Door stop plate

Door-up lock roller

Fig. 4.19 *The Raytheon 1000 main landing-gear (Raytheon)*

Nose gear

The tricycle landing-gear has dual wheels on each leg. The hydraulically operated nose gear retracts forward into a well beneath the forward equipment bay. Hinged nose-wheel doors, normally closed, are sequenced to open when lowering or retracting the nose gear. The advantage of the doors being normally closed is twofold. Firstly, the undercarriage is protected from spray on take-off and landing, and secondly there is a reduction in drag. A small panel on the leg completes enclosure on retraction and a mechanical indicator on the flight-deck shows locking of the gear.

Main gear

The main gear is also hydraulically operated and retracts inwards into wheel bays. Once retracted the main units are fully enclosed by means of fairings attached to the legs and by hydraulically operated doors. Each unit is operated by a single jack and a mechanical linkage maintains the gear in the locked position without hydraulic assistance. The main wheel door's jacks are controlled by a sequencing mechanism that closes the doors when the gear is fully extended or retracted. Figure 4.20 shows the landing-gear sequence for the BAE SYSTEMS 146 aircraft and also shows the clean lines of the nose wheel bay with the doors shut.

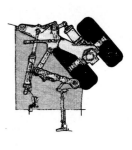

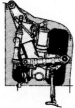

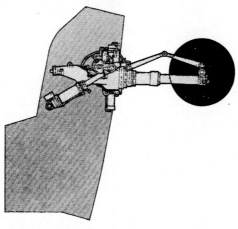

Landing gear extension sequence (front view, left leg). A single hydraulic jack effects extension/retraction of each leg; there is thus no jack sequencing. When viewed from rear, locking/bracing geometry and ease of oleo removal are clear.

Fig. 4.20 The 146 landing-gear sequence (BAE SYSTEMS)

As with main gear, nose gear actuation is by a single jack with mechanically linked doors. Photo shows how closed doors keep nosewheel bay clean.

Steering

A hydraulically operated steering mechanism turns the nose wheel up to 45 degrees from centre. The steering motor responds to demands from the rudder pedals when nose wheel steering is selected.

Braking and anti-skid

Stopping an aircraft safely at high landing speeds on a variety of runway surfaces and temperatures, and under all weather conditions demands an effective braking system. Its design must take into account tyre to ground and brake friction, the brake pressure/volume characteristics, and the response of the aircraft hydraulic system and the aircraft structural and dynamic characteristics. Simple systems are available which provide reasonable performance at appropriate initial and maintenance costs. More complex systems are available to provide minimum stopping distance performance with features such as auto-braking during landing and rejected take-off, additional redundancy and self-test.

One of the simplest and most widely known anti-skid systems is the Dunlop Maxaret unit which consists of a hydraulic valve assembly regulated by the dynamics of a spring-loaded g sensitive flywheel. Figure 4.21 shows an axle-mounted Maxaret together with a modulator.

Rotation of the flywheel is by means of a self-aligning drive from the hub of the wheel, allowing the entire unit to be housed within the axle and protecting the unit from the effects of weather and stones thrown up by the aircraft wheels. Skid conditions are detected by the over-run of the flywheel which opens the Maxaret valve to allow hydraulic pressure to dissipate. A combination of flow-sensitive hydraulic units and switches in the oleo leg provide modulation of pressure for optimum braking force and protection against inadvertent application of the brakes prior to touch-down. This ensures that the aircraft does not land with the brakes applied by only allowing the braking system to become active after the oleo switches have sensed that the oleo is compressed. This condition is known as 'weight-on-wheels'. Without this protection the effect of landing with full braking applied could lead to loss of control of the aircraft; at a minimum a set of burst tyres.

Electronic control

Electronic control of braking and anti-skid systems has been introduced in various forms to provide different features. An electronic anti-skid system with adaptive pressure control is shown in Fig. 4.22.

In this system the electronic control box contains individual wheel deceleration rate skid detection circuits with cross-reference between wheels and changeover circuits to couple the control valve across the aircraft should the loss of a wheel speed signal occur. If a skid develops the system disconnects braking momentarily and the adaptive pressure co-ordination valve ensures that brake pressure is reapplied at a lower pressure after the skid than the level which allowed the skid to occur. A progressive increase in brake pressure between skids attempts to maintain a high level of pressure and braking efficiency.

The adaptive pressure control valve dumps hydraulic pressure from the brake when its first-stage solenoid valve is energized by the commencement of a skid signal. On

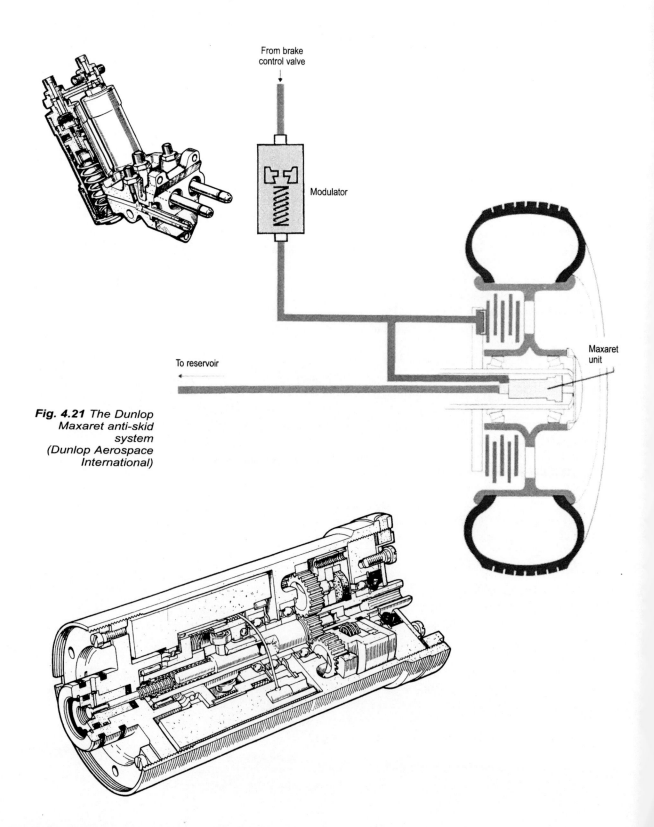

From brake
control valve

Modulator

To reservoir

Maxaret
unit

Fig. 4.21 The Dunlop
Maxaret anti-skid
system
(Dunlop Aerospace
International)

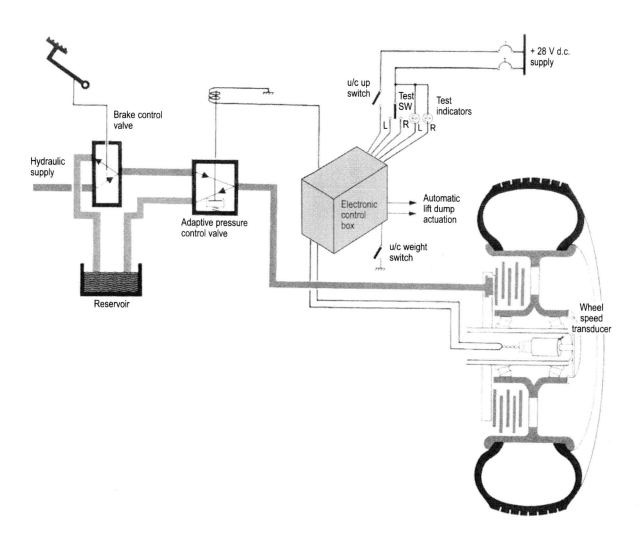

Fig. 4.22 Electronic anti-skid with adaptive pressure control (Dunlop Aerospace International)

wheel speed recovery the solenoid is de-energized and the brake pressure reapplied at a reduced pressure level, depending on the time interval of the skid. Brake pressure then rises at a controlled rate in search of the maximum braking level, until the next incipient skid signal occurs.

Automatic braking

A more comprehensive system is the Dunlop automatic brake control system illustrated in Fig. 4.23 which allows an aircraft to be landed and stopped without pilot braking intervention. During automatic braking a two-position three-way solenoid valve is energized following wheel spin-up to feed system pressure via shuttle valves directly to the anti-skid valves where it is modulated and passed to the brakes. Signals from the auto-braking circuit are responsible for this modulation of pressure at the brake to match a preselected deceleration. However, pilot intervention in the anti-skid control circuit or anti-skid operation will override auto-brake at all times to cater for variations in runway conditions.

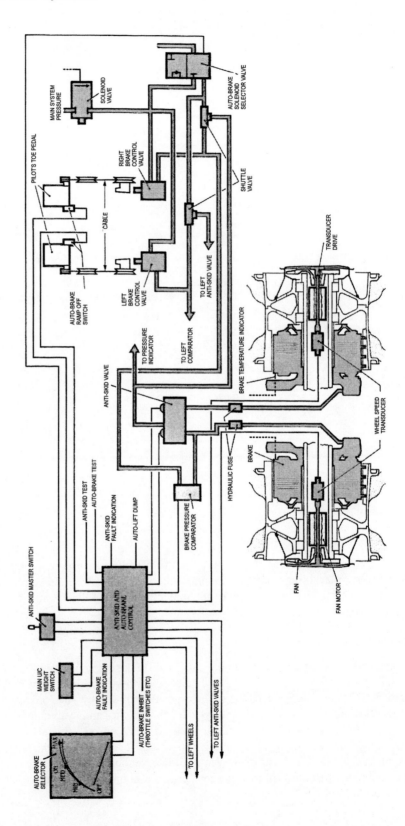

Fig. 4.23 An automatic brake control system (Dunlop Aerospace International)

In the interest of safety a number of prerequisites must be satisfied before auto-braking is initiated:

- Auto-brake switch must be ON and required deceleration selected.
- Anti-skid switch must be ON and operative.
- Throttle must be correctly positioned.
- Hydraulic pressure must be available.
- Brake pedals must not be depressed.
- Wheels must be spun up.

With all these conditions satisfied auto-braking will be operational and will retard the aircraft at a predetermined rate unless overridden by anti-skid activity. At any time during the landing roll the auto-braking may be overridden by the pilot by advancing the throttle levers for go-around, or by normal application of the brakes.

Multi-wheel systems

The systems described thus far apply to most aircraft braking systems. However large aircraft have multi-wheel bogies and sometimes more than two main gears. The Boeing 747-400 has four main oleos, each with a total of four wheels. The Boeing 777 has two main bogies with six wheels each. These systems tend to be more complex and utilize multi-lane dual redundant control. The Boeing 777 main gear shown in Fig. 4.24 is an example.

Fig. 4.24 Simplified Boeing 777 braking configuration

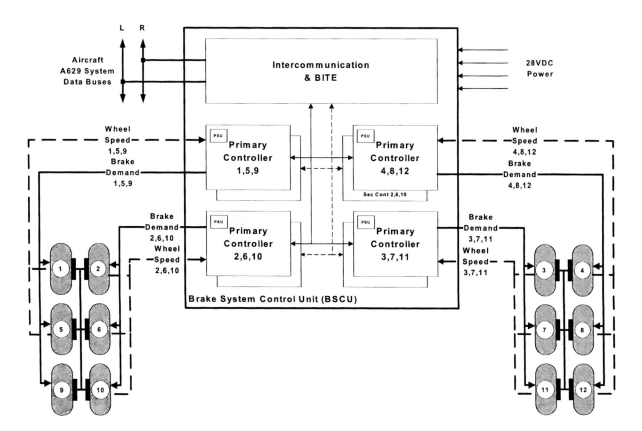

For control purposes the wheels are grouped in four lines of three wheels, each corresponding to an independent control channel as shown in the figure. Each of the lines of three wheels – 1,5,9; 2,6,10 and so on – is controlled by a dual-redundant controller located in the Brake System Control Unit (BSCU). Brake demands and wheel speed sensor readings are grouped by each channel and interfaced with the respective channel control. Control channels have individual power supplies to maintain channel segregation and integrity. The BSCU interfaces with the rest of the aircraft by means of left and right A629 aircraft systems data buses. This system is supplied by the Hydro-Aire division, part of Crane Aerospace, and is indicative of the sophistication which modern brake systems offer for larger systems.

References

(1)   SAE Aerospace Information Report (AIR) 5005, Aerospace – Commercial Aircraft Hydraulic Systems, issued March 2000.

CHAPTER 5

Electrical Systems

Introduction

Electrical systems have made significant advances over the years as aircraft have become more dependent upon electrically powered services. A typical electrical power system of the 1940s and 1950s was the twin 28 VDC system. This system was used a great deal on twin-engined aircraft; each engine powered a 28 VDC generator which could employ load sharing with its contemporary if required. One or two DC batteries were also fitted and an inverter was provided to supply 115 VAC to the flight instruments.

The advent of the V-bombers changed this situation radically due to the much greater power requirements – one, the Vickers Valiant, incorporated electrically actuated landing-gear. They were fitted with four 115 VAC generators, one being driven by each engine. To provide the advantages of no-break power these generators were paralleled which increased the amount of control and protection circuitry. The V-bombers had to power high loads such as radar and electronic warfare jamming equipment. However, examination of the Nimrod maritime patrol aircraft (derived from the de Havilland Comet) shows many similarities. As a yardstick of the rated power generated: the Victor (see Fig. 5.1) was fitted with four 73 kVA AC generators while the Nimrod was fitted with four 60 kVA generators.

Aircraft such as the McDonnell Douglas F-4 Phantom introduced high power AC generation systems to a fighter application. In order to generate constant frequency 115 VAC at 400 Hz, a Constant Speed Drive or CSD is required to negate the aircraft engine speed variation over approximately 2:1 speed range (full power speed: flight idle speed). These are complex hydromechanical devices which by their very nature are not highly reliable. Therefore the introduction of constant frequency AC generation systems was not without accompanying reliability problems, particularly on fighter aircraft where engine throttle settings are changed very frequently throughout the mission.

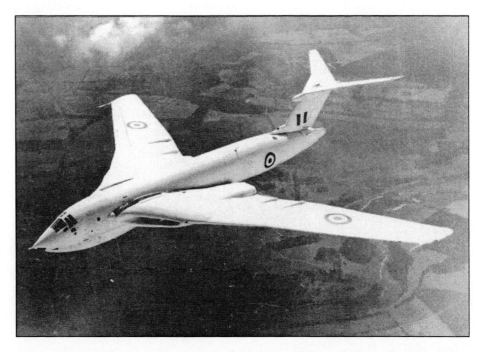

The advances in high-power solid-state switching technology together with enhancements in the necessary control electronics have made Variable-Speed/Constant Frequency (VSCF) systems a viable proposition in the last decade. The VSCF system removed the unreliable CSD portion; the variable frequency or frequency-wild power from the AC generator being converted to 400 Hz constant frequency 115 VAC power by means of a solid-state VSCF converter. VSCF systems are now becoming more commonplace: the F-18 fighter uses such a system and some versions of the Boeing 737-500 did use such a system, not with a lot of success in that particular case. In addition, the Boeing 777 airliner utilizes a VSCF system for back-up AC power generation.

In US military circles great emphasis is being placed by the US Air Force and the US Navy into the development of 270 VDC systems. In these systems high-power generators derive 270 VDC power, some of which is then converted into 115 VAC 400 Hz or 28 VDC required to power specific equipments and loads. This is claimed to be more efficient than conventional methods of power generation and the amount of power conversion required is reduced with accompanying weight savings. These developments are allied to the 'more-electric aircraft' concept where it is intended to ascribe more aircraft power system activities to electrical means rather than use hydraulic or high-pressure bleed air which is presently the case. The fighter aircraft of tomorrow will therefore need to generate much higher levels of electrical power than at present. Schemes for the use of 270 VDC are envisaging power of 250–300 kW and possibly as much as 500 kW per channel: up to ten times the typical level of 50 kVA per channel of today.

At the component level, advances in the development of high-power contactors and solid-state power-switching devices are improving the way in which aircraft primary and secondary power loads are switched and protected. These advances are being married to micro-electronic developments to enable the implementation of new concepts for electrical power management system distribution, protection and load switching. The use of electrical power has progressed to the point where the generation, distribution and protection of electrical power to the aircraft electrical services or loads now comprises one of the most complex aircraft systems. This situation was not always so.

The move towards the higher AC voltage is really driven by the amount of power the electrical channel is required to produce. The sensible limit for DC systems has been found to be around 400 A due to the limitations of feeder size and high-power protection switchgear, known as contactors. Therefore for a 28 VDC system delivering 400 A, the maximum power the channel may deliver is about 12 kW, well below the requirements of most aircraft today. This level of power is sufficient for General Aviation (GA) aircraft and some of the smaller business jets. However, the requirements for aircraft power in business jets, regional aircraft and larger transport aircraft is usually in the range 20–90 kVA per channel and higher. This requirement for more power has been matched in the military aircraft arena.

Aircraft electrical system characteristics

The generic parts of a typical alternating current (AC) aircraft electrical system are shown in Fig. 5.2 and comprise the following:

- Power generation
- Primary power distribution and protection
- Power conversion and energy storage
- Secondary power distribution and protection

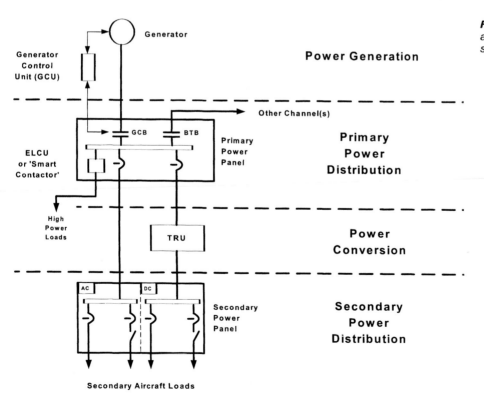

Fig. 5.2 *Generic aircraft AC electrical system*

At this stage it is worth outlining the major differences between AC and DC power generation. Later in the chapter more emphasis is placed upon more recent AC power generation systems.

Power generation

DC power generation

DC systems use generators to develop a DC voltage to supply aircraft system loads; usually the voltage is 28 VDC but there are 270 VDC systems in being which will be described later in the chapter. The generator is controlled – the technical term is regulated – to supply 28 VDC at all times to the aircraft loads such that any tendencies for the voltage to vary or fluctuate are overcome. DC generators are self-exciting, in that they contain rotating electromagnets that generate the electrical power. The conversion to DC power is achieved by using a device called a commutator which enables the output voltage, which would appear as a simple sine wave output, to be effectively half-wave rectified and smoothed to present a steady DC voltage with a ripple imposed.

In aircraft applications the generators are typically shunt-wound in which the high resistance field coils are connected in parallel with the armature as shown in Fig. 5.3.

***Fig. 5.3** Shunt-wound DC generator*

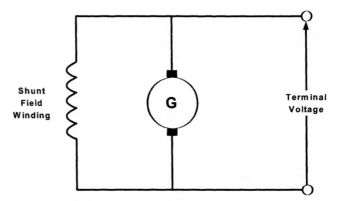

The natural load characteristic of the shunt-wound generator is for the voltage to 'droop' with the increasing load current, whereas the desired characteristic is to control the output at a constant voltage – nominally 28 VDC. For this purpose a voltage regulator is used which modifies the field current to ensure that terminal voltage is maintained while the aircraft engine speed and generator loads vary. The principle of operation of the DC voltage regulator is shown in Fig. 5.4 and it is described later in the chapter.

AC power generation

An AC system uses a generator to generate a sine wave of a given voltage and, in most cases, of a constant frequency. The construction of the alternator is simpler than that of the DC generator in that no commutator is required. Early AC generators used slip rings to pass current to/from the rotor windings; however these suffered from abrasion and pitting, especially when passing high currents at altitude. Modern AC generators work on the principle shown in Fig. 5.5.

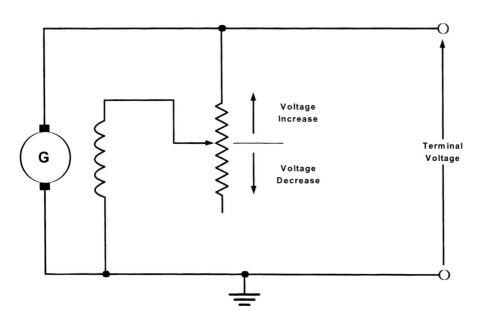

Fig. 5.4 DC voltage regulator

Fig. 5.5 Principle of operation of modern AC generator

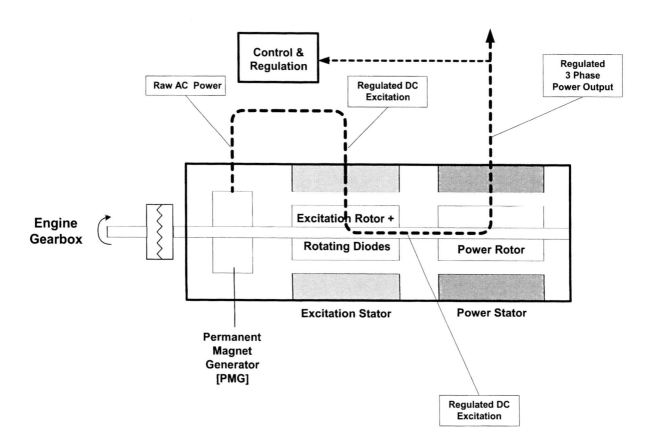

This AC generator may be regarded as several machines sharing the same shaft. From right to left as viewed on the diagram they comprise:

- A Permanent Magnet Generator (PMG)
- An excitation stator surrounding an excitation rotor containing rotating diodes
- A power rotor encompassed by a power stator

The flow of power through this generator is highlighted by the dashed line. The PMG generates 'raw' (variable frequency, variable voltage) power sensed by the control and regulation section that is part of the generator controller. This modulates the flow of DC current into the excitation stator windings and therefore controls the voltage generated by the excitation rotor. The rotation of the excitation rotor within the field produced by the excitation stator windings is rectified by means of diodes contained within the rotor and supplies a regulated and controlled DC voltage to excite the power rotor windings. The rotating field generated by the power rotor induces an AC voltage in the power stator that may be protected and supplied to the aircraft systems.

Most AC systems used on aircraft use a three-phase system, that is the alternator generates three sine waves, each phase positioned 120 degrees out of phase with the others. These phases are most often connected in a star configuration with one end of each of the phases connected to a neutral point as shown in Fig. 5.6. In this layout the phase voltage of a standard aircraft system is 115 VAC, whereas the line voltage measured between lines is 200 VAC. The standard for aircraft frequency-controlled systems is 400 cycles/sec or 400 Hz.

The descriptions given above outline the two primary methods of power generation used on aircraft for many years. The main advantage of AC power is that it operates at a higher voltage; 115 VAC rather than 28 VDC for the DC system. The use of a higher voltage is not an advantage in itself, in fact higher voltages require better standards of insulation. It is in the transmission of power that the advantage of higher voltage is most apparent. For a given amount of power transmission a higher voltage relates to an equivalent lower current. The lower the current the lower are losses such as voltage

Fig. 5.6 *Star connected 3-phase AC generator*

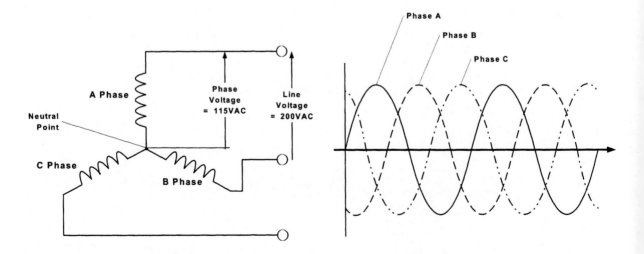

drops (proportional to current) and power losses (proportional to current squared). Also as current conductors are generally heavy is can be seen that the reduction in current also saves weight; a very important consideration for aircraft systems.

Power generation control

The primary elements of power system control are:

- DC systems
  - Voltage regulation
  - Parallel operation
  - Protection functions

- AC systems
  - Voltage regulation
  - Parallel operation
  - Supervisory functions

DC system generation control

Voltage regulation

DC generation is by means of shunt-wound self-exciting machines as briefly outlined above. The principle of voltage regulation is outlined in Fig. 5.4. This shows a variable resistor in series with the field winding such that variation of the resistor alters the resistance of the field winding; hence the field current and output voltage may be varied. In actual fact the regulation is required to be an automatic function that takes account of load and engine speed. The voltage regulation needs to be in accordance with the standard used to specify aircraft power generation systems, namely MIL-STD-704D. This standard specifies the voltage at the point of regulation and the nature of the acceptable voltage drops throughout the aircraft distribution, protection and wiring system. DC systems are limited to around 400 A or 12 kW per channel maximum for two reasons:

(1) The size of conductors and switchgear to carry the necessary current becomes prohibitive.
(2) The brush wear on brushed DC generators becomes excessive with resulting maintenance costs if these levels are exceeded.

Parallel operation

In multi-engined aircraft each engine will be driving its own generator and in this situation it is desirable that 'no-break' or uninterrupted power is provided in cases of engine or generator failure. A number of sensitive aircraft instruments and navigation devices which comprise some of the electrical loads may be disturbed and may need to be restarted or re-initialized following a power interruption. In order to satisfy this requirement generators may be paralleled to carry an equal proportion of the electrical load between them. Individual generators are controlled by means of voltage regulators that automatically compensate for variations. In the case of parallel generator operation there is a need to interlink the voltage regulators such that any unequal loading of the

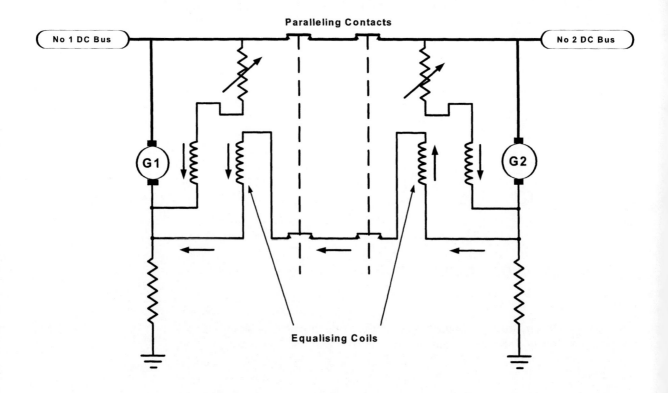

Fig. 5.7 *DC generator parallel operation*

generators can be adjusted by means of corresponding alterations in field current. This paralleling feature is more often known as an equalizing circuit and therefore provides 'no-break' power in the event of a major system failure. A simplified diagram showing the main elements of DC parallel operation is at Fig. 5.7.

Protection functions

The primary conditions for which protection needs to be considered in a DC system are as follows.

- Reverse current. In a DC system it is evident that the current should flow from the generator to the busbars and distribution systems. In a fault situation it is possible for current to flow in the reverse direction and the primary system components need to be protected from this eventuality. This is usually achieved by means of reverse current circuit-breakers or relays. These devices effectively sense reverse current and switch the generator out of circuit thus preventing any ensuing damage.

- Overvoltage protection. Faults in the field excitation circuit can cause the generator to overexcite and thereby regulate the supply voltage to an erroneous overvoltage condition. This could then result in the electrical loads being subject to conditions that could cause permanent damage. Overvoltage protection senses these failure conditions and opens the line contactor taking the generator off-line.

- Undervoltage protection. In a single generator system undervoltage is a similar fault condition as the reverse current situation already described. However, in a multi-generator configuration with paralleling by means of an equalizing circuit, the situation is different. Here an undervoltage protection capability is essential as the equalizing circuit is always trying to raise the output of a lagging generator; in this situation the undervoltage protection is an integral part of the parallel load-sharing function.

AC power generation control

Voltage regulation

As has already been described, AC generators differ from DC machines in that they require a separate source of DC excitation for the field windings although the system described earlier does allow the generator to bootstrap the generation circuits. The subject of AC generator excitation is a complex topic for which the technical solutions vary according to whether the generator is variable frequency or constant frequency. Some of these solutions comprise sophisticated control loops with error detectors, pre-amplifiers and power amplifiers.

Parallel operation

In the same way that DC generators are operated in parallel to provide 'no-break' power, AC generators may also be controlled in a similar fashion. This technique only applies to constant frequency AC generation as it is impossible to parallel frequency-wild or Variable Frequency (VF) AC generators. In fact many of the aircraft loads such as anti/de-icing heating elements driven by VF generators are relatively frequency-insensitive and the need for 'no-break' power is not nearly so important. To parallel AC machines the control task is more complex as both real and reactive (imaginary) load components have to be synchronized for effective load sharing.

The sharing of real load depends upon the relative rotational speeds and hence the relative phasing of the generator voltages. Constant speed or constant frequency AC generation depends upon the tracking accuracy of the constant speed drives of the generators involved. In practice real load sharing is achieved by control laws which measure the degree of load imbalance by using current transformers and error detection circuitry thereby trimming the constant speed drives such that the torques applied by all generators are equal.

The sharing of reactive load between the generators is a function of the voltage generated by each generator as for the DC parallel operation case. The generator output voltages depend upon the relevant performance of the voltage regulators and field excitation circuitry. To accomplish reactive load sharing requires the use of special transformers called mutual reactors, error detection circuitry and pre-amplifiers/power amplifiers to adjust the field excitation current. Therefore by using a combination of trimming the speed of the constant speed drives and balancing the field excitation to the generators, real and reactive load components may be shared equally between the generators. Refer to Fig. 5.8. This has the effect of providing a powerful single vector AC power supply to the aircraft AC system providing a very 'stiff' supply in periods of high power demand. Perhaps the biggest single advantage of paralleled operation is that all the generators are operating in phase synchronism, therefore in the event of a failure there are no change-over transients.

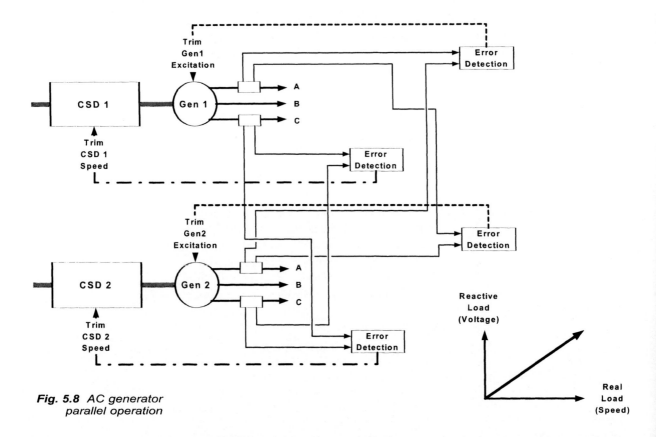

Fig. 5.8 *AC generator*
parallel operation

Supervisory and protection functions

Typical supervisory or protection functions undertaken by a typical AC Generator
Control Unit or GCU are listed below:

- Overvoltage
- Undervoltage
- Under/over-excitation
- Under/over-frequency
- Differential current protection

The overvoltage, undervoltage and under/over-excitation functions are similar to the
corresponding functions described for DC generation control. Under/over-frequency
protection is effectively executed by the real load-sharing function already described
above for AC parallel operation. Differential current protection is designed to detect a
short-circuit busbar or feeder line fault which could impose a very high current demand
on the short-circuited phase. Differential current transformers sense the individual
phase currents at differing parts of the system. These are connected so that detection
circuitry will sense any gross difference in phase current (say in excess of 30 amps per
phase) resulting from a phase imbalance and disconnect the generator from the busbar
by tripping the Generator Control Breaker (GCB).

Modern electrical power generation types

So far basic DC and AC power generating systems have been described. The DC system is limited by currents greater than 400 A and the constant frequency AC method using an Integrated Drive Generator (IDG) has been mentioned. In fact there are many more power generation types in use today. A number of recent papers have identified the issues and projected the growth in aircraft electric power requirements in a civil aircraft setting, even without the advent of more-electric systems. However not only are aircraft electrical system power levels increasing but the diversity of primary power generation types is increasing.

Fig. 5.9 Electrical power generation types

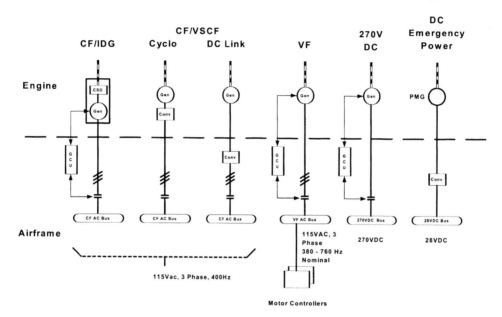

The different types of electrical power generation currently being considered are shown in Fig. 5.9. The Constant Frequency (CF) 115 VAC, three-phase, 400 Hz options are typified by the Integrated Drive Generator (IDG), variable speed constant frequency (VSCF) cycloconverter and DC link options. Variable frequency (VF) 115 VAC, three-phase power generation – sometimes termed 'frequency-wild' – is also a more recent contender, and although a relatively inexpensive form of power generation, it has the disadvantage that some motor loads may require motor controllers. Military aircraft in the US are inclining toward 270 VDC systems. Permanent Magnet Generators (PMGs) are used to generate 28 VDC emergency electrical power for high-integrity systems.

Figure 5.9 is also interesting in that it shows the disposition between generation system components located on the engine and those within the airframe. Without being drawn into the partisan arguments regarding the pros and cons of the major types of power generation in use or being introduced today it is worth examining the main contenders:

- Constant frequency using an IDG
- Variable frequency
- Variable Speed Constant Frequency (VSCF) options

Constant frequency/IDG

Fig. 5.10 *Constant
frequency/IDG
generation*

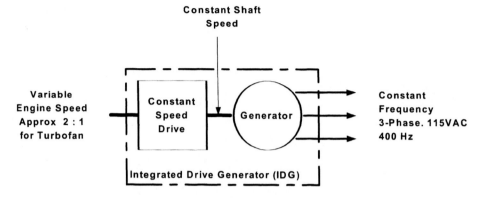

Features:

Constant frequency AC power is most commonly used on turbofan aircraft today

System is expensive to purchase & maintain; primarily due to complexity of Constant Speed Drive (CSD)

Single company monopoly on supply of CSD/IDG

Alternate methods of power generation are under consideration

The main features of CF/IDG power are shown in Fig. 5.10. In common with all the other power generation types this has to cater for a 2:1 ratio in engine speed between maximum power and ground idle. The Constant Speed Drive (CSD) in effect acts as an automatic gearbox, maintaining the generator shaft speed at a constant rpm which results in a constant frequency output of 400 Hz, usually within approximately 10 Hz or less. The drawback of the hydromechanical CSD is that it needs to be correctly maintained in terms of oil charge level and oil cleanliness. Also to maintain high reliability frequent overhauls may be necessary.

That said, the IDG is used to power the majority of civil transport aircraft today as shown in Table 5.1.

Variable frequency

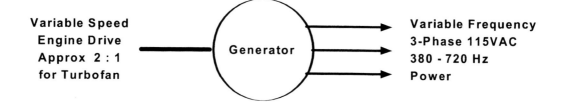

Variable Speed
Engine Drive
Approx 2 : 1
for Turbofan

Generator

Variable Frequency
3-Phase 115VAC
380 - 720 Hz
Power

Features:

Simplest form of generating power, cheapest and most reliable

Variable frequency has impact upon other aircraft subsystems

Motor controllers may be needed for certain aircraft loads

Beginning to be adopted for new programmes: gains outweigh disadvantages

Variable frequency (VF) power generation as shown in Fig. 5.11. is the simplest and most reliable form of power generation. In this technique no attempt is made to nullify the effects of the 2:1 engine speed ratio and the power output, though regulated to 115 VAC, suffers a frequency variation typically from 380 to 720 Hz. This wide band VF power has an effect on frequency-sensitive aircraft loads, the most obvious being the effect on AC electric motors that are used in many aircraft systems. There can therefore be a penalty to be paid in the performance of other aircraft systems such as fuel, ECS and hydraulics. In many cases variations in motor/pump performance may be accommodated but in the worst cases a motor controller may be needed to restore an easier control situation.

Fig. 5.11 Variable frequency power generation

VF is being widely adopted in the business jet community as their power requirements take them above the 28 VDC/12 kW limit of twin 28 VDC systems. Aircraft such as Global Express had VF designed in from the beginning. Other VF power users are the Boeing X-32A/B/C JSF contender and VF power generation has been established as baseline for the Airbus A380 project.

VSCF

Figure 5.12 shows the concept of the VSCF converter. In this technique the variable frequency power produced by the generator is electronically converted by solid-state

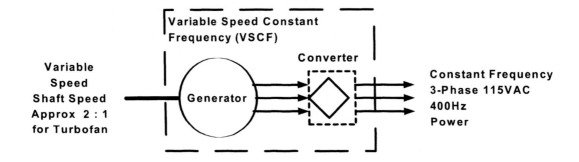

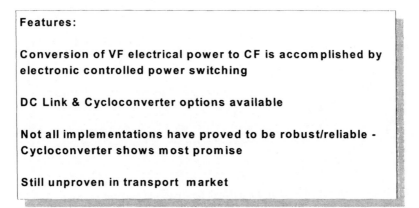

Features:

Conversion of VF electrical power to CF is accomplished by electronic controlled power switching

DC Link & Cycloconverter options available

Not all implementations have proved to be robust/reliable - Cycloconverter shows most promise

Still unproven in transport market

Fig. 5.12 VSCF power generation

power-switching devices to constant frequency 400 Hz, 115 VAC power. Two options exist:

DC link: In the DC link the raw power is converted to an intermediate DC power stage – the DC link – before being electronically converted to three-phase AC power. DC link technology has been used on the B737, MD-90 and B777 but has yet to rival the reliability of CF or VF power generation.

Cycloconverter: The cycloconverter uses a different principle. Six phases are generated at relatively high frequencies in excess of 1,600 Hz and the solid-state devices switch between these multiple phases in a predetermined and carefully controlled manner. The effect is to electronically commutate the input and provide three phases of constant frequency 400 Hz power. Though this appears to be a complex technique it is in fact quite elegant and cycloconverter systems have been successfully used on military aircraft in the US: F-18, U-2 and the F-117A stealth fighter. As yet no civil applications have been used. The cycloconverter concept is revisited later in the chapter.

As suggested earlier in Fig. 5.9 each of these techniques may locate the power conversion section on the engine or in the airframe. Reference (1) examines the implications of moving the VSCF converter from the engine to the airframe in a civil aircraft context.

Table 5.1   *Recent civil and military aircraft power system developments*

Generation type	Civil application		Military application	
IDG/CF [115 VAC/400Hz]	B777 A340 B737NG MD-12 B747-X B717 B767-400 Do728	2 x 120 kVA 4 x 90 kVA 2 x 90 kVA 4 x 120 kVA 4 x 120 kVA 2 x 40 kVA 2 x 120 kVA 2 x 40 kVA		
VSCF (Cycloconverter) [115 V AC/400 Hz]			F-18E/F	2 x 60/65 kVA
VSCF (DC link) [115 VAC/400 Hz]	B777 (Backup) MD-90	2 x 20 kVA 2 x 75 kVA		
VF [115 VAC/380 – 760Hz typical]	Global Ex Horizon A380	4 x 40 kVA 2 x 20/25 kVA 4 x 150 kVA	Boeing JSF [X-32A/B/C]	2 x 50 kVA
270 VDC			F-22 Raptor Lockheed-Martin JSF [X-35A/B/C]	2 x 70 kVA 2 x 50 kVA

Table 5.1 lists the power generation types developed and proposed for civil and military (fighter) aircraft platforms throughout the 1990s. Not only are the electrical power levels increasing in this generation of aircraft but the diversity of electrical power generation methods introduce new aircraft system issues which need to be addressed. For example the Boeing 777 stand-by VSCF and the MD-90 VCSF converters, being located in the airframe, increase the ECS requirements since waste heat is dissipated in the airframe whereas the previous IDG solution rejected heat into the engine oil system. Similarly the adoption of variable frequency (VF) can complicate motor load and power conversion requirements. The adoption of 270 VDC systems by the US military has necessitated the development of a family of 270 VDC protection devices since conventional circuit-breakers cannot be used at such high voltages.

Primary power distribution

The primary power distribution system consolidates the aircraft electrical power inputs. In the case of a typical civil airliner the aircraft may accept power from the following sources.

- Main aircraft generator; by means of a Generator Control Breaker (GCB) under the control of the GCU.
- Alternate aircraft generator – in the event of generator failure; by means of a Bus Tie Breaker (BTB) under the control of a Bus Power Control Unit (BPCU).
- APU generator; by means of an APU GCB under the control of the BPCU.
- Ground power; by means of an External Power Contactor (EPC) under the control of a ground power monitor.

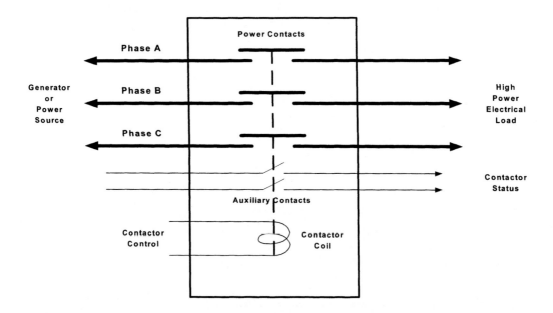

Fig. 5.13(a) Power contactor

- Back-up converter; by means of a Converter Control Breaker (CCB) under the control of the VSCF Converter (Boeing 777 only).
- RAT generator when deployed by the emergency electrical system.

The power switching used in these cases is a power contactor or breaker. These are special high-power switches that usually switch power in excess of 20 A per phase. As well as the power-switching contacts auxiliary contacts are included to provide contactor status – 'Open' or 'Closed – to other aircraft systems.

Higher power aircraft loads are increasingly switched from the primary aircraft bus-bars by using Electronic Load Control Units (ELCUs) or 'smart contactors' for load protection. Like contactors these are used where normal rated currents are greater than 20 A per phase, i.e. for loads of around 7 kVA or greater. Figure 5.13(a) shows the comparison of a line contactor such as a GCB with an ELCU or 'smart contactor' in Fig. 5.13(b). The latter has in-built current sensing coils that enable the current of all three phases to be measured. Associated electronics allow the device trip characteristics to be more closely matched to those of the load. Typical protection characteristics embodied within the electronics are I^2t, modified I^2t and differential current protection. For a paper explaining more about 'smart contactors' refer to reference (2).

Power conversion and energy storage

This chapter so far has addressed the primary generation of electrical power and primary power distribution and protection. There are however many occasions within an aircraft electrical system where it is required to convert power from one form to another. Typical examples of power conversion are:

- Conversion from DC to AC power. This conversion uses units called inverters to convert 28 VDC to 115 VAC single-phase or three-phase power.
- Conversion from 115 VAC to 28 VDC power. This is a much used conversion using units called Transformer Rectifier Units (TRUs).

Fig. 5.13(b) *ELCU or 'smart contactor'*

- Conversion from one AC voltage level to another; a typical conversion would be from 115 VAC to 26 VAC.
- Battery charging. As previously outlined it is necessary to maintain the state of charge of the aircraft battery by converting 115 VAC to a 28 VDC battery charge voltage.

Inverters

Inverters convert 28 VDC power into 115 VAC single-phase electrical power. This is usually required in a civil application to supply Captains or First Officers instruments following an AC failure. Alternatively, under certain specific flight conditions, such as autoland, the inverter may be required to provide an alternative source of power to the flight instruments in the event of a power failure occurring during the critical autoland phase. Some years ago the inverter would have been a rotary machine with a DC motor harnessed in tandem with an AC generator. More recently the power conversion is likely to be accomplished by means of a static inverter where the use of high-power, rapid-switching, Silicon-Controlled Rectifiers (SCRs) will synthesize the AC waveform from the DC input. Inverters are therefore a minor though essential part of many aircraft electrical systems.

Transformer Rectifier Units (TRUs)

TRUs are probably the most frequently used method of power conversion on modern aircraft electrical systems. Most aircraft have a significant 115 VAC three-phase AC power generation capability inherent within the electrical system and it is usual to convert a significant portion of this to 28 VDC by the use of TRUs. TRUs comprise star primary and dual star/delta secondary transformer windings together with three-phase full wave rectification and smoothing to provide the desired 115 VAC/28 VDC conversion. A typical TRU will convert a large amount of power, for example the

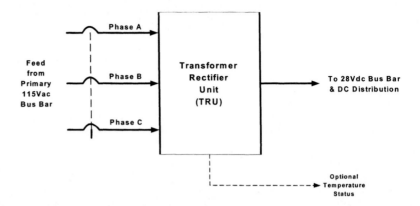

Fig. 5.14 *Transformer rectifier unit (TRU)*

Boeing 767 uses two TRUs, each of which supply a rated load of up to 120 A (continuous) with a five minute rating of 180 A. TRUs dissipate a lot of heat and are therefore forced air cooled. The Boeing 767 unit is packaged in a 6 MCU ARINC 600 case and weighs around 24 lb. Figure 5.14 shows a typical TRU.

TRUs are usually simple, unregulated units, that is the voltage is not controlled to 28 VDC as load is increased and accordingly the load characteristic tends to 'droop'. In some specialist military applications this feature is not desirable and regulated TRUs are used. TRUs are usually operated in isolation, however when regulated they may also be configured to operate in parallel in a similar way to the parallel operation of DC generators. Reference (3) is a paper relating to the development of a regulated TRU.

Auto-transformers

In certain parts of an electrical system simple auto-transformers may be used to provide a simple voltage step-up or step-down conversion. An example of this is the 115 V/26 VAC transformation used to provide 26 VAC aircraft lighting supplies direct from main 115 VAC busbars in the easiest way.

Battery chargers

Battery chargers share many of the attributes of TRUs and are in fact dedicated units whose function is purely that of charging the aircraft battery. In some systems the charger may also act as a stand-by TRU providing a boosted source of DC power to the battery in certain system modes of operation. Usually, the task of the battery charger is to provide a controlled charge to the battery without overheating and for this reason battery temperature is usually closely monitored.

Batteries

The majority of this section has described power generation systems, both DC and AC. However it neglects an omnipresent element – the battery. This effectively provides an electrical storage medium independent of the primary generation sources. Its main purposes are:

- To assist in damping transient loads in the DC system.
- To provide power in system startup modes when no other power source is available.

- To provide a short-term high-integrity source during emergency conditions while alternative/back-up sources of power are being brought on-line.

The capacity of the aircraft battery is limited and is measured in terms of ampere-hours. This parameter effectively describes a current/time capability or storage capacity. Thus a 40 ampere-hour battery when fully charged would have the theoretical capacity of feeding a 1 ampere load for 40 hours or a 40 ampere load for 1 hour. In fact the capacity of the battery depends upon the charge sustained at the beginning of the discharge and this is a notoriously difficult parameter to quantify. Most modern aircraft systems utilize battery chargers to maintain the battery charge at moderately high levels during normal system operation, thereby assuring a reasonable state of charge should solo battery usage be required.

The battery most commonly used is the nickel-cadmium (Ni-Cd) type which depends upon the reaction between nickel oxides for the anode, cadmium for the cathode and operating in a potassium hydroxide electrolyte. Lead-acid batteries are not favoured in modern applications due to corrosive effects. To preserve battery health it is usual to monitor its temperature which gives a useful indication of overcharging and if thermal runaway is likely to occur.

Secondary power distribution

Power switching

In order to reconfigure or to change the state of a system it is necessary to switch power at various levels within the system. At the high-power levels that prevail at the primary power part of the system, power switching is accomplished by high-power electromagnetic devices called contactors. These devices can switch hundreds of amps and are used to switch generator power on to the primary busbars in both DC and AC systems. The devices may be arranged so that they magnetically latch, that is they are magnetically held in a preferred state or position until a signal is applied to change the state. In other situations a signal may be continuously applied to the contactor to hold the contacts closed, and removal of the signal causes the contacts to open. Primary power contactors and ELCUs have been described earlier in the chapter.

For switching currents below 20 A or so, relays are generally used. These operate in a similar fashion to contactors but are lighter, simpler and less expensive. Relays may be used at certain places in the primary electrical system. However relays are more likely to be employed for switching of medium- and high-power secondary aircraft loads or services.

For lower currents still, where the indication of device status is required, simple switches can be employed. These switches may be manually operated by the crew or they may be operated by other physical means as part of the aircraft operation. Such switches are travel limit switches, pressure switches, temperature switches and so on.

Load protection

Circuit-breakers

Circuit-breakers perform the function of protecting a circuit in the event of an electrical overload. Circuit-breakers serve the same purpose as fuses or current limiters. A circuit-breaker comprises a set of contacts which are closed during normal circuit operation. The device has a mechanical trip mechanism which is activated by means of a bi-metallic element. When an overload current flows, the bi-metallic element causes the trip mechanism to activate, thereby opening the contacts and removing power from the circuit. A push-button on the front of the unit protrudes showing that the device has tripped. Pushing in the push-button resets the breaker but if the fault condition still exists the breaker will trip again. Physically pulling the button outwards can also allow the circuit-breaker to break the circuit, perhaps for equipment isolation or aircraft maintenance reasons. Circuit-breakers are rated at different current values for use in differing current-carrying circuits. This enables the trip characteristic to be matched to each circuit. The trip characteristic also has to be selected to co-ordinate with the feeder trip device upstream. Circuit-breakers are used literally by the hundred in aircraft distribution systems; it is not unusual to find 500–600 or more devices throughout a typical aircraft system. Figure 5.15 shows a circuit-breaker and a typical trip characteristic.

Fig. 5.15 *Typical circuit-breaker and trip characteristic*

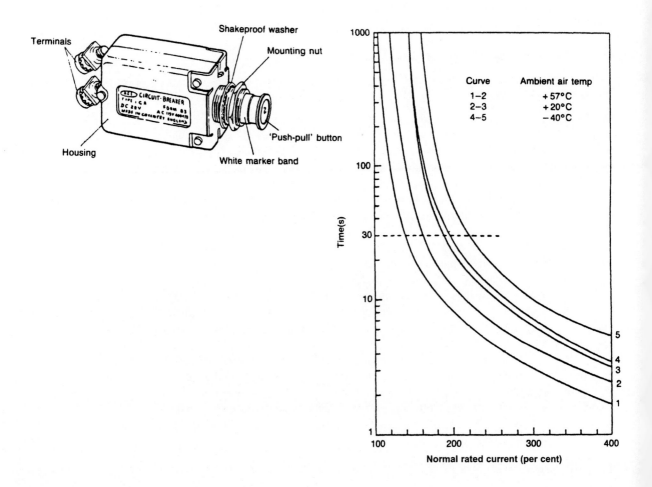

Solid-state power controllers

The availability of high-power solid-state switching devices has been steadily increasing for a number of years, both in terms of variety and rating. More recent developments have led to the availability of solid-state power switching devices which provide a protection capability as well as switching power. These devices known as Solid-State Power Controllers or SSPCs effectively combine the function of a relay or switch and a circuit-breaker. There are disadvantages with the devices available at present; they are readily available up to a rating of 22.5 A for use with DC loads; however the switching of AC loads may only be carried out at lower ratings and with a generally unacceptable power dissipation. Another disadvantage of SSPCs is that they are expensive and costwise may not be comparable with the relay/circuit-breaker combination they replace. They are however predicted to be more reliable than conventional means of switching and protecting small- and medium-sized electrical loads and are likely to become far more prevalent in use in some of the aircraft electrical systems presently under development. SSPCs are also advantageous when utilized in high-duty cycle applications where a relay may wear out.

Present devices are rated at 5, 7.5, 12.5 and 22.5 A and are available to switch 28 VDC and 270 VDC. A recent paper, summarizing the development and capabilities of SSPCs and power management units embodying SSPCs to date, is at reference (4).

Typical aircraft DC system

A generic distribution system is shown in Fig. 5.16. In this case a twin 28 VDC system is shown which might be typical for a twin-engine commuter aircraft requiring less than approximately 12 kW per channel.

The main elements of this electrical system are:

- Two 28 VDC generators operating in parallel to supply No. 1 and No. 2 main DC busbars. These busbars feed the non-essential DC services.
- Two inverters operate, one off each of the DC busbars to provide 115 VAC 400 Hz to non-essential AC services.
- Both No. 1 and No. 2 busbars feed power to a centre or essential busbar which provides DC power for the aircraft essential DC services. An inverter powered off this busbar feeds essential 115 VAC loads. A 28 VDC external power source may also feed this busbar when the aircraft is on the ground without the engines running.
- The aircraft battery feeds the battery busbar from which are fed vital services. The battery may also be connected to the DC essential busbar if required.

To enable a system such as this is to be afforded suitable protection requires several levels of power switching and protection:

- Primary power generation protection of the type described earlier and which includes reverse current and under/over-voltage protection under the control of the voltage regulator. This controls the generator feed contactors which switch the generator output on to the No. 1/No. 2 DC busbars.
- The protection of feeds from the main buses, i.e. the protection of the feeds to the essential busbar. This may be provided by a circuit-breaker or a 'smart' contactor

Fig. 5.16 *Typical twin 28 VDC system*

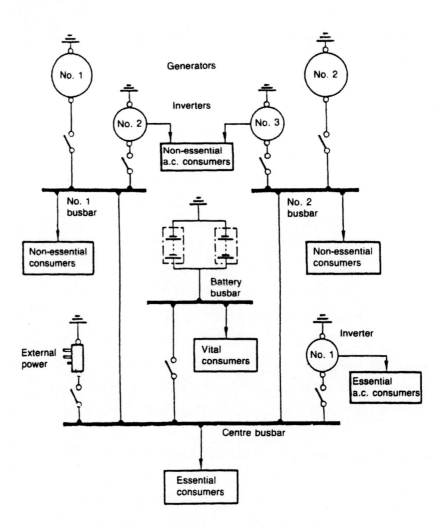

may be used to provide the protection. (Note: The operation of 'smart' contactors will be described later in the chapter).

- The use of circuit-breakers to protect individual loads or groups of loads fed from the supply or feeder busbars.

The cardinal principle is that fault conditions should be contained with the minimum of disruption to the electrical system. Furthermore, faults that cause a load circuit-breaker to trip should not cause the next level of protection to trip also which would be a cascade failure. Thus the trip characteristics of all protection devices should be co-ordinated to ensure that this does not occur.

Typical civil transport aircraft system

A typical civil transport electrical power system is shown in Fig. 5.17. This is a simplified representation of the Boeing 767 aircraft electrical power system that is described in detail in reference (5).

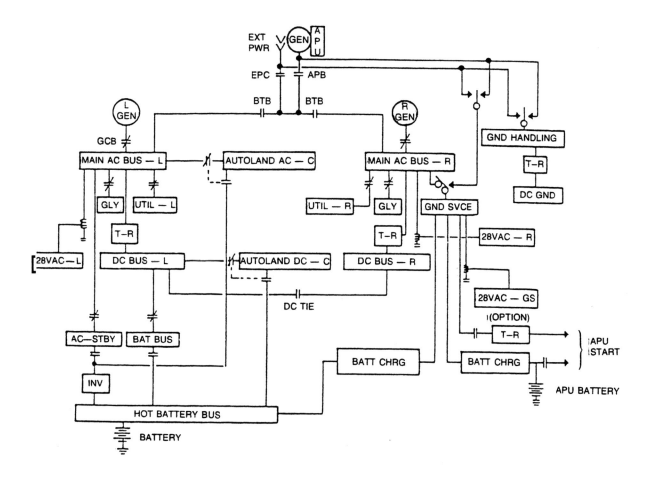

Fig. 5.17 *Simplified Boeing 767 electrical power system (Boeing)*

The primary AC system comprises identical left and right channels. Each channel has an Integrated Drive Generator (IDG) driven from the accessory gearbox of the respective engine. Each AC generator is a three-phase 115 VAC 400 Hz machine producing 90 kVA and is controlled by its own Generator Control Unit (GCU). The GCU controls the operation of the GCB, closing the GCB when all operating parameters are satisfactory and opening the GCB when fault conditions prevail. Two Bus Tie Breakers (BTBs) may be closed to tie both buses together in the event that either generating source is lost. The BTBs can also operate in conjunction with the external power contactor (EPC) or the auxiliary power breaker (APB) to supply both main AC buses power or the 90 kVA APU generator may also feed the ground-handling and ground-servicing buses by means of changeover contactors. The control of the BTBs, EPC, and the ground-handling/servicing contactors is carried out by a unit called a bus power control unit (BPCU). The APU may also be used as a primary power source in flight on certain aircraft in the event that either left or right IDG is lost.

Each of the main AC buses feeds a number of sub-buses or power conversion equipment. TRUs convert 115 VAC to 28 VDC to feed the left and right DC buses respectively. In the event that either main AC bus or TRU should fail, a DC bus-tie contactor closes to tie the left and right DC buses together. The main AC buses also feed the aircraft galleys (a major electrical load) by means of 'smart' contactors. The utility buses are also fed via contactors from each of the main AC buses. In the event of a major electrical system failure the galley loads and non-essential utility bus loads may be shed under the supervision of the BPCU. Both main AC buses feed 26 VAC buses via auto-transformers. Other specific feeds from the left main AC bus are: a switched feed to the autoland AC bus (interlocked with a switched feed from the stand-by inverter); and a switched feed to the AC stand-by bus. Dedicated feeds from the right main AC bus are: via the air/ground changeover contactor to the ground services bus feeding the APU TRU and battery charger; and via the main battery charger to the hot battery bus. The left DC bus also supplies a switched feed to the autoland DC bus (interlocked with a switched feed from the hot battery bus). The hot battery bus also has the capability of feeding the autoland AC bus via the stand-by inverter.

To the uninitiated this may appear to be overly complex; however the reason for this architecture is to provide three independent lanes of AC and DC power for use during autoland conditions. These are:

- Left main AC bus (disconnected from the autoland AC bus) via the left TRU to the left DC bus (which in this situation will be disconnected from the autoland DC bus).
- Right main AC bus via the right TRU, to the right DC bus.
- Right main AC bus via the ground services bus and main battery charger to the hot battery bus and thence to the autoland DC bus (now disconnected from the left DC bus). Also from the hot battery bus via the stand-by inverter to the autoland AC bus (now disconnected from the left main AC bus).

This provides the three independent lanes of electrical power required. It might be argued that two lanes are initially derived from the right main AC bus and therefore the segregation requirements are not fully satisfied. In fact, as the hot battery is fed from the main aircraft battery, this represents an independent source of stored electricity, provided that an acceptable level of charge is maintained. This latter condition is satisfied as the battery charger is fed at all times the aircraft is electrically powered from the ground services bus from either an air or ground source. The battery capacity is such that all stand-by loads may be powered for 30 minutes following primary power loss.

Electrical loads

Once the aircraft electrical power has been generated and distributed then it is available to the aircraft services. These electrical services cover a range of functions spread geographically throughout the aircraft depending upon their task. While the number of electrical services is legion they may be broadly subdivided into the following categories:

- Motors and actuation
- Lighting services
- Heating services
- Subsystem controllers and avionics systems.

Motors and actuation

Motors are obviously used where motive force is needed to drive a valve or an actuator from one position to another depending upon the requirements of the appropriate aircraft system. Typical uses for motors are:

- Linear actuation: electrical position actuators for engine control; trim actuators for flight control systems.
- Rotary actuation: electrical position actuators via screw jacks for flap/slat operation.
- Control valve operation: electrical operation of fuel control valves; hydraulic control valves; air control valves; control valves for ancillary systems.
- Starter motors: provision of starting for engine, APU and other systems that require assistance to reach self-sustaining operation.
- Pumps: provision of motive force for fuel pumps, hydraulic pumps; pumping for auxiliary systems.
- Gyroscope motors: provision of power to run gyroscopes for flight instruments and autopilots.
- Fan motors: provision of power to run cooling fans for the provision of air to passengers or equipment.

Many of the applications for which electric motors are used are not continuously rated; that is, the motor can only be expected to run for a small proportion of the time. Others such as the gyroscope and cooling fan motors may be run continuously throughout the period of operation of the aircraft and the sizing/rating of the motor has to be chosen accordingly. The following categorizes the characteristics of the DC and AC motor types commonly used for aircraft applications.

DC motors

A DC motor is the inverse of the DC generator described earlier in this chapter. It comprises armature field windings and commutator/brushgear and is similarly self-excited. The main elements of importance in relation to motors are the speed and torque characteristics, i.e. the variations of speed and torque with load respectively. Motors are categorised by their field winding configuration (as for generators) and typical examples are series-wound, shunt-wound and compound-wound (a combination of series- and shunt-wound). Each of these types of motor offers differing performance characteristics that may be matched to the application for which they are intended.

A specialized form of series motor is the split-field motor where two sets of series windings of opposite polarity are each used in series with the armature but parallel with each other.

Either one set of field windings or the other may receive power at any one time and therefore the motor may run bi-directionally depending upon which winding is energized. When used in conjunction with suitable switches or relays this type of motor is particularly useful for powering loads such as fuel system valves where there may be a requirement to change the position of various valves several times during flight. Limit switches at the end of the actuator travel prevent the motor/actuator from over-running once the desired position has been reached. Split-field motors are commonly used for linear and rotary position actuators when used in conjunction with the necessary position feedback control.

DC motors are most likely to be used for linear and rotary actuation, fuel valve actuation and starter functions.

AC motors

AC motors used for aircraft applications are most commonly of the 'induction motor' type. An induction motor operates upon the principle that a rotating magnetic field is set up by the AC field current supplied to two or more stator windings (usually three-phase). A simple rotor, sometimes called a 'squirrel cage', will rotate under the effects of this rotating magnetic field without the need for brushgear or slip rings; the motor is therefore simple in construction and reliable. The speed of rotation of an induction motor depends upon the frequency of the applied voltage and the number of pairs of poles used. The advantage of the induction motor for airborne uses is that there is always a source of constant frequency AC power available and for constant rated applications it offers a very cost-effective solution. Single-phase induction motors also exist, however these require a second set of phase windings to be switched in during the start phase, as single-phase windings can merely sustain and not start synchronous running.

AC motors are most likely to be used for continuous operation, i.e. those applications where motors are continuously operating during flight, such as fuel booster pumps, flight instrument gyroscopes and air-conditioning cooling fans.

Lighting

Lighting systems represent an important element of the aircraft electrical services. A large proportion of modern aircraft operating time occurs during night or low-visibility conditions. The availability of adequate lighting is essential to the safe operation of the aircraft. Lighting systems may be categorized as follows:

External lighting systems:
- Navigation lights
- Strobe lights
- Landing/taxi lights
- Formation lights
- Inspection lights (wing/empennage/engine anti-ice)
- Emergency evacuation lights
- Logo lights
- Searchlights (for Search and Rescue or police aircraft)

Internal lighting systems:
- Cockpit/flight-deck lighting (general, spot, flood and equipment panel)
- Passenger information lighting
- Passenger cabin general and personal lighting
- Emergency/evacuation lighting
- Bay lighting (cargo or equipment bays for servicing)

Lighting may be powered by 28 VDC or by 26 VAC provided by auto-transformer from the main AC buses and is mainly achieved by means of conventional filament bulbs. These filaments vary from around 600 W for landing lights to a few watts for minor internal illumination uses. Some aircraft instrument panels or signs may use electroluminescent lighting which is a phosphor layer sandwiched between two electrodes; the phosphor glows when supplied with AC power.

Heating

The use of electrical power for heating purposes on aircraft can be extensive. The highest power usage relates to electrically powered anti-icing or de-icing systems which can consume many tens of kVAs. This power does not have to be frequency-stable and can be variable-frequency and therefore much easier and cheaper to generate. Anti/de-icing elements are frequently used on the tailplane and fin leading edges, intake cowls, propellers and spinners. The precise mix of electrical and hot air (using bleed air from the engines) anti/de-icing methods varies from aircraft to aircraft. Electrical anti/de-icing systems are high current consumers and require controllers to time, cycle and switch the heating current between heater elements to ensure optimum use of the heating capability and to avoid local overheating.

Windscreen heating is another important electrical heating service. In this system the heating element and the controlling thermostat are embedded in the windscreen itself. A dedicated controller maintains the temperature of the element at a predetermined value which ensures that the windscreen is kept free of ice at all times.

Subsystem controllers and avionics systems

As aircraft have become increasingly complex, the sophistication of the aircraft subsystems has increased. Many have dedicated controllers for specific system control functions. For many years the aircraft avionics systems, embracing display, communication and navigation functions, have been packaged into line replaceable units (LRUs) which permit rapid removal should a fault occur. Many of the aircraft subsystem controllers are now packaged into similar LRUs due to increased complexity and functionality and for the same reasons of rapid replacement following a failure. These LRUs may require DC or AC power depending upon their function and modes of operation. Many may utilize dedicated internal power supply units to convert the aircraft power to levels better suited to the electronics that require ±15 VDC and +5 VDC. Therefore these LRUs represent fairly straightforward and, for the most part, fairly low power loads. However there are many of them and a significant proportion may be critical to the safe operation of the aircraft. Therefore two important factors arise: firstly, the need to provide independence of function by distribution of critical LRUs across several aircraft busbars, powered by both DC and AC supplies. Secondly, the need to provide adequate sources of emergency power such that, should a dire emergency occur, the aircraft has sufficient power to supply critical services to support a safe return and landing.

Ground power

For much of the period of aircraft operation on the ground a supply of power is needed. Ground power may be generated by means of a motor-generator set where a prime motor drives a dedicated generator supplying electrical power to the aircraft power receptacle.

The usual standard for ground power is 115 VAC three-phase 400 Hz, that is the same as the aircraft AC generators. In some cases, and this is more the case at major airports, an electrical conversion set adjacent to the aircraft gate supplies 115 VAC three-phase power that has been derived from the national electricity grid. The description given earlier in this chapter of the Boeing 767 system explained how ground power could be applied to the aircraft by closing the EPC.

The aircraft system is protected from sub-standard ground power supplies by means of a ground power monitor. This ensures that certain essential parameters are met before enabling the EPC to close. In this way the ground power monitor performs a similar function to a main generator GCU. Typical parameters which are checked are undervoltage, overvoltage, frequency and correct phase rotation.

Emergency power generation

In certain emergency conditions the typical aircraft power generation system already described may not meet all the airworthiness authority requirements and additional sources of power generation may need to be used to power the aircraft systems. The aircraft battery offers a short-term power storage capability, typically up to 30 minutes. However for longer periods of operation the battery is insufficient. The operation of twin-engined passenger aircraft on ETOPS flights now means that the aircraft has to be able to operate on one engine while up to 180 minutes from an alternative or diversion airfield. This has led to modification of some of the primary aircraft systems, including the electrical system, to ensure that sufficient integrity remains to accomplish the 180 minute diversion while still operating with acceptable safety margins. The three standard methods of providing back-up power on civil transport aircraft are:

Fig. 5.18 Ram Air Turbine (RAT)

(1) Ram Air Turbine (RAT)
(2) Back-up Converters
(3) Permanent Magnet Generators (PMGs)

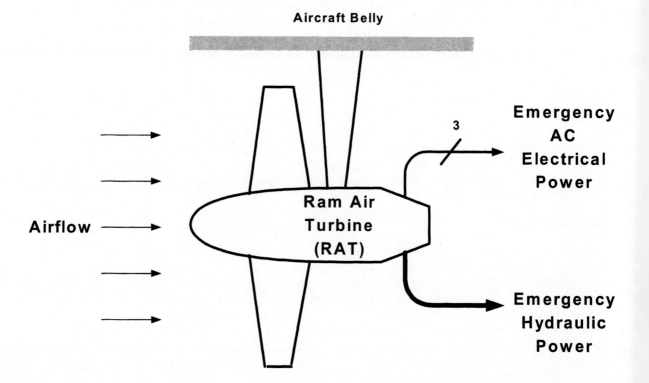

Ram Air Turbine

The Ram Air Turbine or RAT is deployed when most of the conventional power generation system has failed or is unavailable for some reason. The RAT is an air-driven turbine, normally stowed in the aircraft ventral or nose section that is extended either automatically or manually when the emergency commences. The passage of air over the turbine is used to power a small emergency generator of limited capacity, usually enough to power the crew's essential flight instruments and a few other critical services – see Fig. 5.18. Typical RAT generator sizing may vary from 5 to 15 kVA depending upon the aircraft. The RAT also powers a small hydraulic power generator for similar hydraulic system emergency power provision. Once deployed then the RAT remains extended for the duration of the flight and cannot be restowed without maintenance action on the ground. The RAT is intended to furnish the crew with sufficient power to fly the aircraft while attempting to restore the primary generators or carry out a diversion to the nearest airfield. It is not intended to provide significant amounts of power for a lengthy period of operation.

Back-up converters

The requirements for ETOPS have led to the need for an additional method of back-up power supply, short of deploying the RAT that should occur in only the direst emergency. The use of back-up converters satisfies this requirement and is used on the Boeing 777. Back-up generators are driven by the same engine accessory gearbox but are quite independent of the main IDGs. Refer to Fig. 5.19.

The back-up generators are VF and therefore experience significant frequency variation as engine speed varies. The VF supply is fed into a back-up converter which,

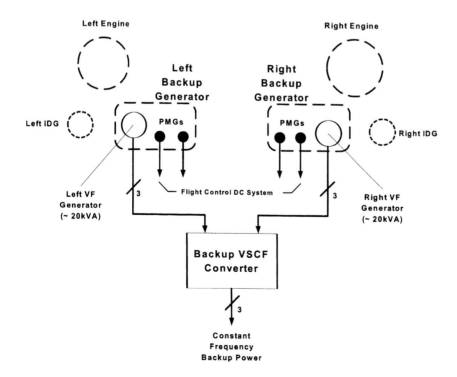

Fig. 5.19 *Simplified back-up VSCF converter system*

using the DC link technique, first converts the AC power to DC by means of rectification. The converter then synthesizes three-phase 115 VAC 400 Hz power by means of sophisticated solid-state power-switching techniques. The outcome is an alternative means of AC power generation which may power some of the aircraft AC busbars; typically the 115 VAC transfer buses in the case of the Boeing 777. In this way substantial portions of the aircraft electrical system may remain powered even though some of the more sizeable loads such as the galleys and other non-essential loads may need to be shed by the Electrical Load Management System (ELMS).

A paper presenting the entire Boeing 777 electrical system may be found at reference (6).

Permanent Magnet Generators (PMGs)

The use of PMGs to provide emergency power has become prominent over the last decade or so. As can be seen from the description of the back-up converter above, the back-up generator hosts PMGs which may supply several hundred watts of independent generated power to the flight control DC system where the necessary conversion to 28 VDC is undertaken. It was already explained earlier in the chapter that AC generators include a PMG to bootstrap the excitation system. PMGs – also called Permanent Magnet Alternators (PMAs) – are used to provide dual independent on-engine supplies to each lane of the FADEC. As an indication of future trends it can therefore be seen that on an aircraft such as the Boeing 777 there are a total of 13 PMGs/PMAs across the aircraft critical control systems – flight control, engine control and electrical systems. See Fig. 5.20.

Reference (6) is an early paper describing the use of a PMG and reference (7) describes some of the work being undertaken in looking at higher levels of PMG power generation.

Fig. 5.20 Boeing 777 PMG/PMA complement

Some military aircraft use Emergency Power Units (EPUs) for the supply of emergency power.

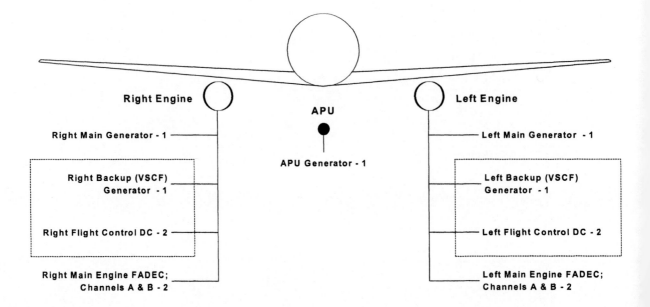

Recent systems developments

In recent years a number of technology advances have taken place in the generation, switching and protection of electrical power. These new developments are beginning to have an impact upon the classic electrical systems that have existed for many years, probably for the first time since WWII. This has resulted in the availability of new devices that in turn have given credibility to new system concepts, or at least provide the means for advanced systems concepts that could not previously be implemented. These techniques and concepts embrace the following:

(1)   Electrical Load Management System (ELMS).
(2)   Variable Speed Constant Frequency (VSCF) – cycloconverter.
(3)   270 VDC systems.
(4)   More electric aircraft.

Electrical Load Management System (ELMS)

The Boeing 777 Electrical Load Management System (ELMS) developed and manufactured by Smiths Industries sets new standards for the industry in terms of electrical load management. The general layout of the ELMS is shown in Fig. 5.21. The system represents the first integrated electrical power distribution and load management system for a civil aircraft.

Fig. 5.21 Boeing 777 Electrical Load Management System (ELMS) (Smiths Industries)

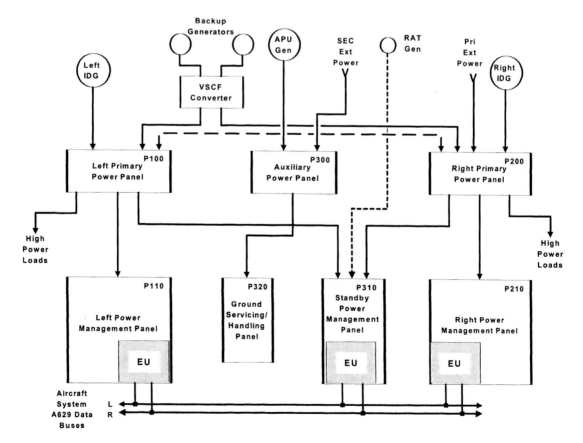

The system comprises seven power panels, three of which are associated with primary power distribution:

- P100 – Left primary power panel distributes and protects the left primary loads.
- P200 – Right primary power panel distributes and protects the right primary loads.
- P300 – Auxiliary power panel distributes and protects the auxiliary primary loads.

The secondary power distribution is undertaken by four secondary power panels:
- P110 – Left power management panel distributes and protects power, and controls loads associated with the left channel.
- P210 – Right power management panel distributes and protects power, and controls loads associated with the right channel.
- P310 – Stand-by power management panel distributes and protects power, and controls loads associated with the stand-by channel.
- P320 – Ground servicing/handling panel distributes and protects power associated with ground handling.

Load management and utilities systems control is exercised by means of Electronic Units (EUs) mounted within the P110, P210 and P310 power management panels. Each of these EUs interfaces with the left and right aircraft systems ARINC 629 digital data buses and contain a dual redundant architecture for reasons of dispatch availability. The EUs contain a modular suite of Line Replaceable Modules (LRMs) that can readily be replaced when the door is open. A total of six module types are utilized to build a system comprising an overall complement of 44 modules across the three EUs. This highly modular construction with multiple use of common modules reduced development risk and resulted in highly accelerated module maturity at a very early stage of airline service. LRMs typically have mature in-service mean time between failures (MTBF) of approximately 200,000 hours as reported by reference (8). See Fig 5.22 for a diagrammatic portrayal of the modular concept.

The load management and utilities control features provided by ELMS are far in advance of any equivalent system in airline service today. Approximately 17 to 19 Electrical Load Control Units (ELCUs) – depending upon aircraft configuration – supply and control loads directly from the aircraft main AC buses. These loads can be controlled by the intelligence embedded within the ELMS EUs. A major advance is the sophisticated load shed/load optimization function which closely controls the availability of functions should a major electrical power source fail or become unavailable. The system is able to reconfigure the loads to give the optimum distribution of the available power. In the event that electrical power is restored, the system is able to re-instate loads according to a number of different schedules. The system is therefore able to make the optimum use of power at all times rather than merely shed loads in an emergency.

The benefits conferred by ELMS have proved to be significant with significant reduction in volume, wiring and connectors, weight, relays and circuit-breakers. Due to the in-built intelligence, use of digital data buses, maintainability features and extensive system built-in test (BIT), the system build and on-aircraft test time turned out to be approximately 30 per cent of that experienced by contemporary systems.

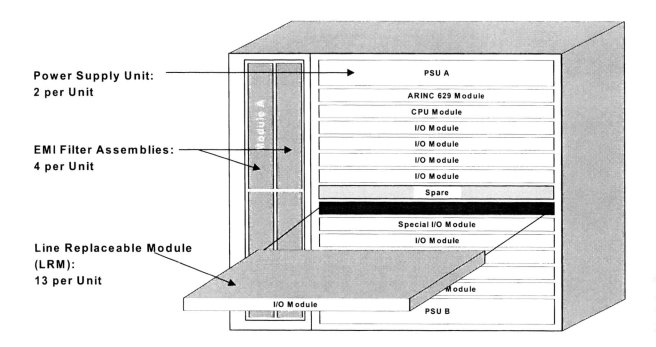

Power Supply Unit:
2 per Unit

EMI Filter Assemblies:
4 per Unit

Line Replaceable Module
(LRM):
13 per Unit

Module A

PSU A

ARINC 629 Module

CPU Module

I/O Module

I/O Module

I/O Module

I/O Module

Spare

Special I/O Module

I/O Module

I/O Module

PSU B

Fig. 5.22 *Boeing 777 ELMS EU concept (Smiths Industries)*

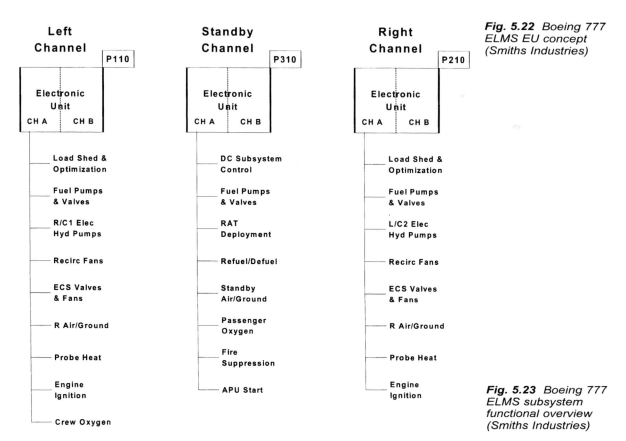

Left Channel	**Standby Channel**	**Right Channel**
P110	P310	P210
Electronic Unit	Electronic Unit	Electronic Unit
CH A CH B	CH A CH B	CH A CH B
Load Shed & Optimization	DC Subsystem Control	Load Shed & Optimization
Fuel Pumps & Valves	Fuel Pumps & Valves	Fuel Pumps & Valves
R/C1 Elec Hyd Pumps	RAT Deployment	L/C2 Elec Hyd Pumps
Recirc Fans	Refuel/Defuel	Recirc Fans
ECS Valves & Fans	Standby Air/Ground	ECS Valves & Fans
R Air/Ground	Passenger Oxygen	R Air/Ground
Probe Heat	Fire Suppression	Probe Heat
Engine Ignition	APU Start	Engine Ignition
Crew Oxygen		

Fig. 5.23 *Boeing 777 ELMS subsystem functional overview (Smiths Industries)*

A large number of utilities management functions are embedded in the system making it a true load management rather than merely an electrical power distribution system. Key functions are the load optimization function already described, fuel jettison, automatic RAT deployment and many others. Figure 5.23 gives an overview of some of the more important functions.

Variable Speed/Constant Frequency (VSCF)

The principle of VSCF has already been outlined in the back-up converter description earlier in the chapter. There are considerable benefits to be accrued by dispensing with the conventional AC power generation techniques using IDGs to produce large quantities of constant frequency 400 Hz 115 VAC power. The constant speed element of the IDG is generally fairly unreliable compared to the remainder of the generation system. The techniques are now available through the use of VSCF to produce significant quantities of primary AC power by means of constant frequency power generation accompanied by suitable power conversion. In particular, the VSCF cycloconverter version developed by Leland Electrosystems, a part of Smiths Industries, is a mature technology. Over 4,000 cycloconverter systems are in service with the US Military: F-18C/D, F-117A, TR-1 and U-2 and the later versions will be fitted to the F-18E/F and V-22 Tilt Rotor.

Theory of VSCF cycloconverter system operation

The VSCF system consists of a brushless generator and a solid-state frequency converter. The converter assembly also has a filter capacitor assembly and control and protection circuit. A simplified block diagram for the VSCF system is shown in Fig. 5.24. The generator is driven by the accessory gearbox and produces AC output voltage at variable frequency proportional to the gearbox speed. The converter converts the variable frequency into constant frequency 400 Hz, three-phase power by using an SCR based cycloconverter. The filter assembly filters out high frequency ripple in the output voltage. The GCU function regulates the output voltage and provides protection to the system.

Generator operation

The function of the generator is to convert mechanical power from the aircraft turbine engine to electrical power suitable for electronic conversion. The electronic converter

Fig. 5.24 Simplified VSCF system diagram

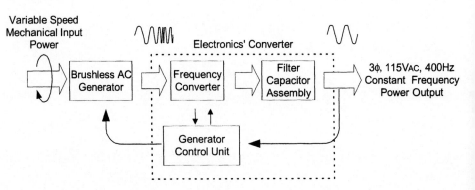

processes the generator output electrical power into high quality 400 Hz electrical power. The brushless, self-excited generator is comprised of three AC machines:

- The permanent magnet generator
- The exciter generator
- The main generator

The Permanent Magnet Generator (PMG) provides electrical power for all control circuitry and the exciter field as soon as the rotor is rotating at minimum speed. The PMG also provides raw electrical power for the Main Line Contactors (MLC). The integral PMG makes the generator self-contained; thus, it does not require any external power for excitation. The PMG is a synchronous machine with flux excitation provided by the permanent magnets contained inside the rotor assembly. The PMG stator contains two separate and electrically isolated windings in a laminated, slotted, magnetic steel core. AC voltages are induced in the stator windings as the flux provided by the PM rotor sweeps past the stator. The PM rotor is driven directly by the gearbox output shaft.

The output of one of the single-phase windings of the PMG stator is fed into the generator voltage regulator. The generator voltage regulator rectifies and modulates the PMG output. This output provides proper current for the exciter field winding, allowing generation of AC voltage on the exciter rotor. The output of the second single-phase winding is used for the converter power supply.

The exciter is a brushless synchronous machine with a DC-excited stator and a three-phase wound rotor. The exciter stator winding receives controlled DC current from the rectified PMG output through the generator voltage regulator. This in turn develops the AC power in the three-phase rotor windings as they rotate past the exciter generator stator winding, inducing an AC voltage in the three-phase windings of the exciter's rotor. The magnitude of this rectified AC voltage is proportional to the speed of the shaft and to the DC excitation current on the exciter's stator winding. The rotor output is rectified with three silicon rectifiers mounted inside the rotor shaft. The exciter and rectifiers are used to eliminate brushes anywhere in the generator. The rectified exciter output supplies field current for the main generator.

The main generator is a wound rotor, synchronous machine with a 16-pole rotor and a six-phase stator. The connections between the exciter rotor windings, three rectifier diodes and the main rotor field winding are all on the rotor. The six-phase stator output winding is star connected. All six-phase leads and the neutral connection are brought out to the terminal block. The wound rotor, when excited with DC current supplied by the exciter, establishes magnetic flux in the air gap between the rotor and the stator. This magnetic flux, when driven by the gearbox's shaft, induces alternating voltage into the six-phase windings of the stator. The magnitude of this AC stator voltage is proportional to the speed of the rotor and the DC current supplied by the exciter rotor. The magnitude of the rotor DC current in turn depends upon the excitation current provided by the generator voltage regulator to the exciter's stator. Therefore, the magnitude of the exciter's stator current determines the magnitude of the main generator stator's AC voltage output. The frequency of the main generator's output is dependent upon the shaft speed. With 16 poles, the frequency of the main generator varies from 1,660–3,500 Hz as the input speed is varied from 12,450–26,250 rpm. The main generator output supplies a variable frequency, six-phase AC power to the cycloconverter for further processing.

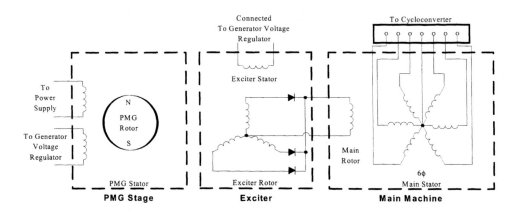

Fig. 5.25 *Generator electrical schematic (Leland Electrosystems/Smiths Industries)*

The neutral ends of each of the six stator windings are connected to the neutral through Current Transformers (CTs). The CTs sense the current in each winding and compare it with the current in each phase in the converter. If any current differential is detected in the zone between the generator neutral and the converter, the system de-energizes quickly by means of the High-Frequency Differential Protection (HFDP) circuit, preventing damage to any of the generator windings.

All connections between the generator and frequency converter are internal to the VSCF package so the converter cannot be subjected to abnormal phase rotation unless the generator rotation is reversed. The Generator Over-Current (GOC) protection will de-energize the system in the event of reversed generator rotation.

The electrical schematic for the generator is shown Fig. 5.25.

Fig. 5.26 *VSCF converter*

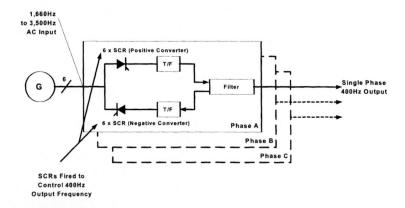

Converter operation

This section describes the cycloconverter design and operation as configured for a 30/40 kVA rating. This review concentrates on the most critical aspects of a Variable Speed, Constant Frequency (VSCF) system, i.e., the power flow section and switch module control circuits.

The frequency conversion system consists of three frequency converters, one for each phase (Fig. 5.26). The generator delivers six-phase, variable frequency power to each converter. Each frequency converter consists of a cycloconverter (12 SCRs) and its associated control circuits: modulators, mixer, firing wave generator, reference wave generator, feedback control circuit, and low-pass filter. The SCRs are controlled by the modulators. They compare the cosine firing wave with the processed reference wave to generate appropriately timed SCR gating signals. The low-pass output filter attenuates the ripple frequency components.

Negative feedback is used to improve the linearity of the cycloconverter and to reduce the output impedance. Thus, the cycloconverter is a high-power amplifier producing an output wave that is a replica of the reference sine wave. The actual feedback loop has multiple feedback paths to improve the waveform, reduce the DC content, and lower the output impedance. The mixer amplifier adds the feedback signals in the correct proportions.

The 400 Hz output voltage is regulated with individual phase voltage regulators that adjust the 400 Hz reference wave amplitudes. Consequently, the voltage unbalance in the line-to-neutral output voltages is negligible even with large unbalanced loads.

The unfiltered output of the two rectifier banks – solid jagged lines in Fig. 5.27 shows the conduction period where the rectifiers are connected to the generator lines. The heavy, smooth lines are the filtered output of the cycloconverter. Both rectifier banks are programmed to operate over the entire 360 degree of the output wave, and each bank can supply either voltage polarity. The positive half of the output voltage wave is formed by operating either the positive bank in the rectifying mode or the negative bank in the inverting mode.

Fig. 5.27 *VSCF 400 Hz waveform formulation (Leland Electrosystems/Smiths Industries)*

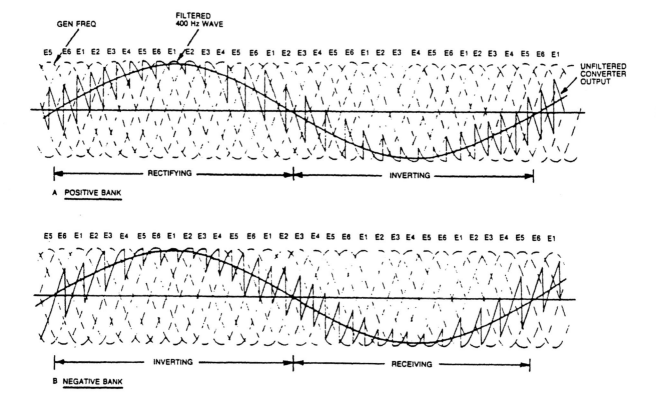

Fig. 5.28 Leland
VSCF cycloconverter
(Leland
Electrosystems/Smiths
Industries)

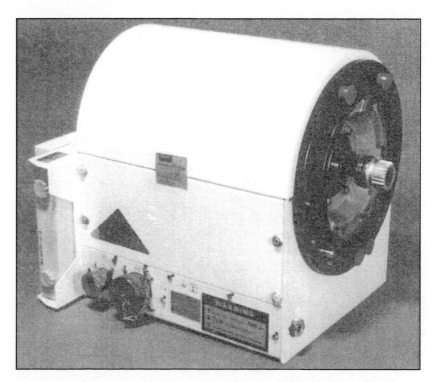

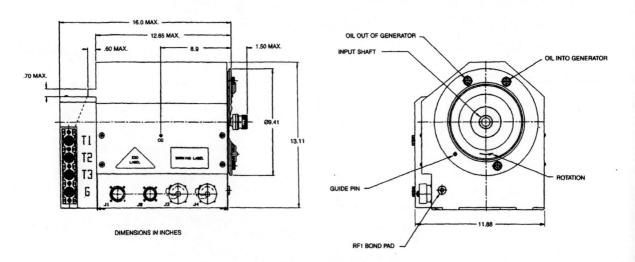

Fig. 5.29 Leland
VSCF cycloconverter
dimensions (Leland
Electrosystems/Smiths
Industries)
The negative half of the output wave is formed in reverse fashion. The rectifying and inverting modes define the direction of power flow; toward the load in the rectifying mode and toward the source in the inverting mode.

Some of the physical attributes of the 60/65 kVA machine are shown in Figs 5.28 and 5.29. This particular version also embodies PMGs capable of supplying three independent channels of 28 VDC regulated power to feed flight control and other essential loads. A simplified version of the F-18E/F electrical system is shown in Fig. 5.30.

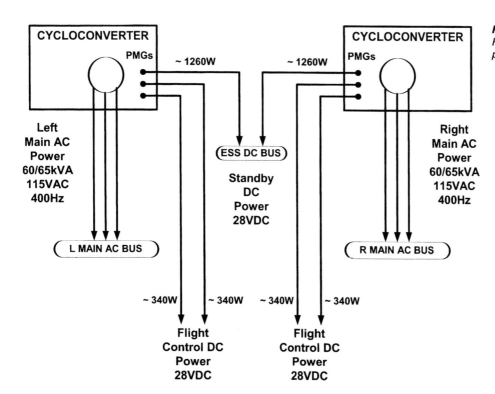

Fig. 5.30 Simplified F-18E/F electrical power system

270 VDC systems

An initiative which has been under way for a number of years in the US military development agencies is the 270 VDC system. The US Navy has championed this concept and the technology has developed to the point that some of the next generation of US combat aircraft will have this system imposed as a tri-Service requirement. The aircraft involved are the US Air Force Advanced Tactical Fighter (ATF) – now the Lockheed Martin F-22 Raptor, the former US Navy Advanced Tactical Aircraft (ATA) or A-12, and the US Army Light Helicopter (LHX or LH) – now known as RAH-66 Comanche. More recent projects noted in Table 5.1 include the Joint Strike Fighter (JSF) offerings from Lockheed Martin (X-35A/B/C) and the Boeing (X-32A/B/C), although the latter is reportedly a predominantly VF 115 VAC system with some power conversion for 270 VDC loads.

The use of 270 VDC is an extension of the rationale for moving from 28 VDC to 115 VAC: reduction in the size of current-carrying conductors thereby minimizing weight, voltage drop and power dissipation. There are however a number of disadvantages associated with the use of 270 VDC. 270 VDC components are by no means commonplace; certainly were not so at the beginning of development and even now are not inexpensive. Also, a significant number of aircraft services will still require 28 VDC or 115 VAC supplies and the use of higher voltages places greater reliance on insulation techniques to avoid voltage breakdown. The US military addressed these technical issues by a wide range of funded technology development and demonstrator programmes. Some of these are also directed at the greater use of electrical power on the combat aircraft, possibly to supplant conventional secondary power and hydraulic power systems or at least to augment them to a substantial degree. The term for these developments is the more-electric aircraft (MEA), implying a much greater if not total use of electrical power for aircraft systems.

The high DC voltage poses a risk in military aircraft of increased possibility of fire resulting from battle damage in carbon-fibre composite aircraft. Care must be taken to reduce the risk of arcing at high altitudes or in humid salt-laden air conditions such as tropical or maritime environments. There is also a potential lethal hazard to ground crew during servicing operations. All these must be taken into account in design.

One of the problems in moving to 270 VDC is that there is still a need for the conventional 115 VAC and 28 VDC voltages for some equipment as mentioned above. The 270 VDC aircraft therefore becomes a somewhat hybrid system as shown in Fig. 5.31 that may lose some of the original 270 VDC advantages.

Fig. 5.31 Simplified F-22 electrical system

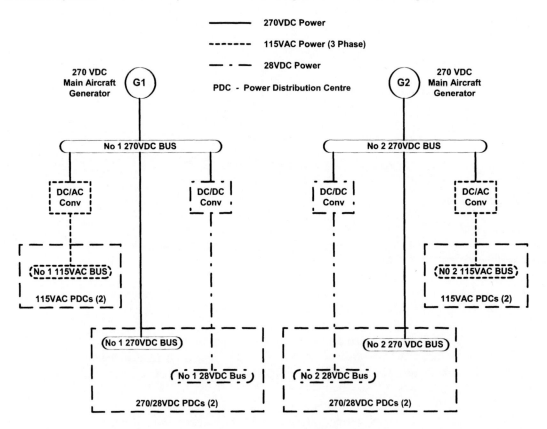

More-electric Aircraft (MEA)

For at least the last ten years a number of studies have been under way in the US which have examined the all-electric aircraft. As stated earlier, aircraft developed in the UK in the late 1940s/early 1950s, such as the V-Bombers, utilized electric power to a greater extent than present-day aircraft. In the 1980s, a number of studies promoted by NASA, the US Navy, and US Air Force development agencies, and undertaken by Lockheed and Boeing, addressed the concept in detail. The topic is covered in this book under Chapter

10, Advanced Systems, since the implications of the MEA are more embracing than merely organizing the aircraft electrical system in a different manner. The concept addresses more energy-efficient ways of converting and utilizing aircraft power in the broadest sense and therefore has a far-reaching effect upon overall aircraft performance.

Electrical system displays

The normal method of displaying electrical power system parameters to the flight crew has been via dedicated control and display panels. On a fighter or twin-engined commuter aircraft the associated panel is likely to be fairly small. On a large transport aircraft the electrical systems control and display would have been achieved by a large systems panel forming a large portion of the flight engineer's panel showing the status of all the major generation and power conversion equipment. With the advent of two crew flight-deck operations, of which the Boeing 757, 767, 747-400 and Airbus A320, and indeed most modern aircraft, are typical examples, the electrical system selection panel was moved into the flight crew overhead panel. EICAS or ECAM systems now permit the display of a significant amount of information by the use of:

- Synoptic displays
- Status pages
- Maintenance pages

These displays show in graphic form the system operating configuration together with the status of major system components, key system operating parameters, and any degraded or failure conditions which apply. The maximum use of colour will greatly aid the flight crew in assimilating the information displayed. The overall effect is vastly to improve the flight crew/system interface giving the pilots a better understanding of the system operation while reducing the crew workload.

References

(1) **Bonneau, V.** Dual-use of VSCF cycloconverter, FITEC'98, London.
(2) **Boyce, J.W.** An introduction to smart relays, Paper presented at the SAE AE-4 Symposium.
(3) **Johnson, W., Casimir, B., Hanson, R., Fitzpatrick, J.,** and **Pusey, G.** Development of 200 ampere regulated transformer rectifier, SAE, Mesa
(4) **Wall, M.B.** Electrical power system of the Boeing 767 airplane.
(5) **Layton, S.G.** Solid state power control, ERA Avionics Conference, London, Heathrow.
(6) **Tenning, C.** Boeing 777 Electrical system, RAeS Conference, London.
(7) **Rinaldi, M.R.** A highly reliable DC power source for avionics subsystems, SAE Conference.
(8) **Haller, J.P., Weale, D.V.,** and **Loveday, R.G.** Integrated utilities control for civil aircraft, FITEC'98, London.

CHAPTER 6

Pneumatic Systems

Introduction

The modern turbofan engine is a very effective gas generator and this has led to the use of engine bleed air for a number of aircraft systems, either for reasons of heating, provision of motive power or as a source of air for cabin conditioning and pressurization systems. Bleed air is extracted from the engine compressor and after cooling and pressure reduction/regulation it is used for a variety of functions.

On the engine, high-pressure bleed air is used as the motive power – sometimes called 'muscle power' – for many of the valves associated with the bleed air extraction function. Medium-pressure bleed air is used to start the engine in many cases, either using air from a ground power unit, APU or cross-bled from another engine on the aircraft which is already running. Bleed air is also used to provide anti-ice protection by heating the engine intake cowling and it is also used as the motive power for the engine thrust reversers.

On the aircraft, bleed air tapped from the engine is used to provide air to pressurize the cabin and provide the source of air to the cabin-conditioning environmental control system. A proportion of bleed air is fed into air-conditioning packs which cool the air, dumping excess heat overboard; this cool air is mixed with the remaining warm air by the cabin temperature control system such that the passengers are kept in a comfortable environment. Bleed air is also used to provide main wing anti-ice protection.

Bleed air is also used for a number of ancillary functions around the aircraft: pressurizing hydraulic reservoirs, providing hot air for rain dispersal from the aircraft windscreen, pressurizing the water and waste system and so on. In some aircraft Air-Driven Pumps (ADPs) are used as additional means of providing aircraft hydraulic power.

Pitot static systems are also addressed in the pneumatic chapter, as although this is a sensing system associated with measuring and providing essential air data parameters for safe aircraft flight, it nonetheless operates on pneumatic principles. Pitot systems have been used since the earliest days of flight using pneumatic, capsule-based mechanical flight instruments. The advent of avionics technology led first to centralized Air Data Computers (ADCs) and eventually on to the more integrated solutions of today such as the Air Data and Inertial Reference System (ADIRS).

Pneumatic power is the use of medium-pressure air to perform certain functions within the aircraft. While the use of pneumatic power has been ever present since aircraft became more complex, the evolution of the modern turbojet engine has lent itself to the use of pneumatic power, particularly on the civil airliner.

The easy availability of high-pressure air from the modern engine is key to the use of pneumatic power as a means of transferring energy or providing motive power on the aircraft. The turbojet engine is in effect a gas generator where the primary aim is to provide thrust to keep the aircraft in the air. As part of the turbojet combustion cycle, air is compressed in two- or three-stage compressor sections before fuel is injected in an atomized form and then ignited to perform the combustion process. The resulting expanding hot gases are passed over turbine blades at the rear of the engine to rotate the turbines and provide shaft power to drive the LP fan and compressor sections. When the engine reaches self-sustaining speed the turbine is producing sufficient shaft power to equal the LP fan/compressor requirements and the engine achieves a stable condition – on the ground this equates to the ground idle condition. The availability of high-pressure, high-temperature air bled from the compressor section of the engine lends itself readily to the ability to provide pneumatic power for actuation, air-conditioning or heating functions for other aircraft subsystems.

Other areas of the aircraft use pneumatic principles for sensing the atmosphere surrounding the aircraft for instrumentation purposes. The sensing of air data is key to ensuring the safe passage of the aircraft in flight.

Use of bleed air

The use of the aircraft engines as a source of high-pressure, high-temperature air can be understood by examining the characteristics of the turbojet, or turbofan engine as it should more correctly be described. Modern engines 'bypass' a significant portion of the mass flow past the engine and increasingly a small portion of the mass flow passes through the engine core or gas generation section. The ratio of bypass air to engine core air is called the bypass ratio and this is usually 4:1 to 5:1 for modern civil engines.

The characteristics of a modern turbofan engine are shown in Fig. 6.1. This figure shows the pressure (in psi) and the temperature (in degrees Centigrade) at various points throughout the engine for three engine conditions: ground idle, take-off power and in the cruise condition.

It can be seen that in the least stressful condition – ground idle – the engine is in a state of equilibrium but that even at this low level the compressor air pressure is 50 psi and the temperature 180 °C. At take-off conditions the compressor air soars to 410 psi/540 °C. In the cruise condition the compressor air is at 150 psi/400 °C. The engine is therefore a source of high-pressure and high-temperature air that can be 'bled' from the engine to perform various functions around the aircraft. The fact that there are such considerable variations in air pressure and temperature for various engine conditions places an imposing control task upon the pneumatic system. Also the variation in engine characteristics between similarly rated engines of different manufacturers poses additional design constraints. Some aircraft, such as the Boeing 777, offer three engine

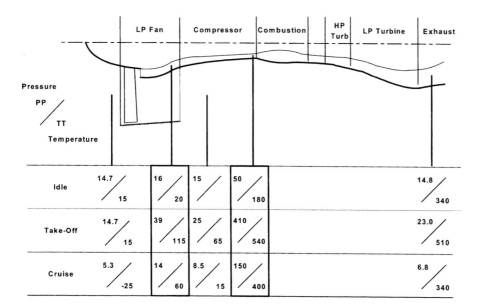

Fig. 6.1
Characteristics of a modern turbofan engine

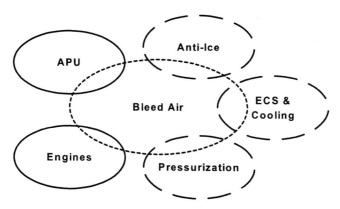

Fig. 6.2 *Relationship of bleed air with major aircraft systems*

choices: Pratt & Whitney, General Electric and Rolls-Royce and each of these engines has to be separately matched to the aircraft systems, the loads of which may differ as a result of operator-specified configurations.

As well as the main aircraft engines the Auxiliary Power Unit (APU) is also a source of high-pressure bleed air. The APU is in itself a small turbojet engine, designed more from the viewpoint of an energy and power generator than a thrust provider which is the case for the main engines. The APU is primarily designed to provide electrical and pneumatic power while the aircraft is on the ground, although it can be used as a back-up provider of power while airborne. Some aircraft designs are actively considering the use of in-flight operable APUs to assist in in-flight engine relighting and to relieve the engines of off-take load in certain areas of the flight envelope.

It is also usual for the aircraft to be designed to accept high-pressure air from a ground power cart, for aircraft engine starting.

These three sources of pneumatic power provide the muscle or means by which the pneumatic system is able to satisfy the aircraft demands. In a simplified form the pneumatic system may be represented by the interrelationships shown in Fig. 6.2.

This simplified drawing – the ground air power source is omitted – shows how the aircraft high-pressure air sources provide bleed air which forms the primary source for the three major aircraft air-related systems.

- Ice protection: the provision of hot air to provide anti-icing of engine nacelles and the wing, tailplane or fin leading edges; or to dislodge ice that has formed on the surfaces.
- ECS and cooling: the provision of the main air source for environmental temperature control and cooling.
- Pressurization: the provision of a means by which the aircraft may be pressurized, giving the crew and passengers a more comfortable operating environment.

A simplified representation of this relationship is shown in Fig. 6.3. This example shows a twin-engine configuration typical of many business jets and regional jet transport aircraft.

Bleed air from the engines is passed through a Pressure-Reducing Shut-Off Valve (PRSOV) which serves the function of controlling and, when required, shutting off the engine bleed air supply. Air downstream of the PRSOV may be used in a number of ways.

- By means of a cross-flow Shut-Off Valve (SOV) the system may supply air to the opposite side of the aircraft during engine start or if the opposite engine is inoperative for any reason.

Fig. 6.3 *Simplified bleed air system and associated aircraft systems*

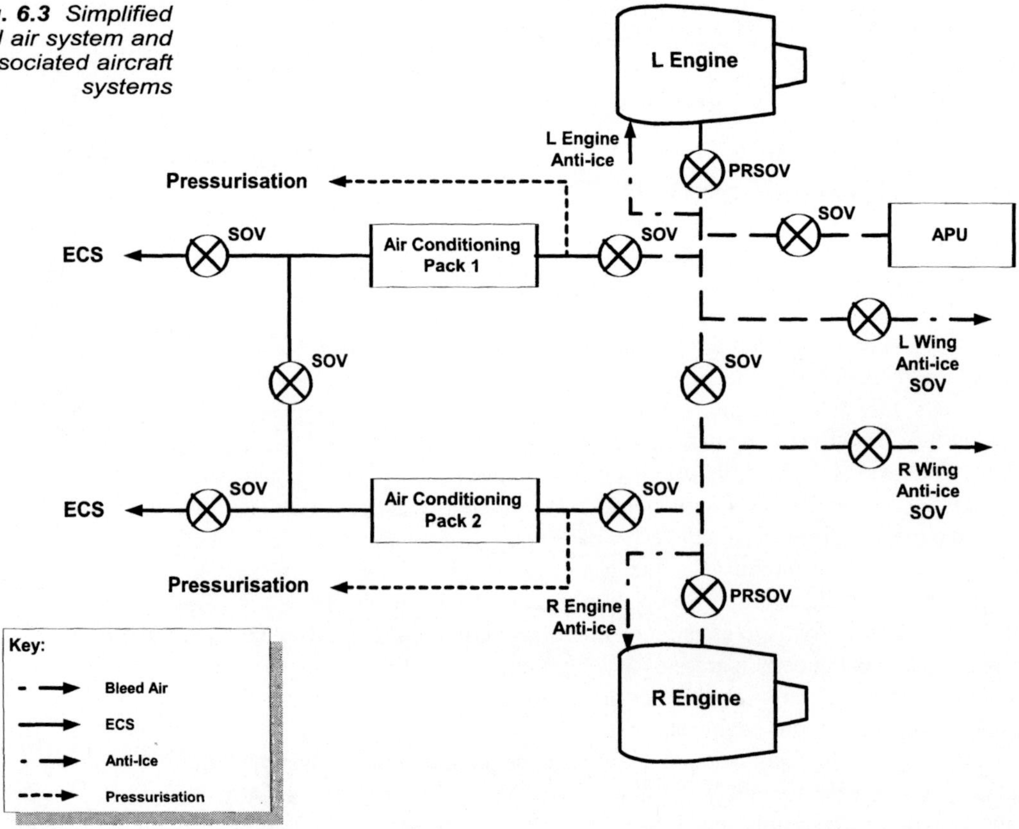

- A SOV from the APU may be used to isolate the APU air supply.
- SOVs provide isolation as appropriate to the left and right air-conditioning packs and pressurization systems.
- Additional SOVs provide the means by which the supply to left and right wing anti-icing systems may be shut off in the event that these functions are not required.

This is a simplified model of the use of engine bleed air in pneumatic systems. A more comprehensive list of those aircraft systems with which bleed air is associated are listed as follows with the accompanying civil Air Transport Association (ATA) chapter classification.

- Air conditioning (ATA Chapter 21)
- Cargo compartment heating (ATA Chapter 21)
- Wing and engine anti-icing (ATA Chapter 30)
- Engine start (ATA Chapter 80)
- Thrust reverser (ATA Chapter 78)
- Hydraulic reservoir pressurization (ATA Chapter 29)
- Rain-repellent nozzles – aircraft windscreen (ATA Chapter 30)
- Water tank pressurization and toilet waste (ATA Chapter 38)
- Air-driven hydraulic pump (ADP) (ATA Chapter 29)

Several examples will be examined within this pneumatic systems chapter. However, before describing the pneumatically activated systems it is necessary to examine the extraction of bleed air from the engine in more detail.

Engine bleed air control

Figure 6.4 gives a more detailed portrayal of the left-hand side of the aircraft bleed air system, the left side being an identical mirror image of the right-hand side.

Fig. 6.4 *Typical aircraft bleed air system – left-hand side*

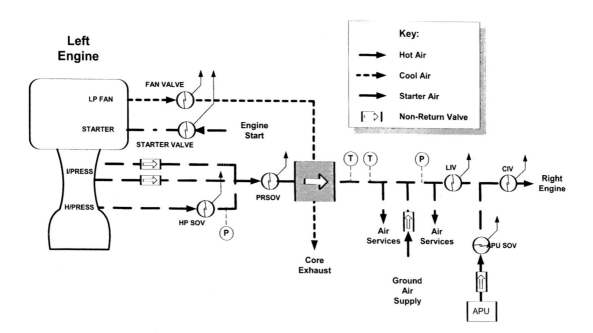

Fig. 6.5(a) *Typical pressure-reducing shut-off valve (PRSOV)*

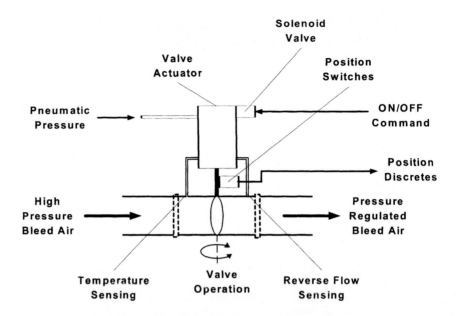

Air is taken from an intermediate stage or high-pressure stage of the engine compressor depending upon the engine power setting. At lower power settings, air is extracted from the high-pressure section of the compressor while at higher power settings the air is extracted from the intermediate compressor stage. This ameliorates to some degree the large variations in engine compressor air pressure and temperature for differing throttle settings as already shown in Fig. 6.1. A pneumatically controlled High-Pressure Shut-Off Valve (HP SOV) regulates the pressure of air in the engine manifold system to around 100 psi and also controls the supply of bleed air from the engine.

The Pressure-Reducing Shut-Off Valve (PRSOV) regulates the supply of the outlet air to around 40 psi before entry into the pre-cooler. Flow of cooling air through the pre-cooler is regulated by the fan valve which controls the temperature of the LP fan air and therefore of the bleed air entering the aircraft system. Appropriately located pressure and temperature sensors allow the engine bleed air temperature and pressure to be monitored and controlled within specified limits.

A typical PRSOV is shown in Fig. 6.5(a); an example of a Harrier II valve which is solenoid-controlled and pneumatically-operated and which controls temperature, flow and pressure is shown in Fig. 6.5(b).

The PRSOV performs the following functions:

- ON/OFF control of the engine bleed system.
- Pressure regulation of the engine supply air by means of a butterfly valve actuated by pneumatic pressure.
- Engine bleed air temperature protection and reverse flow protection.
- Ability to be selected during maintenance operations in order to test reverse thrust operation.

The PRSOV is pneumatically operated and electrically controlled. Operation of the solenoid valve from the appropriate controller enables the valve to pneumatically control the downstream pressure to approximately 40 psi within predetermined limits.

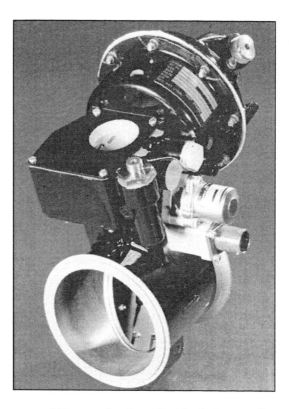

Fig. 6.5(b) *Harrier II pneumatic valve (Honeywell Normalair-Garret Ltd)*

The valve position is signalled by means of discrete signals to the bleed air controller and pressure switches provide over- and under-pressure warnings. The various pressure, flow and discrete signals enable the bleed air controller built-in test (BIT) to confirm the correct operation of the PRSOV and fan control valve combination. This ensures that medium-pressure air (approximately 40 psi) of the correct pressure and temperature is delivered to the pre-cooler and thence downstream to the pneumatic and air distribution system.

Downstream of the PRSOV and pre-cooler, the air is available for the user subsystems, a number of which are described below.

A number of isolation valves or SOVs are located in the bleed air distribution system. These valves are usually electrically-initiated, pneumatically-operated solenoid valves taking 28 VDC electrical power for ON/OFF commands and indication. A typical isolation valve is shown in Fig. 6.6. The valve shaft runs almost vertically across the duct as shown in the diagram and the valve mechanism and solenoid valve is located on the top of the valve.

Bleed air system indications

It is common philosophy in civil aircraft bleed air systems, in common with other major aircraft subsystems, to display system synoptic and status data to the flight crew on the Electronic Flight Instrument System (EFIS) displays. In the case of Boeing aircraft the synoptics are shown on the Engine Indication and Crew Alerting System (EICAS) display whereas for Airbus aircraft the Electronic Checkout And Monitoring (ECAM) displays are used. Both philosophies display system data on the colour displays located

Fig. 6.6 *Bleed air system isolation valve*

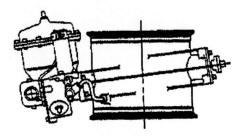

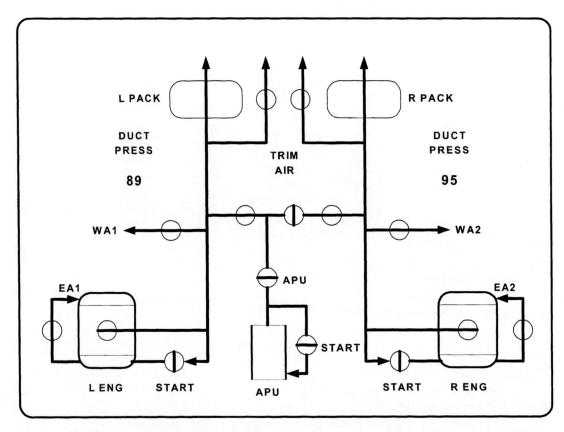

Fig. 6.7 *Typical bleed air system synoptic display*

on the central display console where they may be easily viewed by both Captain and First Officer. A typical bleed air system synoptic is shown in Fig. 6.7.

The synoptic display as shown portrays sufficient information in a pictorial form to graphically show the flight crew the operating status of the system. In the example, both main engines are supplying bleed air normally but the APU is isolated. The cross-flow valve is shut, as are both engine start valves. The wing and engine anti-ice valves are open allowing hot bleed air to be fed to the engines and wing leading edge to prevent any ice accretion.

Bleed air system users

The largest subsystem user of bleed air is the air system. Bleed air is used as the primary source of air into the cabin and fulfils the following functions:

- Cabin environmental control – cooling and heating.
- Cabin pressurization.
- Cargo bay heating.
- Fuel system pressurization.

The environmental control chapter – Chapter 7 – addresses the air systems. However there are other subsystems where the use of engine bleed air is key. These subsystems are:

- Wing and engine anti-ice protection
- Engine start
- Thrust reverser actuation
- Hydraulic system

Wing and engine anti-ice

The protection of the aircraft from the effects of aircraft icing represents one of the greatest and flight-critical challenges which confront the aircraft. Wing leading edges and engine intake cowlings need to be kept free of ice accumulation at all times. In the case of the wings, the gathering of ice can degrade the aerodynamic performance of the wing, leading to an increased stalling speed with the accompanying hazard of possible loss of aircraft control. Ice that accumulates on the engine intake and then breaks free entering the engine can cause substantial engine damage with similar catastrophic results. Considerable effort is also made to ensure that the aircraft windscreens are kept clear of ice by the use of window heating so that the flight crew has an unimpeded view ahead. Finally, the aircraft air data sensors are heated to ensure that they do not ice up and result in a total loss of air data information that could cause a hazardous situation or the aircraft to crash. The prevention of ice build-up on the windscreen and air data system probes is achieved by means of electric heating elements. In the case of the wing and engine anti-icing the heating is provided by hot engine bleed air which prevents ice forming while the system is activated.

The principles of wing anti-ice control are shown in Fig. 6.8. The flow of hot air to the outer wing leading edges is controlled by the Wing Anti-ice (WA) valve. The air flow is modulated by the electrically enabled anti-icing controller, this allows air to pass down the leading edge heating duct. This duct can take the form of a pipe with holes appropriately sized to allow a flow of air on to the inner surface of the leading edge – sometimes known as a 'piccolo tube'. The pressure of air in the ducting is controlled to about 20–25 psi. Telescopic ducting is utilized where the ducting moves from fixed wing to movable slat structure and flexible couplings are used between adjacent slat sections. These devices accommodate the movement of the slat sections relative to the main wing structure as the slats are activated. The air is bled out into the leading-edge slat section to heat the structure before being dumped overboard. A pressure switch and an overheat switch protect the ducting downstream of the wing anti-ice valve from over-pressure and over-temperature conditions.

Engine anti-icing is similarly achieved. An Engine Anti-Ice (EAI) valve on the engine fan casing controls the supply of bleed air to the fan cowl in order to protect against the formation of ice. As in the case of the wing anti-ice function, activation of the engine anti-icing system is confirmed to the flight crew by means of the closure of a pressure switch that provides an indication to the display system.

The presence of hot-air ducting throughout the airframe in the engine nacelles and wing leading edges poses an additional problem; that is to safeguard against the possibility of hot-air duct leaks causing an overheat hazard. Accordingly, overheat detection loops are provided in sensitive areas to provide the crew with a warning in the event of a hot-gas leak occurring. An overheat detection system will have elements adjacent to the air-conditioning packs, wing leading edge and engine nacelle areas to warn the crew of an overheat hazard – a typical system is shown in Fig. 6.9.

Fig. 6.8 Wing anti-ice control

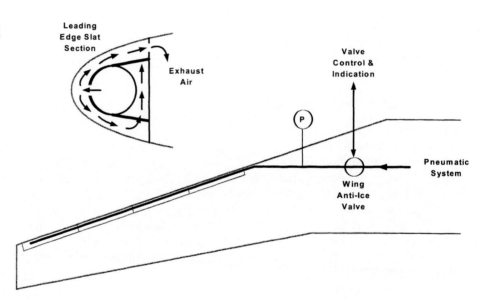

Fig. 6.9 Typical overheat warning system

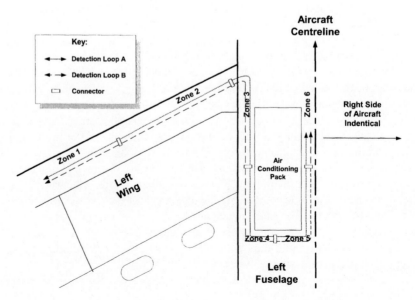

The operation of fire detection elements is described in Chapter 8 – Emergency Systems. In a civil airliner the hazardous areas are split into zones as shown in the figure. Each zone is served by two overheat detection loops – Loop A and Loop B. Modern technology is capable not just of locating an overheat situation but locating the point of detection downstream to within about one foot, thereby giving more information as to where the leak has actually occurred. Civil systems employ a dual system to aid dispatch. It is possible to dispatch the aircraft with one loop inoperative for a specific operating period provided that assurance is given that the remaining loop is operating correctly. This feature would allow the aircraft to recover to main base in order to have corrective maintenance action carried out

A number of low-speed commercial aircraft, employ a method of de-icing based on a flexible rubber leading-edge 'boot' that is inflated by air pressure to dislodge ice built up on the surface. The system is operated manually or in response to an ice detector input. The Advanced Turbo Prop (ATP) wing, tailplane and fin leading edges are protected by pneumatic rubber boots actuated by low-pressure engine compressor air. A cycling system is used to reduce the amount of air required. The ice is removed by successive inflation and deflation cycles of the boots. The crew is able to select light or heavy ice removal modes.

Engine start

The availability of high-pressure air throughout the bleed air system lends itself readily to the provision of motive power to crank the engine during the engine start cycle. As can be seen from earlier figures, a start valve is incorporated which can be activated to supply bleed air to the engine starter. On the ground the engines may be started in a number of ways:

- By use of a ground air supply cart
- By using air from the APU – probably the preferred means
- By using air from another engine which is already running

The supply of air activates a pneumatic starter motor located on the engine accessory gearbox. The engine start cycle selection enables a supply of fuel to the engine and provision of electrical power to the ignition circuits. The pneumatic starter cranks the engine to approximately 15–20 per cent of full speed by which time engine ignition is established and the engine will pick up and stabilize at the ground-idle rpm.

Thrust reversers

Engine thrust reversers are commonly used to deflect engine thrust forward during the landing roll-out to slow the aircraft and preserve the brakes. Thrust reversers are commonly used in conjunction with a lift dump function, whereby all the spoilers are simultaneously fully deployed, slowing the aircraft by providing additional aerodynamic drag while also dispensing lift. Thrust reversers deploy two buckets, one on each side of the engine, which are pneumatically operated by means of air turbine motor actuators to deflect the fan flow forward thereby achieving the necessary braking effect when the aircraft has a 'weight-on-wheels' condition. The air turbine motor has an advantage in that it is robust enough to operate in the harsh temperature and acoustic noise environment associated with engine exhaust, where hydraulic or electrical motors would not be sufficiently reliable.

Interlock mechanisms are provided which prevent inadvertent operation of the thrust reversers in flight. The Tornado thrust reversers are selected by rocking the throttle levers outboard in flight. On touchdown a signal is sent by the engine control systems to an air turbine motor connected to a Bowden cable and a screw jack mechanism to deploy the buckets.

Hydraulic system

Pneumatic pressure is commonly used to pressurize the aircraft hydraulic reservoirs. Some Boeing aircraft – usually the wide-bodies also use pneumatic power or air-driven hydraulic pumps to augment the normal Engine-Driven Pumps (EDPs) and AC Motor Pumps (ACMPs) for certain phases of flight. Figure 6.10 shows a typical centre hydraulic power channel as implemented by the Boeing philosophy – this is shown in a hydraulic system context in Chapter 4, Hydraulic Systems.

The hydraulic reservoir is pressurized using regulated bleed air from the pneumatic/bleed air system. Supply hydraulic fluid may be pressurized by the two alternate pumps:

- By means of the ACMP powered by three-phase 115 VAC electrical power
- By means of the Air-Driven Pump (ADP) using pneumatic power as the source.

Either pump in this hydraulic channel is able to deliver hydraulic pressure to the system services downstream; it is however more usual for the ACMP to be used as the primary source of power with the ADP providing supplementary or demand power for specific high-demand phases of flight. The ACMP may be activated by supplying a command to a high-power electrical contactor, or Electrical Load Management Unit (ELCU), as described in Chapter 5, Electrical Systems. The pneumatic pressure driving the ADP is controlled by means of a 28 VDC powered, solenoid-controlled, Modulating Shut-Off Valve (MSOV) upstream of the ADP. Hydraulic fluid temperature and pressure is monitored at various points in the system and the system information displayed on system synoptic or status pages as appropriate.

Fig. 6.10 Simplified
pneumatic system –
hydraulic system
interaction

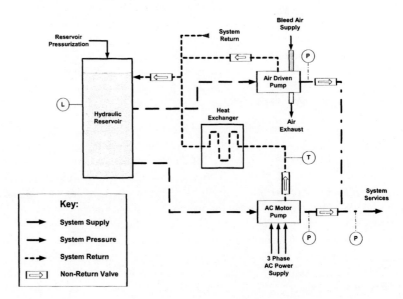

Pitot-static systems

By contrast with the bleed air system already described which provides energy or power for a number of diverse aircraft systems, the pitot-static system is an instrumentation system used to sense air data parameters of the air through which the aircraft is flying. Without the reliable provision of air data the aircraft is unable to safely continue flight. The pitot-static system is therefore a high-integrity system with high levels of redundancy. There are two key parameters which the pitot static system senses:

- Total pressure P_t, is the sum of local static pressure and the pressure caused by the forward flight of the aircraft. The pressure related to the forward motion of the aircraft is given by the following formula:

  Pressure $= \frac{1}{2} \rho V^2$ where ρ is the air density of the surrounding air and V is the velocity

- Static pressure or P_s is the local pressure surrounding the aircraft and varies with altitude

  Therefore total pressure, $P_t = P_s + \frac{1}{2} \rho V^2$

The forward speed of the aircraft is calculated by taking the difference between P_t and P_s.

An aircraft will have three or more independent pitot and static sensors. Figure 6.11 shows the principle of operation of pitot and static sensors.

The pitot probe shown in the top diagram is situated such that it faces in the direction of the air flow, thereby being able to sense the variation in aircraft speed using the formula quoted above. The sensing portion of the pitot probe stands proud from the aircraft skin to minimize the effect of laminar air flow. Pitot pressure is required at all stages throughout flight and a heater element is incorporated to prevent the formation of ice that could block the sensor or create an erroneous reading. The pitot heating

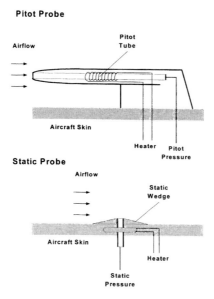

Fig. 6.11 *Pitot and static sensors*

Fig. 6.12 *Typical pitot and static probe installation*

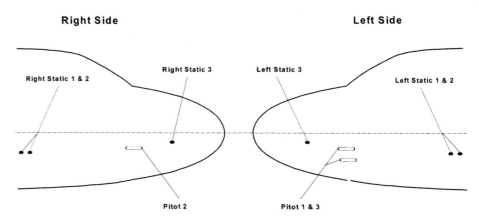

element is active throughout the entire flight.

The static probe shown in the lower diagram is located perpendicular to the air flow and so is able to sense the static pressure surrounding the aircraft. Like the pitot probe the static probe is provided with a heater element that continuously heats the sensor and prevents the formation of ice.

On some aircraft the pitot and static sensing functions are combined to give a pitot-static probe capable of measuring both dynamic and static pressures. A typical installation on a civil transport aircraft is depicted in Fig. 6.12.

This shows a configuration where three pitot probes are used; pitot 2 on the right side and pitot 1 and pitot 3 on the left side of the aircraft nose. Three static probes are located on the left and right sides of the aircraft. Pitot and static probes are placed carefully towards the nose of the aircraft such that the sensitive air data measurements are unaffected by other probes or radio antennae. Residual instrumentation errors due to probe location or installation are measured during the aircraft development phase and the necessary corrections applied further downstream in the system.

Fine-bore tubing carries the sensed air data pressure – pitot and static – to the aircraft instruments or the air data suite. Due to the sensitivity of the sensed data, water drain traps are provided so that extraneous moisture such as condensation may be extracted from the pitot-static lines. Also, following the replacement of any part of the pipework or the destination instrument, leak checks have to be carried out to ensure pipework integrity.

The way in which the air data is used to portray meaningful data to the crew by means of the aircraft instruments is shown in Fig. 6.13.

Three major parameters may be calculated from the pitot-static pressure information sensed by the pitot and static probes or by a combined pitot-static probe as shown in the diagram:

- Air speed may be calculated from the deflection in the left-hand instrument where Pt and Ps are differentially sensed. Air speed is proportionate to Pt - Ps and therefore the mechanical deflection may be sensed and air speed deduced. This may be converted into a meaningful display to the flight crew in a mechanical instrument by the mechanical gearing between capsule and instrument dial.

- Altitude may be calculated by the deflection of the static capsule in the centre instrument. Again in a mechanical instrument the instrument linkage provides the

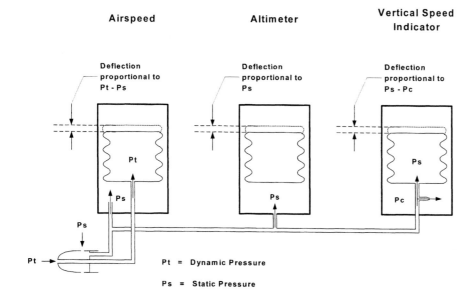

Fig. 6.13 *Use of air data to drive flight instruments*

mechanical scaling to transform the data into a meaningful display.

- Vertical speed may be deduced in the right-hand instrument where the capsule deflection is proportional to the rate of change of static pressure with reference to a case pressure, Pc. Therefore the vertical speed is zero when the carefully-sized bleed orifice between capsule inlet and case allows these pressures to equalize.

The examples given above are typical for aircraft instruments used up to about 40 years ago. There are three methods of converting air data into useful aircraft-related parameters etc. that the aircraft systems may use:

- On older aircraft conventional mechanical flight instruments may be used; these tend to be relatively unreliable, expensive to repair, and are limited in the information they can provide to an integrated system. Mechanical instruments are also widely used to provide stand-by or back-up instrumentation.
- On some integrated systems the pitot-static sensed pressures are fed into centralized Air Data Computers (ADCs). This allows centralization of the air data calculations into dedicated units with computational power located in electrical bay racks. The ADCs can provide more accurate air data calculations more directly aligned to the requirements of a modern integrated avionics system. When combined with digital computation techniques within the ADC and the use of modern data buses such as Mil-Std-1553B, ARINC 429 and ARINC 629 to communicate with other aircraft systems, higher degrees of accuracy can be achieved and the overall aircraft system performance improved.
- More modern civil aircraft developed in the late 1980s and beyond use Air Data Modules (ADMs) located at appropriate places in the aircraft to sense the pitot and static information as appropriate. This has the advantage that pitot-static lines can be kept to a minimum length reducing installation costs and the subsequent maintenance burden. By carefully selecting an appropriate architecture greater redundancy and improved fault tolerance may be designed in at an early stage, improving the aircraft dispatch availability.

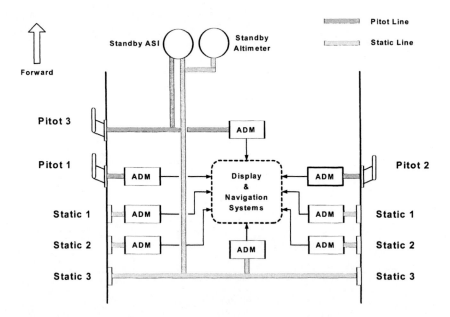

Fig. 6.14 *Air data system using ADMs*

An example of a modern air data system using ADMs is shown in Fig. 6.14. This architecture equates to the probe configuration installation shown in Fig. 6.12 namely, three pitot probes and a total of six static probes, three each on the left- and right-hand side of the aircraft.

Figure 6.14 shows how these probes are connected to ADMs and the degree of redundancy that can be achieved.

- Each pitot probe is connected to an individual ADM so there is triple redundancy of pitot pressure sensing. Pitot probe 3 also connects to the mechanical stand-by air speed indicator (ASI) that operates as shown in Fig. 6.13.
- The four static probes represented by static probes 1 and 2, left and right are connected to individual ADMs effectively giving quadruple redundancy of static pressure. Static probes left and right are physically interconnected and linked to a further ADM while also providing the static pressure sensing for the mechanical stand-by ASI and stand-by altimeter – see Fig. 6.13.
- Each of the eight ADMs shown in this architecture can be identical, since each is merely sensing an air data pressure parameter – pitot or static. The use of pin-programming techniques in the aircraft wiring means that an ADM may be installed in any location and will automatically adopt the personality required for that location.
- The ADMs interconnect to the aircraft display and navigation systems by means of ARINC 429 data buses as shown in Fig. 6.14.

CHAPTER 7

Environmental Control Systems

Introduction

Throughout the operation of an aircraft, whether on the ground or in the air, the crew and passengers must be kept in comfortable conditions. They must be neither too hot nor too cold, they must have air to breathe and they must be kept in comfortable atmospheric pressure conditions. This is by no means easy.

A military aircraft may have only a small crew, but the aircraft may be designed to perform in climatic extremes ranging from Arctic to full desert sunlight. A commercial aircraft may carry over 300 fare-paying passengers. In neither case can the human cargo be subjected to extremes of discomfort – passengers will go to another airline and the military crew will not perform at their most effective.

The environmental control system must cope with widely differing temperature conditions, must extract moisture and provide air with optimum humidity, and must ensure that the air in the aircraft always contains a sufficient concentration of oxygen.

Modern systems do this and more, for the term 'environmental control' also includes the provision of suitable conditions for the avionic, fuel and hydraulic systems by allowing heat loads to be transferred from one medium to another. In addition to these essentially comfort-related tasks, environmental control systems provide de-misting, anti-icing, anti-g and rain dispersal services.

The need for a controlled environment for crew, passengers and equipment

In the early days of flight, pilots and passengers were prepared to brave the elements for the thrill of flying. However, as aircraft performance has improved and the operational role of both civil and military aircraft has developed, requirements for Environmental Control Systems (ECS) have arisen. They provide a favourable environment for the instruments and equipment to operate accurately and efficiently, to enable the pilot and crew to work comfortably, and to provide safe and comfortable conditions for the fare-paying passengers.

In the past large heating systems were necessary at low speeds to make up for the losses to the cold air outside the aircraft. With today's aircraft operating at high subsonic or supersonic speeds, the emphasis is more towards the provision of cooling systems, although heating is still required, for example on cold night flights and for rapid warm-up of an aircraft which has been soaked in freezing conditions on the ground for long periods. Providing sufficient heat for the aircraft air-conditioning system is never a problem, since hot air can be bled from the engines to provide the source of conditioning air. The design requirement is to reduce the temperature of the air sufficiently to give adequate conditioning on a hot day. The worst case is that of cooling the pilot and avionics equipment in a high-performance military aircraft (reference (1)). The following heat sources give rise to the cooling problem.

Kinetic heating

Kinetic heating occurs when the aircraft skin heats up due to friction between itself and air molecules. The skin, in turn, heats up the interior of the aircraft such as the cockpit and equipment bays. Skin temperatures can reach up to 100 °C or more in low-level flight at transonic speeds, and even higher temperatures can be reached in supersonic flight at medium and high altitudes. Figure 7.1 shows a typical flight envelope for a high-performance military aircraft.

Note that in some flight cases, for example subsonic cruise at altitude on a cold day, kinetic heat loads can actually be negative. This is when heating is required.

Aircraft leading edges feel the full effect of kinetic heating due to friction and reach what are known as ram temperatures. All other surfaces away from the leading edges are subject to slightly lower temperatures termed recovery temperatures. For design purposes, the following equations can be used to calculate ram and recovery temperatures:

$$T_{rec} = T_{amb}(1 + 0.18 \, M^2)$$

$$T_{ram} = T_{amb}(1 + 0.2 \, M^2)$$

T_{rec} = Recovery air temperature °K

T_{ram} = Ram air temperature °K

T_{amb} = Ambient air temperature °K

M = Mach number

Unconditional equipment bays may reach recovery temperatures during flight.

Solar heating

Solar radiation affects a military aircraft cockpit directly through the windscreen and canopy. Equipment bays and civil aircraft cabins are only affected indirectly. A fighter aircraft is the worst case, since it usually has a large transparent canopy to give the pilot

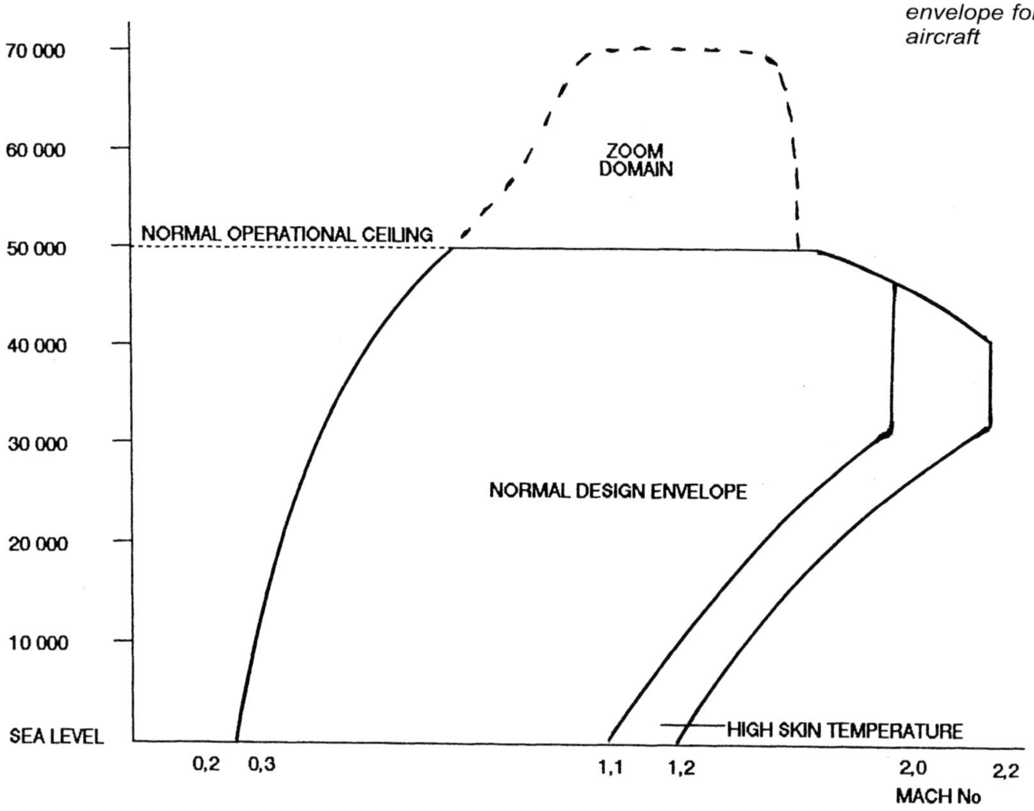

Fig. 7.1 *Typical flight envelope for a combat aircraft*

good all round vision, and can fly typically up to twice the maximum altitude of a civil aircraft. At such altitudes solar radiation intensity is much higher.

Solar heating significantly affects both cabin and equipment bays on ground stand-by, since surfaces exposed to direct solar radiation will typically rise 20 °C above the ambient temperature, depending on the thermal capacity of the surface material. This is of special concern in desert areas of the world where the sun is hot and continuous throughout the day.

Avionics heat loads

While advances in technology have led to reductions in heat dissipation in individual electronic components, the increased use of avionics equipment and the development of high density digital electronics has increased the heat load per unit volume of avionics equipment. This has resulted in an overall increase in heat load.

Airframe system heat loads

Heat is produced by the environmental control system itself, as well as hydraulic systems, electrical generators, engines and fuel systems components. This takes the form of heat produced as radiation from energy-consuming components in the systems such as pumps or motors, or from heat rejected in cooling fluids such as oil.

The need for cabin conditioning

Design considerations for providing air-conditioning in the cockpit of a high-performance fighter are far more demanding than those for a subsonic civil airliner cruising between airports.

The cockpit is affected by the sources of heat described above, but a high-performance fighter is particularly affected by high skin temperatures and the effects of solar radiation through the large transparency. However, in designing a cabin conditioning system for the fighter, consideration must also be taken of what the pilot is wearing. If, for example, he is flying on a mission over the sea, he will be wearing a thick rubber immersion suit which grips firmly at the throat and wrists. In addition, the canopy and windscreen will have hot air blown over the inside surfaces to prevent misting which would affect the temperature of the cabin. Another important factor is pilot workload or high-stress conditions such as may be caused by a failure, or by exposure to combat. All these factors make it very difficult to cool the pilot efficiently so that his body temperature is kept at a level that he can tolerate without appreciable loss of his functional efficiency.

The need for avionics conditioning

Most aircraft equipment which generates heat will operate quite satisfactorily at a much higher ambient air temperature than can be tolerated by a human. The maximum temperatures at which semi-conductor components can safely operate is above 100 °C, although prolonged operation at this level will seriously affect reliability.

Air-conditioning systems are typically designed to provide a maximum conditioned bay temperature of 70 °C, which is considered low enough to avoid significantly affecting the reliability of components. The minimum design equipment operating temperature for worldwide use tends to be about –30 °C. Equipment must also be designed to remain undamaged over a wider temperature range, typically from –40 °C to +90 °C for worldwide use. These figures define the maximum temperature range to which the equipment may be subjected depending on the storage conditions, or in the event that the aircraft is allowed to remain outside for long durations in extreme hot or cold conditions.

The International Standard Atmosphere (ISA)

An international standard atmosphere has been defined for design purposes. Tables of figures can be found in textbooks which show how values of temperature, pressure and air density vary with altitude. At sea level it is defined as follows:

    Air pressure   = 101.3 kPa absolute
    Air temperature = 15 °C
    Air density   = 1.225 kg/m

In addition, maximum and minimum ambient air temperatures have been derived from temperatures which have been recorded over a number of years throughout the world. These figures have been used to define a standard to which aircraft can be designed for worldwide operation. Examples are illustrated in Figs 7.2, 7.3, and 7.4, which are to be considered for design purposes only, and should not be considered as realistic atmospheres which could occur at any time.

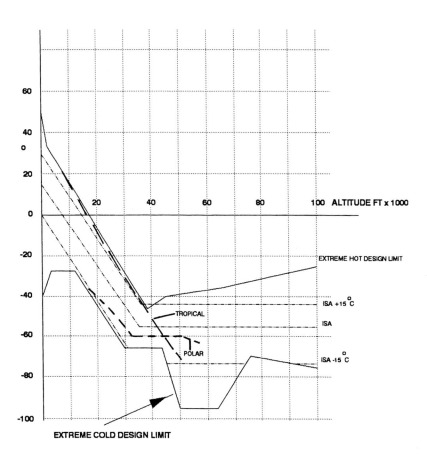

Fig. 7.2 *Ambient temperature versus altitude*

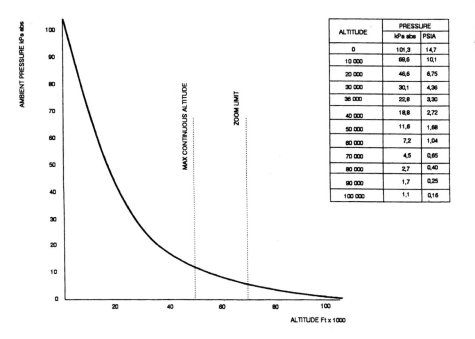

Fig. 7.3 *Ambient pressure variation with altitude*

ALTITUDE	PRESSURE	
	kPa abs	PSIA
0	101,3	14,7
10 000	69,6	10,1
20 000	46,6	6,75
30 000	30,1	4,36
36 000	22,8	3,30
40 000	18,8	2,72
50 000	11,6	1,68
60 000	7,2	1,04
70 000	4,5	0,65
80 000	2,7	0,40
90 000	1,7	0,25
100 000	1,1	0,16

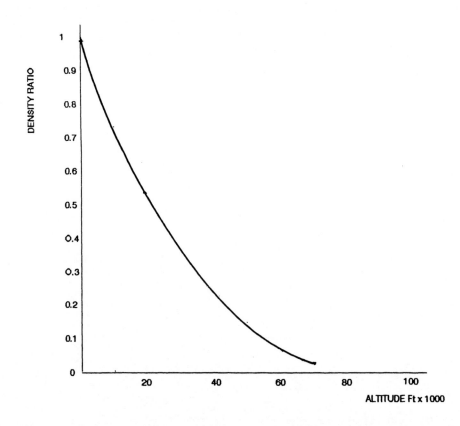

Fig. 7.4 *Air density ratio variation with altitude*

Fig. 7.5 *Typical mid-day world temperatures*

Figure 7.5. shows a distribution of maximum temperatures below and above ISA which are typically encountered throughout the world. These figures are used as a guide for designers of systems which are required to operate in particular areas.

Environmental control system design

This section describes methods of environmental control in common use and, in addition, outlines some recent advances and applications in environmental control system design.

The cooling problem brought about by the heat sources described above must be solved to successfully cool the aircraft systems and passengers in flight. For ground operations some form of ground cooling system is also required.

Heat must be transferred from these sources to a heat sink and rejected from the aircraft. Heat sinks easily available are the outside air and the internal fuel. The outside air is used either directly as ram air, or indirectly as air bled from the engines. Since the available heat sinks are usually at a higher temperature than that required for cooling the systems and passengers, then some form of heat pump is usually necessary.

Ram air cooling

Ram air cooling is the process of rejecting aircraft heat load to the air flowing round the aircraft. This can be achieved by scooping air from the aircraft boundary layer or close to it. The air is forced through a scoop which faces into the external air flow, through a heat exchanger matrix and then rejected overboard by the forward motion of the aircraft. The heat exchanger works just like the radiator of a car.

This system has the disadvantage that it increases the aircraft drag because the resistance of the scoop, pipework and the heat exchanger matrix slows down the ram air flow.

The use of ram air as a cooling medium has its limitations, since ram air temperature increases with air speed and soon exceeds the temperature required for cabin and equipment conditioning. For example, at Mach 0.8 at sea level on a 40 °C day, the ram air temperature is about 80 °C. Ram air is also a source of heating itself as described above (kinetic heating). In addition, at high altitude the air density becomes very low, reducing the ram air mass flow and hence its cooling capacity. In fact, when conditioning is required for systems which require cooling on the ground, then ram air cooling alone is unsuitable.

However, this situation can be improved by the use of a cooling fan, such as used on a civil aircraft, or a jet pump, mainly used on military aircraft, to enhance ram air flow during taxiing or low-speed flight. The jet pump enhances ram air cooling in the heat exchanger by providing moving jets of primary fluid bled from the engines to entrain a secondary fluid, the ram air, and move it downstream as shown in Fig. 7.6.

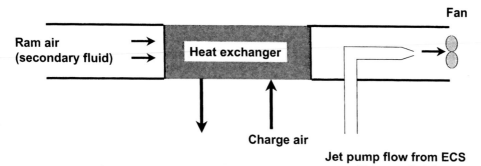

Fan

Ram air (secondary fluid)

Heat exchanger

Charge air

Jet pump flow from ECS (Primary fluid)

Fig. 7.6 The use of cooling fans and jet pumps to improve ram air flow

Fuel cooling

Fuel cooling systems have limited applications on aircraft for the transfer of heat from a heat source into the aircraft fuel. This is mainly due to the fact that fuel flow is variable and is greatly reduced when the engines are throttled back. However, fuel is much better than air as a cooling medium because it has a higher heat capacity and a higher heat transfer coefficient. Fuel is typically used to cool engine oil, hydraulic oil and gearbox oil.

Figure 7.7. shows a typical fuel cooling system. When the fuel flow is low, the fuel temperature will rise significantly, so recirculation lines are used to pipe the hot fuel back into the fuel tank. Ram air cooled fuel coolers often need to be introduced into the recirculation flow lines to prevent a rapid increase in fuel temperatures in the tank when fuel level is low. This can only be brought into effect in low-speed flight when ram temperatures are low enough. This prevents a rapid rise in the tank fuel temperature during the final taxi after landing, when the tanks are most likely to be almost empty.

Fig. 7.7 *Example of fuel cooling system*

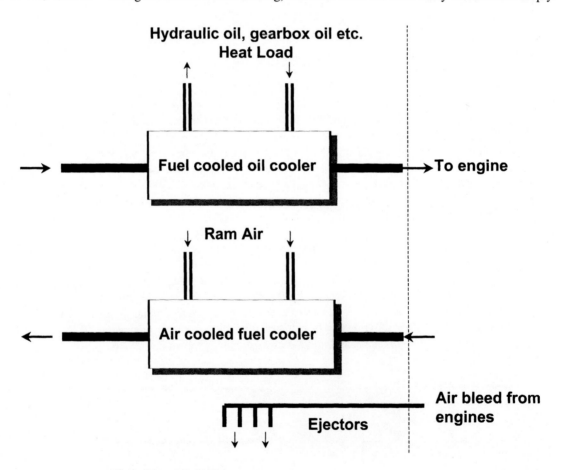

Engine bleed

The main source of conditioning air for both civil and military aircraft is engine bleed from the high-pressure compressors. This provides a source whenever the engines are running. The conditioning air is also used to provide cabin pressurization.

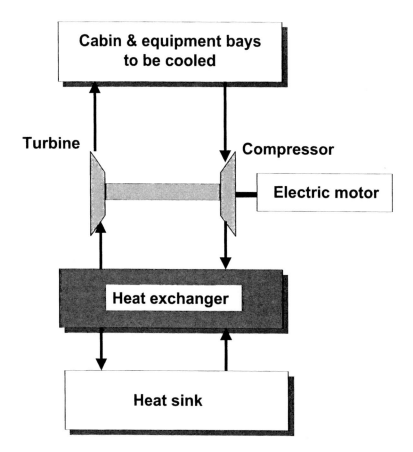

Fig. 7.8 *Closed loop cooling system schematic*

There are two types of bleed air system: open loop and closed loop. Open-loop environmental control systems continually bleed large amounts of air from the engines, refrigerate it, and then use it to cool the passengers and crew, as well as equipment, before dumping the air overboard. Closed-loop systems (Fig. 7.8) collect the air once it has been used for cabin conditioning, refrigerate it and recycle it to be used again. In this way bleed air is used only to provide pressurization, a low venting air supply and sufficient flow to compensate for leaks in the closed-loop system. This means that such a system uses considerably less engine bleed air than an open-loop system and therefore has a correspondingly reduced effect on engine performance. It follows that with a closed-loop system, a military aircraft has more available thrust at its disposal, or that a civil aircraft is able to operate more efficiently, particularly on long flights.

Since only a small amount of air is bled off from the engines, the need for ram air cooling of the bleed air is reduced. However, to recycle conditioning air it is necessary to seal and pressurize the equipment bays. The cooling air is distributed between equipment using cooling trays with fans to draw equipment exhaust air into the recirculation loop.

Closed-loop systems have to date only been used in a few aircraft applications. Not only are there the practical difficulties of collecting and reusing the conditioning air, but closed-loop systems also tend to be heavier and more expensive than equivalent open-loop systems. As a result the latter, using air cycle refrigeration to cool engine bleed air are most commonly used in aircraft applications. However, some recirculation of cabin air has been introduced on civil aircraft to reduce the ECS cooling penalty. The cabin air is drawn into the recirculation line by a jet pump or fan, and then mixed with refrigerated engine bleed air before being supplied to the cabin inlet at the required temperature. The utilization of such a recirculation flow can double the efficiency of the system in some cases.

The above method of reducing bleed flow has limited application on high performance military aircraft because of problems such as the lack of recirculation air available at high altitudes from unpressurized bays and restricted space for ducting.

Therefore, bleed flow reduction on most military aircraft is achieved by modulation of system flow in accordance with demand as described in the following passage.

Bleed flow and temperature control

Typically air at a workable pressure of about 650 kPa absolute (6.5 atm) and a temperature of about 100 °C is needed to provide sufficient system flow and a temperature high enough for such services as rapid demisting and anti-icing. However, the air tapped from the engine high-pressure compressor is often at higher pressures and temperatures than required. For example, in a high-performance fighter aircraft the air can be at pressures as high as 3,700 kPa absolute (37 atm) and temperatures can be over 500 °C: high enough to make pipes manufactured from conventional materials glow red hot. Tapping air at lower pressures and temperatures from a lower compressor stage would be detrimental to engine performance. On many civil aircraft, different bleed tappings can be selected according to engine speed.

The charge air pressure needs to be reduced as soon as possible to the required working pressure for safety reasons and to reduce the complexity of components since there are problems with sealing valves at such high pressures.

A pressure-reducing valve can be used to reduce the pressure of the engine bleed air. This valve controls its downstream pressure to a constant value, no matter what the upstream pressure. The maintenance of this downstream pressure controls the amount of flow from the engines through the environmental control system.

This is acceptable for an aircraft with very few speed variations, such as a civil airliner. However, the faster an aircraft flies the more conditioning air is required, since the greater is the effect of kinetic heating.

In a supersonic aircraft, if the pressure-reducing valve was designed to provide sufficient cooling air at high speeds, there would be an excess of flow at low speed. This is wasteful and degrades engine performance unnecessarily. On the Eurofighter the environmental control system contains a variable Pressure-Reducing Valve (PRV) which automatically controls its downstream pressure and, therefore, the amount of engine bleed, depending on aircraft speed. This means that the effect of engine bleed on engine performance can be kept to a minimum at all times.

Once the air pressure has been reduced to reasonable working values, the air temperature needs to be reduced to about 100 °C for such services as de-icing and demisting. Heat exchangers are used to reject unwanted heat to a cooling medium, generally ram air (Fig. 7.9).

In some flight conditions, particularly on cold days, there is so much relatively cool air that the heat exchanger outlet temperature is much less than the 100 °C required for de-icing or demisting. In such cases the correct proportion of hot air from upstream of the heat exchanger is mixed with heat exchanger outlet flow to maintain at least 100 °C mixed air outlet temperatures.

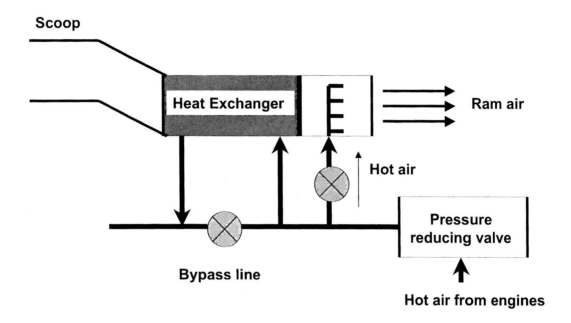

Refrigeration systems

Fig. 7.9 *Mixing hot air with heat exchanger outlet*

There are two main types of refrigeration systems in use:

   (1)   Air cycle refrigeration systems
   (2)   Vapour cycle refrigeration systems

Air cycle refrigeration systems

Air cycle refrigeration systems are used to cool engine bleed air down to temperatures required for cabin and equipment conditioning. Since engine bleed air is generally available, air cycle refrigeration is used because it is the simplest solution to the cooling problem, fulfilling both cooling and cabin pressurization requirements in an integrated system. However, although lighter and more compact than vapour cycle, air cycle systems have their limitations. Very large air flows are required in high heat load applications which require large diameter ducts with the corresponding problems of installation in the limited space on board an aircraft. Large engine bleed flows are detrimental to engine performance and large aircraft drag penalties are incurred due to the need for ram air cooling.

   There are many different types of air cycle systems, but the basic principles remain the same. A source of high pressure air is cooled as much as possible in a heat exchanger using ram air as a coolant. The air is then expanded across a turbine to reduce the temperature and pressure further. The turbine drives a fan or compressor which acts as a brake while doing work forcing ambient air across the heat exchanger. Examples of such systems are described below.

Turbofan system

This will typically be used in a low-speed civil aircraft where ram temperatures will never be very high. A typical turbofan system is illustrated in Fig. 7.10.

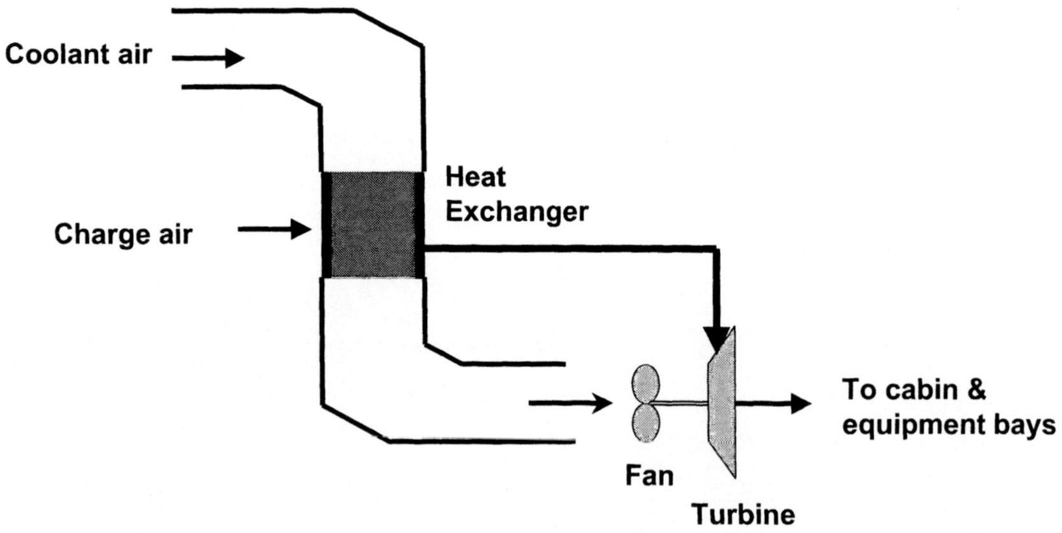

Coolant air
Charge air
Heat
Exchanger
To cabin &
equipment bays
Fan
Turbine

Fig. 7.10 Example of turbofan refrigeration system

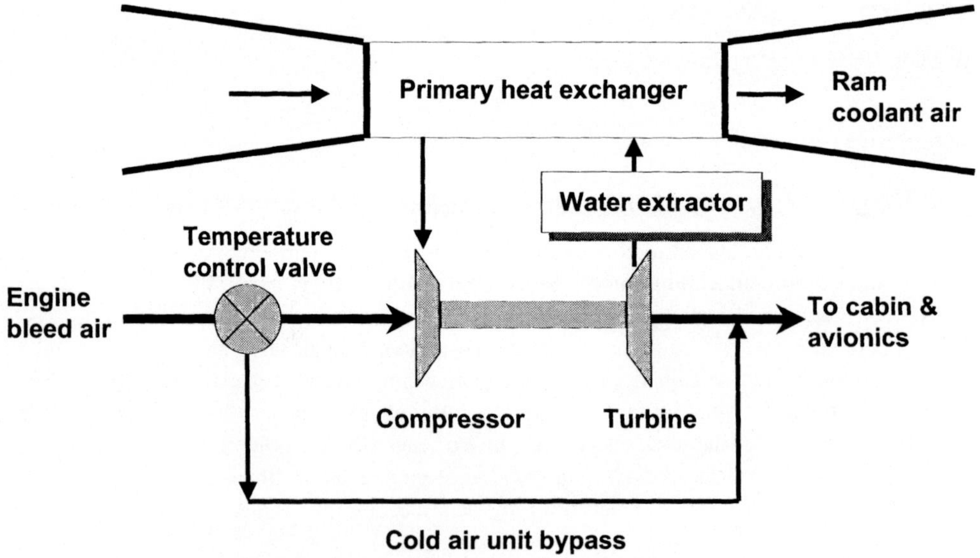

Primary heat exchanger
Ram
coolant air
Water extractor
Temperature
control valve
Engine
bleed air
To cabin &
avionics
Compressor
Turbine
Cold air unit bypass

Fig. 7.11 Example of bootstrap refrigeration system

Bootstrap system

Conventional bootstrap refrigeration is generally used to provide adequate cooling for high ram temperature conditions, for example a high-performance fighter aircraft.

The basic system consists of a cold air unit and a heat exchanger (Fig. 7.11). The turbine of the cold air unit drives a compressor. Both are mounted on a common shaft. This rotating assembly tends to be supported on ball bearings, but the latest technology uses air bearings. This provides a lighter solution which requires less maintenance, for example no oil is required.

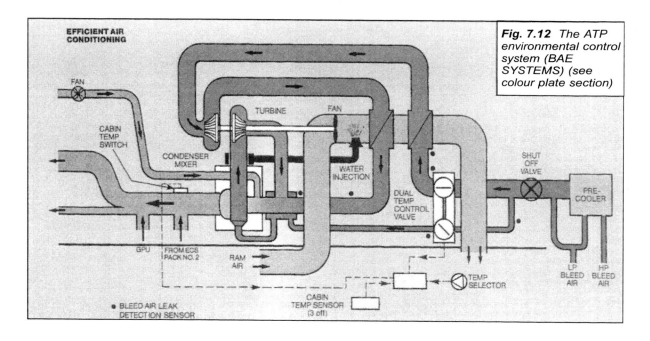

Fig. 7.12 *The ATP environmental control system (BAE SYSTEMS) (see colour plate section)*

Fig. 7.13 *Examples of air cycle machine and air conditioning packs (Honeywell)*

Three-rotor cold air units or air cycle machines can be found on most recently designed large aircraft, incorporating a heat exchanger coolant fan on the same shaft as the compressor and turbine. Military aircraft tend to use the smaller and simpler two-rotor cold air unit using jet pumps to draw coolant air through the heat exchanger when the aircraft is on the ground and in low-speed flight. Figure 7.12 shows the environmental control system of the BAE SYSTEMS Advanced Turbo-Prop (ATP) aircraft as a typical example.

The compressor is used to increase the air pressure with a corresponding increase in temperature. The temperature is then reduced in the ram air cooled heat exchanger.

This reduction in temperature may lead to water being condensed out of the air, especially when the aircraft is operating in a humid climate. Figure 7.12 shows a water extractor at the turbine inlet which will remove most of the free water, helping to prevent freezing of the turbine blades and water being sprayed into the cabin and equipment bays. As the air expands across the turbine the temperature can drop below 0°C in certain flight conditions. Figure 7.12. also shows a cold air unit bypass line which is used to vary turbine outlet temperature to the required value for cabin and equipment cooling. The volume of air flowing round the bypass is varied by a temperature control valve until the air mixture at the turbine outlet is at the required temperature.

Examples of some of the machines presently in use on the Boeing 737 and Boeing 757/Boeing 767 are shown in Fig. 7.13.

Reversed bootstrap

The reversed bootstrap system is so named because the charge air passes through the turbine of the cold air unit before the compressor. Following initial ram air cooling from a primary heat exchanger the air is cooled further in a regenerative heat exchanger and is then expanded across the turbine with a corresponding decrease in temperature. This air can then be used to cool an air or liquid closed-loop system, for radar transmitter cooling for example. The air then passes through the coolant side of the regenerative heat exchanger before being compressed by the compressor and dumped overboard (Fig. 7.14).

Fig. 7.14 Example of a reverse bootstrap refrigeration system

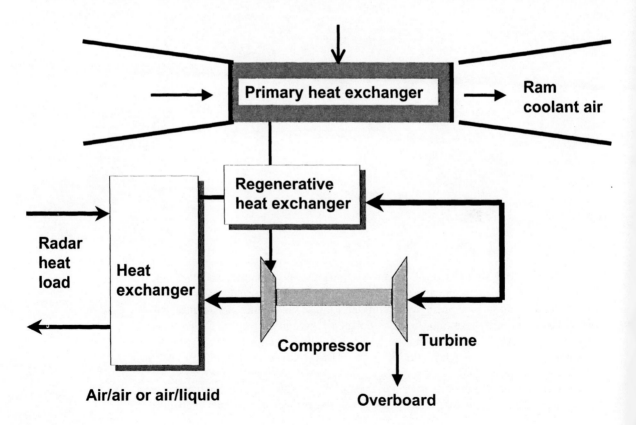

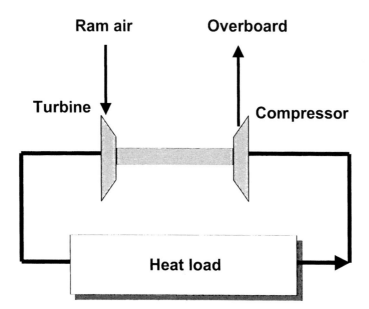

Fig. 7.15 *Example of ram-powered reverse bootstrap*

Ram powered reverse bootstrap

In some cases equipments may be remotely located where it is not practicable to duct an air supply from the main ECS. In such cases a separate cooling package must be employed. This situation is becoming particularly common on military aircraft with equipment mounted in fin tip or under-wing pods, where it is not possible to find a suitable path to install ducting or pipes. A ram-powered reverse bootstrap air-cycle system can be used to meet such 'stand-alone' cooling requirements.

The method increases the capability of a ram air cooled system by expanding the ram air through a turbine, so reducing air temperature as shown in Fig. 7.15. Therefore cooling can be provided up to much higher air speeds than a purely ram air cooled system. However, cooling is still a problem on the ground and in low-speed flight. Therefore this system is typically used as a 'stand-alone' cooling system for equipment which is operated only during flight.

Vapour cycle systems

The vapour cycle system is a closed-loop system where the heat load is absorbed by the evaporation of a liquid refrigerant such as freon in an evaporator. The refrigerant then passes through a compressor with a corresponding increase in pressure and temperature, before being cooled in a condenser where the heat is rejected to a heat sink. The refrigerant flows back to the evaporator via an expansion valve. This system is illustrated in Fig. 7.16.

Although vapour cycle systems are very efficient, with a coefficient of performance typically five times that of a comparable closed-loop air-cycle system. Applications are limited due to problems such as their limited temperature range and heavy weight compared to air-cycle systems. The maximum operating temperatures of many refrigerants are too low, typically between 65 °C and 70 °C, significantly less than the temperatures which are required for worldwide operation.

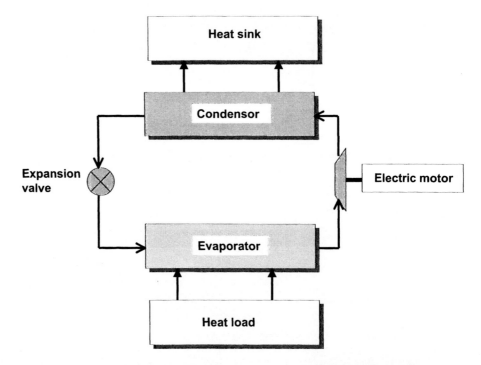

Fig. 7.16 *Basic vapour cycle system*

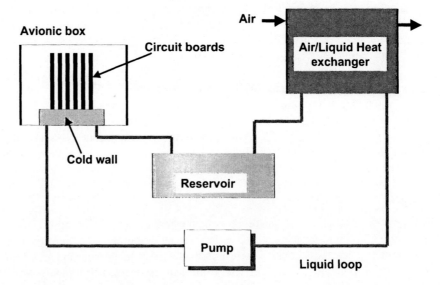

Fig. 7.17 *Example of liquid-cooling system*

Liquid-cooled systems

Liquids such as Coolanol are now more commonly being used to transport the heat away from avionics equipment. Liquid can easily replace air as a transport medium flowing through the cold wall heat exchanger.

A typical liquid loop consists of an air/liquid heat exchanger which is used to dump the heat load being carried by the liquid into the air-conditioning system, a pump and a reservoir as illustrated in Fig. 7.17.

The advantages are that it is a more efficient method of cooling the heat source, and the weight and volume of equipment tends to be less than the air-conditioning equipment which would otherwise be required. The disadvantages are that it is expensive, and the liquid Coolanol is toxic. Self-sealing couplings must be provided to prevent spillage wherever a break in the piping is required for maintenance purposes.

Expendable heat sinks

An expendable coolant, typically water, can be carried to provide a heat sink by exploiting the phenomenon of latent heat of vaporization. A simple system is shown in Fig. 7.18.

The liquid refrigerant is stored in a reservoir which supplies an evaporator where the heat load is cooled. The refrigerant is then discharged overboard. This type of system can only be used to cool small heat loads (or large loads for a short time), otherwise the amount of liquid refrigerant that must be carried on board the aircraft would be too large.

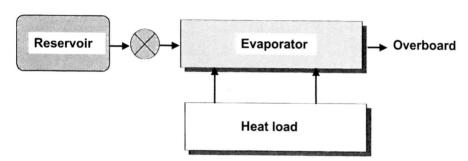

Fig. 7.18 *Simple expendable heat sink system*

Humidity control

Passenger comfort is achieved not only by overcoming the problems of cooling and cabin pressurization, but also by controlling humidity in the passenger cabin. This is only a problem on the ground and at low altitudes, since the amount of moisture in the air decreases with increasing altitude. There is a particular difficulty in hot humid climates. For example in Northern Europe the typical air moisture content can be 10 grams of water per kilogram of air, but in some parts of the Far East moisture contents of more than 30 grams per kilo can be encountered. In a hot, humid climate the cabin inlet air supply temperature needs to be cold to keep the passengers and aircrew comfortable. Without good humidity control this can result in a wet mist being supplied to the cabin.

In addition to the aim of ensuring passenger comfort, humidity levels must be controlled to prevent damage to electrical and electronic equipment due to excessive condensation. Humidity control also reduces the need for windscreen and window de-misting and anti-misting systems.

The fine mist of water droplets in the cold cabin inlet supply must be coalesced into large droplets that can then be trapped and drained away. Two types of water separator are in common use with air cycle refrigeration systems: a centrifugal device and a mechanical device. In the centrifugal devices a turbine is commonly used to swirl the moist air. The relatively heavy water droplets are forced to the sides of a tube, where the water and a small amount of air is trapped and drained away, thus reducing the water content of the air downstream of a water separator.

The mechanical water separator, which consists of a coalescer, a relief valve and a water collector, achieves the same result by forcing the moist air to flow through the coalescer where large droplets are formed and blown on to collector plates. The water runs down the plates and is then drained away. The relief valve opens to allow the air to bypass the water separator if ice forms.

Simple water collection devices can be used in vapour cycle refrigeration systems to reduce humidity levels since the air is cooled to its dew-point as it flows through the evaporator. Water droplets collect on the heat exchanger surfaces and can be simply trapped and drained away.

Chemicals can also be used to reduce moisture content. In civil aircraft the air gaps between two plates of the passenger windows are commonly vented via an absorbent material such as silica gel to prevent condensation of moisture on the window. Moisture is condensed from the air as it flows through the gel, and the latent heat given up by the condensing moisture increases the air temperature.

Molecular sieves can also be used to remove moisture from air. These are absorbent materials which are used to sieve out the large water molecules from the air in the same way as the molecular sieve oxygen concentrators described later in this chapter remove the large gas molecules and impurities from air to leave almost pure oxygen.

The inefficiency of current environmental control systems

In cooling down engine bleed air to temperatures low enough to provide adequate cooling capacity for aircrew, passengers and equipment, a great deal of heat and therefore potentially useful energy is rejected to atmosphere. Typically, the ratio of engine power used to heat load cooled in order to provide sufficient cooling for the total aircraft heat load is 10:1

In addition, further engine power is required to overcome the drag caused by the ram air heat exchangers. This problem becomes worse, particularly on military aircraft which suffer a continually increasing avionics heat load; while the design requirements are to improve engine performance and reduce aircraft weight. The more avionics, the heavier the aircraft, not only due to the avionics equipment weight itself, but also due to the weight of the environmental control system equipment and the air distribution pipework. Furthermore, additional engine bleed air is required as the avionics heat load increases, but bleeding more air off the engines is detrimental to engine performance. More efficient cooling by closed-loop systems would undoubtedly increase equipment reliability.

The increasing avionics heat load on military aircraft may lead to further developments of closed-loop environmental control systems in the future, since there is potential to vastly reduce the amount of engine bleed required, and thus overcome the problems of detrimental effects of open-loop systems on engine performance.

Air distribution systems

Avionics cooling

In civil aircraft the total avionics heat load is low when compared with the many applications which have been, and continually are being found in military aircraft. In civil aircraft it is often sufficient to draw cabin ambient air over the avionics equipment racks using fans. This will have the effect of increasing the overall cabin temperature but, since the total avionics heat load is not massive, the environmental control system has sufficient capacity to maintain cabin temperatures at acceptable levels.

However, on a military aircraft with a high avionics heat load, only a few of the avionics equipment are located in the cabin. The majority are located in either conditioned or non-conditioned equipment bays, an installation decision which is made by taking into consideration such criteria as the effect of temperature on equipment reliability or damage, and the amount of engine bleed available for air-conditioning.

Since the equipment can operate in ambient temperatures higher than humans can tolerate, the air used to condition it tends to be cabin exhaust air. There is usually very little space in equipment bays as they are tightly packed with equipment. There is little space left for the installation of cooling air ducts. Therefore, the equipment racking and air distribution system must be carefully designed to ensure an even temperature distribution.

Unconditioned bays

Unconditioned bays may reach temperatures up to recovery temperature. However, air in these bays is not totally stagnant. The aircraft is usually designed to have a continuous venting flow through each equipment bay; only the pressure cabin is sealed. This ensures that there is no build-up of differential pressure between bays, particularly during rapid climb and descent. The venting flow tends to be the conditioned bay outlet flow.

Conditioned bays

Equipment can be cooled by a variety of methods, including the following:

- Cooling by convection air blown over the outside walls of the equipment boxes. (External air wash.)
- Air blown through the boxes and over the printed-circuit boards. (Direct forced air.)
- Air blown through a cold wall heat exchanger inside the box. (Indirect forced air.)
- Fans installed in the box to draw a supply of cooling air from the box surroundings.

The first method of cooling is adequate for equipment with low heat loads. As the heat load increases it tends to become very inefficient, requiring a lot more cooling air than the other methods to achieve the same degree of cooling. It is very difficult to design an avionics equipment box with a high heat load to enable the efficient dissipation of heat by convection via the box walls. Local 'hot spots' inside the box will lead to component unreliability.

The other three methods of cooling are very much more efficient, but the boxes must have a good thermal design to ensure precious conditioning air is not wasted.

Indirect forced air is often the most efficient way of cooling. The box is thermally designed so that the component heat load is conducted to a cold wall heat sink. The cold wall acts as a small heat exchanger.

The fan cooling method is only acceptable in an equipment bay layout where there is no chance of re-ingestion of hot exhaust air from another box. This is not practical in a closely packed equipment bay. It tends to be selectively used for TRVs.

Fig. 7.19 *Typical air distribution scheme*

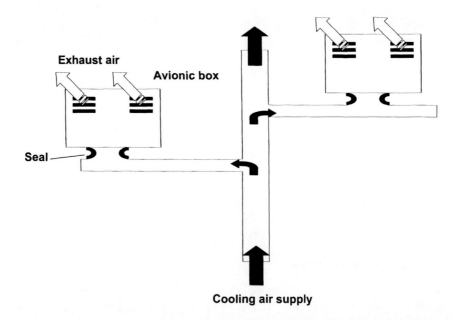

Particular attention must be given to the cooling requirements of equipment whose correct operation is critical to the safety of the aircraft. For example, the Eurofighter is an inherently unstable aircraft which is controlled by the pilot via the flight control computers. These computers must be continuously fully conditioned, since failure of all computers would render the aircraft uncontrollable.

Figure 7.19 shows a typical method of air distribution. The distribution ducting provides a supply of air into a plenum chamber which is built into a shelf on which the equipment is installed. The air is supplied directly into the equipment via orifices in the shelf and the equipment box. It is prevented from leaking away by a soft seal between the shelf and the box. The air exhausts from the box through louvres in the wall.

Fig. 7.20 *Different methods of cooling avionics equipment*

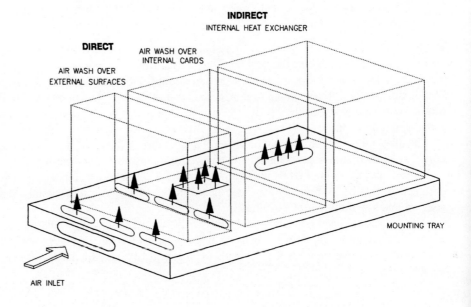

Conditioned bay equipment racking

In a closely packed equipment bay cooling methods (1), (2) and (3) are often used side by side as shown in Fig. 7.20.

Ground cooling

For aircraft with separate equipment bays fans are provided which are often located in the undercarriage bays. These are used to provide ambient cooling air flow for the avionics bays when the aircraft is on the ground, and there is only enough bleed air flow from the engines in this case to provide cabin conditioning. The fans can also be used to cool the equipment if the environmental control system fails.

Cabin distribution systems

Cabin distribution systems on both civil and military aircraft are designed to provide as comfortable an environment as possible. The aircrew and passenger body temperatures should be kept to acceptable levels without hot spots, cold spots or draughts. Civil aircraft are designed to maintain good comfort levels throughout the cabin since passengers are free to move about. On modern aircraft each passenger has personal control of flow and direction of local air from an air vent above his head. There are usually additional vents which blow air into the region of the passengers' feet so that there is no temperature gradient between the head and feet.. Figure 7.21 shows how the air is distributed in the passenger cabin of the BAE SYSTEMS 146. This can be more difficult to achieve for the pilot on a fighter aircraft, where his head receives the full effect of solar radiation through the transparency.

The air velocities must be high and the air temperatures near freezing for the pilot to feel any effect through his clothing (including an immersion suit). The distribution system must also be designed so that the cold air jet picks up as little heat as possible from its surrounding environment before it reaches the subject to be cooled.

Cabin noise

Aircraft are designed aerodynamically or structurally to keep externally generated noise levels to a minimum. At crew stations the noise levels should be such that the aircrew are able to communicate satisfactorily over a radio or intercom, or to operate direct voice input avionics systems. As in any other work environment noise levels must be kept to satisfactory levels to avoid damage to hearing. The noise levels in the passenger cabin of a civil aircraft are kept to a minimum to ensure passenger comfort since fare-paying passengers are free to take their custom elsewhere.

Cabin pressurization

Cabin pressurization is achieved by a cabin pressure control valve which is installed in the cabin wall to control cabin pressure to the required value depending on the aircraft altitude by regulating the flow of air from the cabin.

For aircraft where oxygen is not used routinely, and where the crew and passengers are free to move around, as in a long-range passenger airliner, the cabin will be pressurized so that a cabin altitude of about 8,000 ft is never exceeded. This leads to a high differential pressure between the cabin and the external environment. Typically

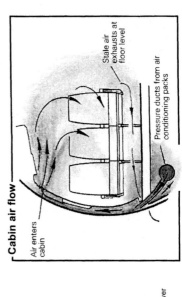

Cabin air flow

Stale air exhausts at floor level

Pressure ducts from air conditioning packs

Air enters cabin

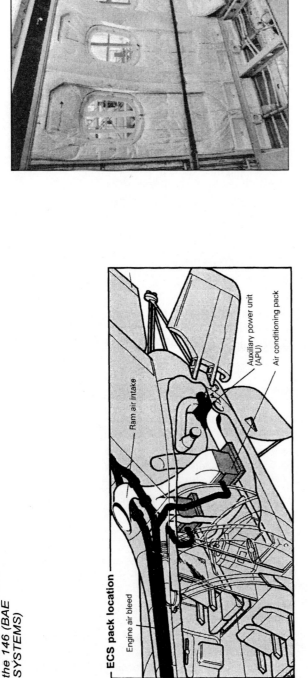

Fig. 7.21 Air distribution system of the 146 (BAE SYSTEMS)

ECS pack location

Ram air intake

Engine air bleed

Auxiliary power unit (APU)

Air conditioning pack

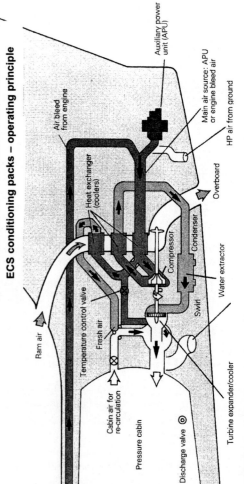

ECS conditioning packs – operating principle

Air bleed from engine

Auxiliary power unit (APU)

Main air source: APU or engine bleed air

HP air from ground

Heat exchanger (coolers)

Overboard

Compressor

Condenser

Ram air

Temperature control valve

Fresh air

Water extractor

Swirl

Cabin air for re-circulation

Pressure cabin

Discharge valve

Turbine expander/cooler

for an airliner cruising at 35,000 ft with a cabin altitude of 8,000 ft there will be a differential pressure of about 50 kPa (0.5 atm) across the cabin wall. The crew is able to select a desired cabin altitude from the cockpit and cabin pressurization will begin when the aircraft reaches this altitude. This will be maintained until the maximum-design cabin differential pressure is reached.

For aircraft with the crew in fixed positions, using oxygen routinely as in a military aircraft, the pressurization system is usually designed so that the cabin altitude does not exceed about 20,000 ft. Figure 7.22 shows a typical fighter aircraft automatic pressurization schedule. The cabin pressure control valve is designed to automatically maintain the cabin altitude to this schedule depending on aircraft altitude without any intervention from the pilot.

The differential pressure is maintained high enough so that if the cabin pressurization fails when the aircraft is at a high altitude there is sufficient time for the pilot to descend. For example, at 50,000 ft; then the pressure will not leak away causing the cabin altitude to exceed a safe value before the pilot has had enough time to descend to a safe altitude.

Therefore, the cabin must be designed as a pressure vessel with minimum leakage. In the event of loss of pressurization the cabin pressure control valve will close and the only leakage will be through the structure. Non-return valves are installed in the air distribution pipes where they pass through the cabin wall, so that when the air supply fails the air already in the cabin cannot leak back out through the pipes. A safety valve is installed in the cabin wall to relieve internal pressure if it increases above a certain value in the event of failure of the pressure control valves.

Following the loss of the cabin pressurization system and descent to a safe altitude, the pilot can select the opening of a valve to enable ram air to be forced into the distribution system by a scoop which faces into the external air flow. This system of purging with ram air can also be selected should the cabin be contaminated by fumes or smoke coming from the main environmental control system air supply.

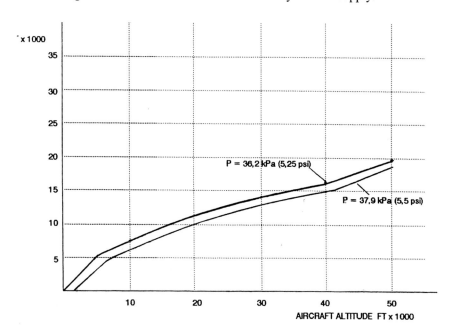

Fig. 7.22 *Fighter aircraft automatic pressurization schedule*

Hypoxia

Oxygen is essential for the maintenance of life. If the oxygen supply to the brain is cut off, unconsciousness soon follows, and brain death is likely to occur within 4–5 min. Breathing air at reduced atmospheric pressure results in a reduction in alveolar oxygen pressure which in turn results in an oxygen supply deficiency to the body and brain tissues. This condition is termed hypoxia.

The effects of hypoxia can be demonstrated by imagining a slow ascent by balloon. Up to 10,000 ft there will be no significant effects of altitude on the body. At 15,000 ft it will be markedly more difficult to perform physical tasks, and the ability to perform skilled tasks will be severely reduced, although this fact will probably go unnoticed. At 20,000 ft the performance of physical tasks will be grossly impaired, thinking will be slow, and calculating ability will be unreliable. However, the occupants of the balloon will be unaware of their deficiencies, and may become light-headed and over-confident. Any physical exertion may cause unconsciousness. Even a highly qualified and experienced pilot will be in a totally unfit state to fly an aircraft. Above 20,000 ft loss of consciousness sets in (reference (2)).

During a rapid climb to altitude in a fighter aircraft, without any protection against hypoxia, rapid and sudden loss of consciousness will result without any of the symptoms of hypoxia appearing. The dangerous effects of breathing air at reduced atmospheric pressures can be alleviated by pressurizing the cabin. Typically, on a civil aircraft with a maximum operating altitude of 25,000 ft, the cabin will be pressurized to maintain a cabin altitude of 8,000 ft.

An alternative method of preventing hypoxia is to increase the concentration of oxygen in the cabin atmosphere. Civil aircraft only supply oxygen in cases of rapid cabin depressurization or contamination of the cabin air by smoke or harmful gases. Emergency procedures require quick action from the pilot and crew, or an automatic system, in the event of a rapid depressurization since hypoxia is much more severe when it is initiated by a sudden exposure to high altitude compared to a more gradual degradation of performance with gradually increasing altitude (reference (3)).

For both civil and military applications, an oxygen regulator is used to control the flow of breathing gas in response to the breathing action of the person requiring the supply of gas. The proportions of air to oxygen mixture supplied can be varied depending on the altitude. A mask is connected by hoses and connectors to the regulator output.

The source of breathing gas will be either from pre-charged or liquid oxygen bottles, or from a Molecular Sieve Oxygen Concentrator (MSOC) which produces breathable gas from engine bleed air.

Molecular sieve oxygen concentrators

Until recently the only practical means of supplying oxygen during flight has been from a cylinder or a liquid oxygen bottle. This has several disadvantages, particularly for military aircraft. It limits sortie duration (fuel may not be the limiting factor if in-flight refuelling is used), the equipment is heavy and the bottles need replenishing frequently.

Molecular Sieve Oxygen Concentrators (MSOC) are currently being developed for military applications. The MSOCs use air taken from the environmental control systems as their source of gas. Most of the gases in air have larger molecules than oxygen. These molecules are sieved out of the air mixture until mostly oxygen remains. This means that a continuous supply of oxygen can be made available without needing to replenish the traditional oxygen storage system after each flight. The residual inert gases can be used for fuel tank pressurization and inerting. A system designed specifically for the production of inert gases is known as the On-Board Inert Gas Generating System (OBIGGS).

However, MSOCs have a major disadvantage. If the environmental air supply from the engines stops then so does the supply of oxygen. Therefore, small back-up oxygen systems are required for emergency situations to enable the pilot to descend to altitudes where oxygen levels are high enough for breathing. Developments of MSOCs are watched with interest, and future systems may be efficient enough to provide oxygen enriched air for civil aircraft cabins.

In military aircraft which are typically designed to fly to altitudes in excess of 50,000 ft, both cabin pressurization and oxygen systems are employed to help alleviate the effects of hypoxia. In cases where aircrew are exposed to altitudes greater than 40,000 ft, either due to cabin depressurization or following escape from their aircraft, then additional protection is required. In the event of cabin depressurization the pilot would normally initiate an emergency descent to a 'safe' altitude. However, short-term protection against the effects of high altitude is still required.

At altitudes up to 33,000 ft, the alveolar oxygen pressure can be increased up to its value at ground level by increasing the concentration of oxygen in the breathing gas. However, even when 100 per cent oxygen is breathed, the alveolar oxygen pressure begins to fall at altitudes above 33,000 ft. It is possible to overcome this problem by increasing the pressure in the lungs above the surrounding environmental pressure. This is called positive pressure breathing. At altitudes above 40,000 ft the rise in pressure in the lungs relative to the pressure external to the body seriously affects blood circulation round the body and makes breathing more difficult. Partial-pressure suits are designed to apply pressure to parts of the body to counter the problems of pressure breathing for short durations above 40,000 ft.

A partial-pressure suit typically includes a pressure helmet and a bladder garment which covers the entire trunk and the upper part of the thighs. The pressure garments are inflated when required by air taken from the environmental control system and are used in conjunction with an inflatable bladder in anti-g trousers which are used primarily to increase the tolerance of the aircrew to the effects of g.

Full-pressure suits can be used to apply an increase in pressure over the entire surface of the body. This increases duration at altitude. For durations exceeding 10 min, however, other problems such as decompression sickness and the effects of exposure to the extremely low temperatures at altitude must be overcome.

g Tolerance

Engineers strive constantly to improve the agility and combat performance of military aircraft. Indeed technology is such that it is now man who is the limiting factor and not the machine. Accelerations occur whenever there is a change in velocity or a change in direction of a body at uniform velocity. For a centripetal acceleration, towards the centre of rotation, a resultant centrifugal force will act to make the body feel heavier than normal, as illustrated in Fig. 7.23.

Forces due to acceleration are measured in g. 1 g is the acceleration due to gravity,

Fig. 7.23 *g forces in a combat aircraft*

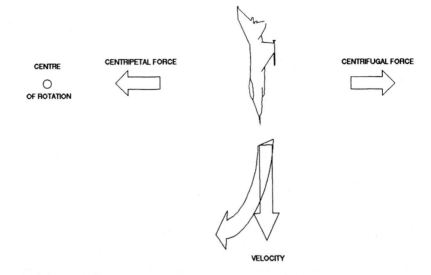

i.e. 9.81 m/s. A typical pilot is capable of performing aircraft manoeuvres up to 3 or 4 g, i.e. until he feels about three or four times his normal body weight. At g levels above this the heart becomes unable to maintain an adequate supply of oxygenated blood to the brain, which will result in blackout. This is a very dangerous condition, particularly in low-flying aircraft (reference (3)).

If the acceleration onset is gradual then the blood supply to the eyes is the first to reduce sufficiently to provide the symptoms of tunnelling of vision, before blackout and loss of consciousness occurs.

Anti-g trousers are used to partially alleviate the effects of excessive g on the body. The trousers consist of inflatable air bladders retained beneath a non-stretch belt and leggings. The trousers are inflated using air from the environmental control system. Inflation and deflation of the trousers is typically controlled by an inertial valve. The valve consists of a weight acting on a spring. At the onset of g, as the pilot is pushed down in his seat, the weight compresses the spring which acts to open the valve, thus allowing a supply of air to inflate the bladders in the trousers. The inflation action acts to restrict the flow of blood away from the brain. Using anti-g trousers a typical pilot can perform manoeuvres up to about 8 g. Positive pressure breathing also increases short-term resistance to g.

Another method of increasing g tolerance is to recline the pilot's seat. This increases the ability of the heart to provide an adequate supply of blood to the brain under high g conditions. However, in practice the seat can only be slightly reclined because of cockpit design problems, pilot visibility and the need to provide a safe ejection pathway to ensure injury-free emergency exit from the cockpit.

Rain dispersal

A pilot must have clear vision through the windscreen under all weather conditions, particularly on approach to landing. The use of windscreen wipers can be effective up to high subsonic speeds, particularly on large screens. As on a car, wipers are used in conjunction with washing fluid to clean the screen of insect debris, dust, dirt and salt spray etc. However, wipers are not suitable for use with plastic windscreens since they tend to scratch the surface. They also have the disadvantage of increasing drag.

Hot-air jets for rain dispersal can be used up to much higher speeds than wipers and are suitable for use on glass and plastic. The air is discharged at high velocity over the outside surface of the screen from a row of nozzles at the base. The air discharged from the nozzles is supplied from the environmental control system at temperatures of at least 100 °C. Such high temperatures are required to evaporate the water. However, the nozzles must be designed so that the windscreen surface temperature is not increased to such an extent that damage occurs. This is particularly a problem with stretched acrylic windscreens which begin to shrink back to the cast acrylic state at temperatures above 120 °C.

Distortion of the surface of the acrylic at the locations where the air jet impinges on the screen has been known to occur.

The system can also be used for anti-icing and demisting.

Anti-misting and demisting

Misting will occur when the surface temperature of the transparency falls below the dew-point of the surrounding air. Misting typically occurs when an aircraft has been cruising at an altitude where air is cold and relatively dry. When the aircraft descends into a warmer and more humid atmosphere, misting will occur on the surfaces which have not had enough time to warm up to a temperature above the dew-point of the air.

An anti-misting system can be provided to keep the surface temperature of the transparency above the dew-point and thus prevent misting. A system of nozzles blowing air at about 100 °C over the canopy from its base can be used, or alternatively an electrically heated gold or metal oxide film can be deposited on the transparency surface or placed between laminations.

A transparency demist system can be provided to clear the transparency of mist should misting occur suddenly, or if the anti-mist system fails. This is particularly important on landing for aircraft where the pilot is tightly strapped into his seat and cannot clear the screen with his hand. The demist system consists of nozzles blowing environmental control system air at high flow rates across the transparency.

References

(1) **Society of Aerospace Engineering (SAE)** (1969) Aerospace Applied Thermodynamics Manual, Developed by SAE Committee, AC-9, Aircraft Environmental Systems.

(2) **Ernsting, J.** (Ed) (1988) *Aviation Medicine*, Second edition, Butterworths, London.

(3) **Ashcroft, Frances** (2000) *Life at the extremes – The Science of Survival*, Harper Collins.

CHAPTER 8

Emergency Systems

Introduction

Despite the best intentions of the designers and the manufacturers of equipment, something will inevitably fail or malfunction. Emergency systems are designed to cater for just such failures. They fall into two major categories:

(1) Those that inform the crew that something is wrong.
(2) Those that allow the crew to perform some corrective action.

The first category usually takes the form of warning devices, either lamps or panels. In the early days of aircraft cockpit design signal lamps with different coloured lenses and an engraved legend were installed on the cockpit panels. Eventually these lamps became grouped into a single area in the form of a panel – variously known as central warning panels, master warning panels or master caution panels. Their aim was to bring the occurrence of a detected failure or malfunction to the attention of the crew quickly and unambiguously. Today these panels are being replaced by areas on the cockpit visual displays units containing warning captions.

Usually the circuit providing the warning signal is independent of the circuit or system providing the controlling action. Typical devices used in independent warning systems are pressure switches, over-temperature detectors or end-of-travel microswitches that are connected directly to a lamp.

The second category covers a wide range of systems to provide alternative means of control or to provide a means of safe evacuation or escape from the aircraft. If motion or energy is needed it is provided by directly operated motors, mechanical levers, hydraulic accumulators, or pyrotechnic devices. In all cases, emergency systems are designed to be foolproof and fail-safe. They will be vital to the continuing safety of the aircraft, passengers and crew.

Fighter aircraft of WW II had a performance that made bailing out of them in flight a difficult proposition. The Spitfire and Hurricane had a simple hood jettison system but the pilot had then to climb out of his seat and jump clear of his aircraft, before he could inflate his parachute. At speeds of 350 mph, pilots were pinned against rear bulkheads of their cockpits and it could take a while for them to extract themselves. Clearly, a means of assisted escape was required. This became more urgent as the birth of the jet engine heralded even greater aircraft speeds.

Several ideas were put forward for providing some impetus to push the pilot from the cockpit. These ranged from huge springs under the crew seat to adaptations of medieval catapult technologies to fling the pilot from the cockpit. The key was to find a reliable source of stored energy that could be released quickly and in a controlled manner. James Martin, chief designer of the Martin Baker Aircraft Company, played the major role in solving the energy storage and release problem. Martin's solution was to use a chemical propellant in a specially designed telescopic ejection gun. The standard ejection seat family using this approach was the Martin Baker MK 4 seat which was fitted to many aircraft types throughout the world.

When the performance limits of the ejection gun were reached, Sir James Martin, as he had by now become, again turned his mind to the problem. The solution was adding a solid fuel rocket motor to his seat to provide a greater impetus which was able to cope with ejection from high-performance jet fighters. Even when the novel escape problems, posed by engine failures during the hover, were introduced with the Harrier, the rocket-assisted seat was able to cope. The definitive rocket-assisted seat family is the Martin Baker Mk 10, which has many variants in current service around the world.

There will always be the risk of failure or accident that impairs the continuing safe operation of the aircraft. Under such circumstances there is the possibility of damage to it, and the risk of injury and death to the occupants or to members of the public on the ground. Although it can never be possible to cover all eventualities and account for them in design, it is possible to predict certain failures or accidents. If the statistical probability of their occurrence is sufficiently high, and the consequences of such occurrences sufficiently severe, then the aircraft design will incorporate emergency systems to improve the survivability of the aircraft.

Because emergency systems may be the final means of survival for the aircraft, crew and passengers, then the integrity of these systems must be high. Hence there is a need to separate them from the aircraft primary systems so that failures are not propagated from the primary systems into the emergency systems. Emphasis is placed on separate sources of power, alternative methods of operation and clear emergency warning indications. This will ensure that the systems can be operated during or after an emergency, and, if necessary, by untrained operators such as passengers or rescuers at a crash site.

Warning systems

Since many systems in a modern aircraft perform their functions automatically and in many instances take full control of the aircraft's flight and propulsion systems, it is essential that any detected malfunctions are instantly signalled to the crew.

In previous generations of aircraft, warnings were presented to the crew as individual warning lights, each with an engraved legend on the lamp lens or on the instrument panel. Such warnings were rarely placed together but tended to be sited on the cockpit panels near to the controls or indicators of the system to which they related, or even wherever there happened to be sufficient space (Fig. 8.1).

Fig. 8.1 *Spitfire cockpit (Gordon G. Bartley)*

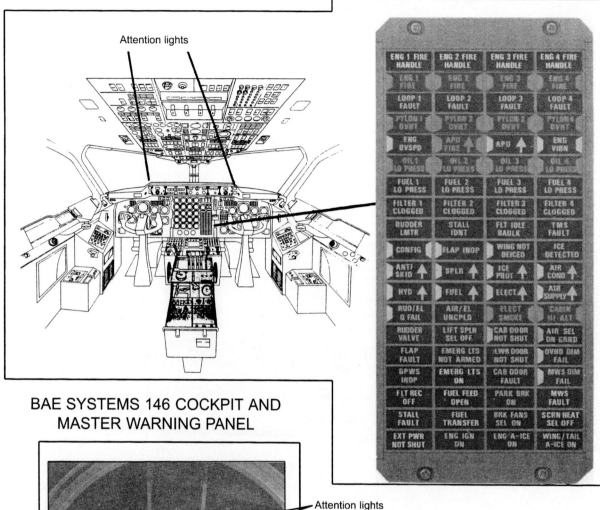

ENG 1 FIRE HANDLE	ENG 2 FIRE HANDLE	ENG 3 FIRE HANDLE	ENG 4 FIRE HANDLE
ENG 1 FIRE	ENG 2 FIRE	ENG 3 FIRE	ENG 4 FIRE
LOOP 1 FAULT	LOOP 2 FAULT	LOOP 3 FAULT	LOOP 4 FAULT
PYLON 1 OVHT	PYLON 2 OVHT	PYLON 3 OVHT	PYLON 4 OVHT
ENG OVSPD	APU FIRE	APU	ENG VIBN
OIL 1 LO PRESS	OIL 2 LO PRESS	OIL 3 LO PRESS	OIL 4 LO PRESS
FUEL 1 LO PRESS	FUEL 2 LO PRESS	FUEL 3 LO PRESS	FUEL 4 LO PRESS
FILTER 1 CLOGGED	FILTER 2 CLOGGED	FILTER 3 CLOGGED	FILTER 4 CLOGGED
RUDDER LMTR	STALL IDNT	FLT IDLE BAULK	TMS FAULT
CONFIG	FLAP INOP	WING NOT DEICED	ICE DETECTED
ANTI SKID	SPLR	ICE PILOT	AIR COND
HYD	FUEL	ELECT	AIR SUPPLY
RUD/EL Q FAIL	AIR/EL UNCPLD	ELECT SMOKE	CABIN HI ALT
RUDDER VALVE	LIFT SPLR SEL OFF	CAB DOOR NOT SHUT	AIR SEL ON GRBD
FLAP FAULT	EMERG LTS NOT ARMED	LWR DOOR NOT SHUT	OVHD DIM FAIL
GPWS INOP	EMERG LTS ON	CAB DOOR FAULT	MWS DIM FAIL
FLT REC OFF	FUEL FEED OPEN	PARK BRK ON	MWS FAULT
STALL FAULT	FUEL TRANSFER	BRK FANS SEL ON	SCRN HEAT SEL OFF
EXT PWR NOT SHUT	ENG IGN ON	ENG A-ICE ON	WING/TAIL A-ICE ON

Attention lights

BAE SYSTEMS 146 COCKPIT AND MASTER WARNING PANEL

Attention lights

Centralized warning panel

HAWK 200 COCKPIT AND CENTRALIZED WARNING PANEL

Fig. 8.2 Examples of central warning panels *(see colour plate section)*

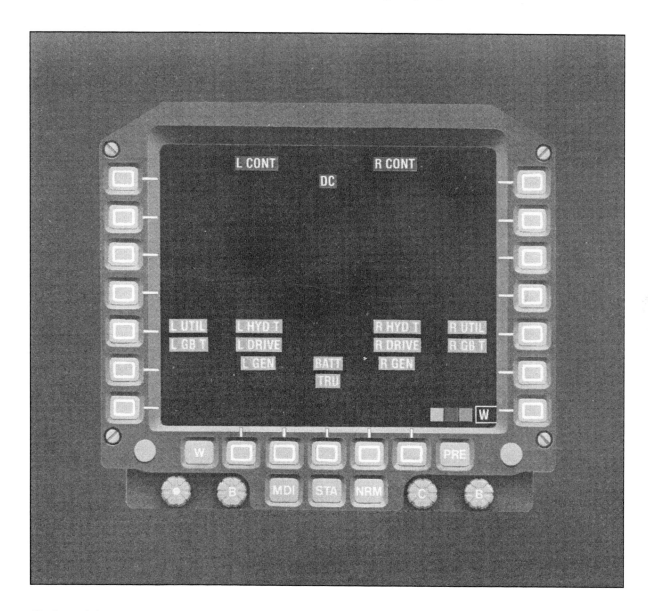

Fig. 8.3 MFD
warning and display
system (Smiths
Industries)

Haphazard though this may seem, a traditional hierarchy of warnings and a philosophy of colour usage emerged. Red was used for failures requiring instant corrective action, amber was used for cautions with less need for an immediate response; blue, green or clear were used as advisory or status indications. This was developed further by grouping together warnings into a single area of the cockpit or flight-deck in the form of a central warning panel or master caution panel (Fig. 8.2).

The attention of the crew to the generation of a warning can be achieved by incorporating a flashing lamp or attention-getter in the direct vision of the pilot, and by using audible tones in the cockpit or on the crew headsets. Bells, buzzers, electronic warbles and tones are in use on many aircraft today. A hierarchy of tones is required to ensure unambiguous attention-getting in circumstances where a number of warnings arise together.

A typical sequence of events for an immediate attention warning is as follows:

(1) A system warning is detected by a sensor or control unit.
(2) A signal is sent to the central warning panel.
(3) The attention-getters flash, an audible tone sounds in the pilot's headset, and a caption on the panel is illuminated.
(4) The pilot presses the attention-getter to stop it flashing and to silence the tone.
(5) The pilot reads the caption and takes the necessary corrective action.

Any further warnings will start the sequence again. To ensure that the pilot takes the correct action, a set of flight reference cards is carried. The cards enable the pilot to locate the caption rapidly and to read from the cards a series of corrective actions.

Aircraft being built today tend to use Multi-Function Displays (MFDs) units for the presentation of aircraft data, and areas on the screen can be reserved for the display of warning messages (Fig. 8.3). The use of voice is available as an alternative to tones; it allows multiple word messages to be generated in response to different failures. An incidental benefit of this method is that such messages will automatically be recorded on the cockpit voice recorder for analysis in the event of a crash.

Multi-word visual and aural messages can be sufficiently explicit about the failure condition and do not leave the crew with the difficult task of trying to decipher a single-word lamp legend together with systems indications in a stressful situation. In fact modern display systems can tell the crew what the system failure is and what actions they should take to recover to a safe condition. This can be used to replace the flight reference cards. Further information on warning systems can be found in reference (1).

Fire detection and suppression

The occurrence of a fire in an aircraft is an extremely serious event, since the structure is unlikely to remain sound in the continued presence of flame or hot gases. The most likely place for a fire to start is the engine compartment. Fires may occur as a result of mechanical damage leading to the engine breaking up or overheating, from pipe or casing ruptures leading to the escape of hot gases which may impinge on the structure, or from escaping fuel coming into contact with hot surfaces.

All the necessary ingredients for starting and maintaining a fire are readily available – plenty of fuel, plenty of air and hot surfaces. Needless to say, everything that can be done to prevent the escape of fluids and to reduce the risk of fire is done. Nevertheless it is prudent to install some means of detecting one and a means of extinguishing it.

Detection systems are usually installed in bays where the main and auxiliary power-plants are located (Fig. 8.4). The intention is to monitor the temperature of the bays and to warn the crew when a predetermined temperature has been exceeded. The system consists of a temperature-measuring mechanism, either discrete or continuous, a control unit and a connection to the aircraft warning displays. The temperature detection mechanism is usually installed in different zones of the engine bay so that fires can be localized to individual areas of the power-plant.

Discrete temperature sensors usually take the form of bi-metallic strips constructed so that a contact is made up to a certain temperature, when the strips part. A number of sensors are placed at strategic locations in the engine compartment, and wired to cause the contacts to open, then the control unit detects the change in resistance of the series wiring and causes a warning to illuminate in the cockpit.

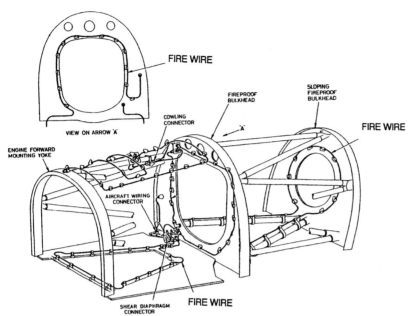

Fig. 8.4 *Fire detection system in an engine bay*

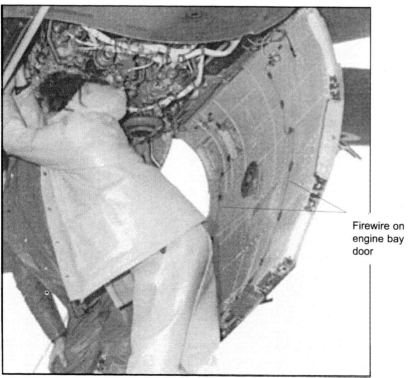

Firewire on engine bay door

An alternative method is a continuous loop of tubular steel coaxial sensor which can be routed around the engine bay. This sensor changes its physical and electrical characteristics when subjected to heat. This change of characteristic is sensed by a control unit which causes a warning to light (Fig. 8.5).

Fig. 8.5 Diagrams of discrete and continuous systems

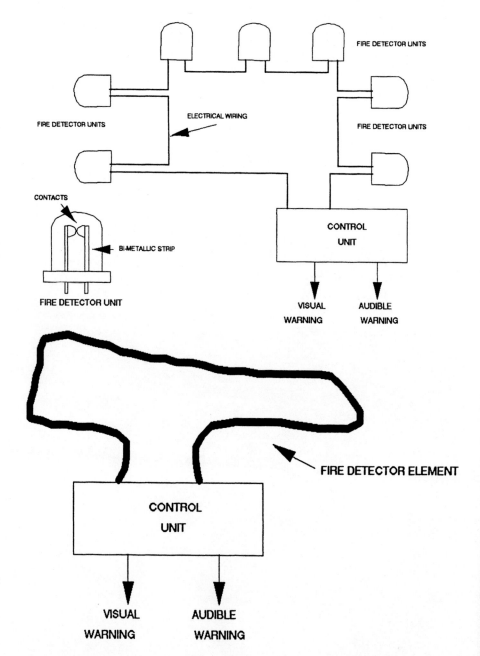

The Graviner FIREWIRE sensing element is a slim stainless steel tube with a centrally located coaxial wire surrounded by a temperature-sensitive, semi-conductive material. This material has a negative temperature coefficient of resistance. The resistance measured between the centre wire and the outer sheath decreases with temperature, and is accompanied by a corresponding increase in capacitance. The resistance and capacitance of a loop is monitored continuously by a control unit. The control unit will provide a warning signal when the resistance reaches a predetermined value, as long as the capacitance is sufficiently high. Monitoring both parameters in this way reduces the

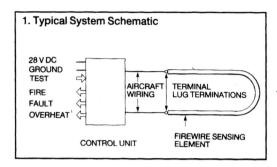

1. Typical System Schematic

28 V DC
GROUND
TEST

FIRE
FAULT
OVERHEAT

AIRCRAFT WIRING

TERMINAL LUG TERMINATIONS

CONTROL UNIT

FIREWIRE SENSING ELEMENT

TYPICAL SYSTEM SCHEMATIC

Fig. 8.6
Construction of continuous firewire system (Kidde-Graviner)

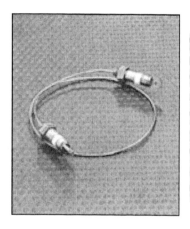

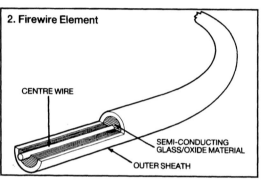

2. Firewire Element

CENTRE WIRE

SEMI-CONDUCTING GLASS/OXIDE MATERIAL

OUTER SHEATH

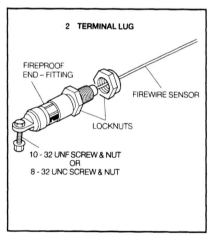

2   TERMINAL LUG

FIREPROOF END – FITTING

FIREWIRE SENSOR

LOCKNUTS

10 - 32 UNF SCREW & NUT
OR
8 - 32 UNC SCREW & NUT

TERMINAL LUG

potential for false recognition of fires resulting from damage or moisture contamination of the element.

The Kidde CFD system, shown in Fig. 8.6, uses a ceramic-like thermistor surrounding two electrical conductors. The thermistor material has a high resistance at normal ambient temperature which reduces rapidly as the sensor is heated. A control unit senses the resistance and signals a warning when the value drops below a pre-set condition. An alternate technology uses a pressurized sensor element to detect severe changes in temperature.

Both discrete and continuous systems work as detectors of overheating or fire, but both are susceptible to damage by the very condition they are monitoring. The fire or jet of hot gas that leads to the temperature rise can easily burn through the wiring or the sensor. The system must be designed so that if this does occur, then the warning is not extinguished. Equally the system must be designed so that no warnings are given when there is no fire. Both these conditions are dangerous. The first because the crew may think that a fire has been extinguished, the second because a system which continually gives spurious warnings may be disregarded when a real fire occurs.

Once a fire warning is observed a formal drill is initiated by the crew to extinguish the fire. This will include shutting down the engine and isolating the fuel system at the engine fire-wall by closing a cock in the fuel system, and then discharging extinguisher fluid into the bay.

This is done by pressing a switch in the cockpit (often a switch built into the fire warning lamp) which fires a cartridge built into a bottle containing a fluid such as BromoChlorodiFluoromethane (BCF). This causes a spray of fluid to be directed into the engine bay. Usually the bottles are single shot. If, after discharging the bottle, the fire warning remains, the crew must decide if the warning is genuine. In a commercial aircraft this can be done by looking out of a window to see if flames can be seen in the engine nacelle, in a military aircraft by asking another aircraft to observe from behind. If a fire is confirmed then the aircraft must be landed as soon as possible, or abandoned.

Modern fluids have been developed, which do not contain environmentally harmful fluoro-carbon materials and are being introduced to meet mandatory regulations (Montreal Protocol).

Emergency power sources

Modern commercial aircraft rely on multiple redundancy to achieve continued safe operation in the presence of single or even multiple failures in critical systems such as electrical or hydraulic power generation, engine or flight control. This redundancy may achieve levels as high as quadruple independent systems.

Military aircraft can rarely go to such levels and it is necessary to provide some form of emergency power source in some types. Notably these are aircraft with full-authority electrical engine and flight control systems in which total power loss would result in loss of the aircraft. Very often this applies only to prototypes and test aircraft which are flown up to and beyond normal flight envelope restrictions.

An aircraft exploring high incidence boundaries is likely to depart into a stall or a spin, which may lead to such a disturbance of the engine intake air flow as to cause all engines to flame out. This will result in a total loss of engine generated power, such as electrical and hydraulic power, both of which may be required to attain the correct flight attitude and forward speed necessary to restart the engines.

Emergency power can be provided by a number of means including an Emergency Power Unit (EPU), an electrohydraulic pump, or a Ram Air Turbine (RAT). These are described in Chapter 5, Electrical Power.

An emergency power unit consists of a turbine which is caused to rotate by the release of energy from a mono-fuel such as hydrazine. The hydrazine is stored in a sealed tank and isolated from the turbine by a shut-off cock. The shut-off cock is opened in emergency conditions, either manually by a pilot-operated switch or automatically by a sensor which detects that the aircraft is in flight and that all engines

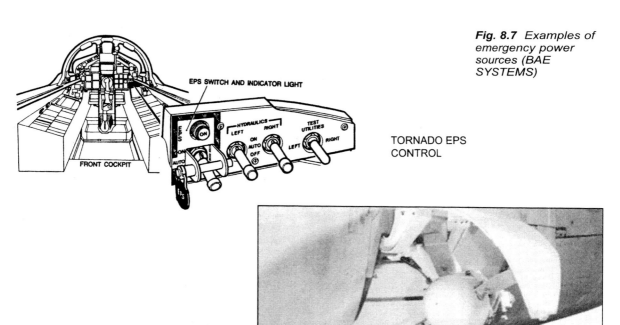

Fig. 8.7 *Examples of emergency power sources (BAE SYSTEMS)*

EPS SWITCH AND INDICATOR LIGHT

TORNADO EPS CONTROL

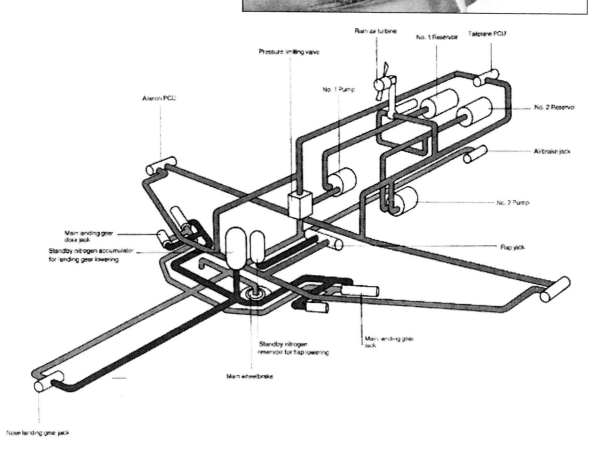

are below a predetermined speed of rotation. The rotating turbine drives an aircraft gearbox which enables at least one hydraulic pump and one generator to be energized. A hydrazine EPU was used in Concorde prototypes and some Tornado prototypes.

An electrohydraulic pump can be used to provide hydraulic power for aircraft in which the flight control system can be used without the need for electrical control. A manual or automatic operation can be used to initiate a one-shot or thermal battery to drive a hydraulic pump. This will provide power for a limited duration, sufficient to recover the aircraft and start the engines. Such a unit was used on the Jaguar prototype for spinning trials.

The Tornado GRI emergency power system (EPS) provides hydraulic power following a double engine flame-out, a double generator failure or a double transformer rectifier unit failure. In this system a single-shot battery is activated by an explosive device. This activation can be automatic or initiated by the pilot. As well as an hydraulic pump, the system also drives a fuel pump which can be supplied with power for up to 13 min as long as hydraulic demands are minimized. The cockpit controls are shown in Fig. 8.7.

A ram air turbine does not require a source of power other than that provided by forward movement of the aircraft. It is limited in the amount of power that can be provided. The multi-bladed unit drops from a stowed position in the aircraft and provides electrical power. The Tornado F3 is fitted with a RAT which is deployed automatically when both engine speeds fall below a prescribed level. The RAT maintains sufficient pressure in the No. 1 hydraulic system to provide adequate taileron control during engine re-light. The Tornado RAT is shown in Fig. 8.7.

The Hawk aircraft also uses a RAT which extends into the airstream from the top fuselage following an engine failure, thereby providing power to the flying controls down to landing speed. The position of the RAT in the Hawk hydraulic circuit is shown in Fig. 8.7.

Explosion suppression

The aircraft fuel tanks are a potential explosion hazard when partially full since the volume above the fuel fills with fuel vapour. If air is present in sufficient concentration then the resulting mixture is combustible. This presents a potential hazard to military aircraft which may be involved in combat, since penetration of the tank by particles of munitions or by tracer or incendiary rounds may result in an explosion. A commercial aircraft is at risk following an emergency landing. To reduce the risk of this happening the tank may be pressurized with an inert gas such as nitrogen or filled with a reticulating foam. Nitrogen for this purpose can be obtained from an On-Board Inert Gas Generating System (OBIGGS). This system uses a molecular sieve to extract nitrogen from the air in a similar fashion to OBOGS; in fact nitrogen and other inert gases are a waste by product of OBOGS.

Emergency oxygen

Commercial aircraft operating above 10,000 ft pressurize the fuselage to an altitude condition that is comfortable for the crew and passengers. If there is any failure of the cabin pressurization system then oxygen must be provided for the occupants. The aircrew are provided with face masks which they can fit rapidly to obtain oxygen from a pressurized bottle. Face masks for passengers are normally stowed in the racks above

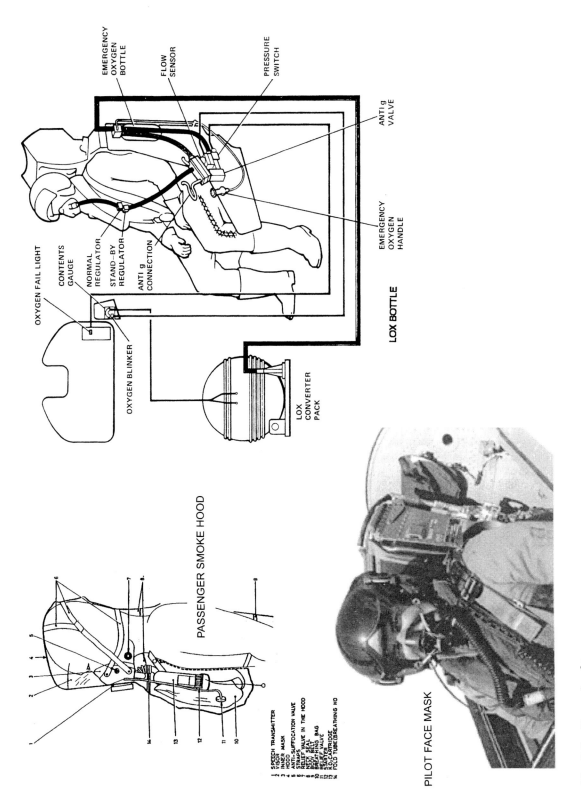

EMERGENCY OXYGEN BOTTLE

FLOW SENSOR

PRESSURE SWITCH

ANTI g VALVE

EMERGENCY OXYGEN HANDLE

LOX BOTTLE

OXYGEN FAIL LIGHT

CONTENTS GAUGE

NORMAL REGULATOR

STAND–BY REGULATOR

ANTI g CONNECTION

OXYGEN BLINKER

LOX CONVERTER PACK

PASSENGER SMOKE HOOD

1 SPEECH TRANSMITTER
2 VISOR
3 INNER MASK
4 HOOD
5 ANTI-SUFFOCATION VALVE
6 STRAPS
7 RELIEF VALVE IN THE HOOD
8 NECK SEAL
9 BODY BELT
10 BREATHING BAG
11 RELIEF VALVE
12 STARTER
13 O₂ CARTRIDGE
14 FOLD TUBE (BREATHING HO

PILOT FACE MASK

Fig. 8.8 *Example of a face mask, passenger hood and LOX bottle*

the seats, and fall automatically on depressurization. Before each flight the cabin crew will brief passengers and demonstrate the use of the masks. The aircraft descends to an altitude where oxygen concentration in the air is sufficiently high to allow normal air breathing.

Most combat aircraft crews breathe oxygen throughout the flight using a face mask supplied with oxygen from a liquid oxygen (LOX) container which can be charged before each flight. One or two wire-wound cylinders are provided in the aircraft. The gas flows through a pressure regulating valve, and a regulator enables the pilot to select an oxygen-air mixture or pure oxygen. Two 1,400 l bottles provide sufficient oxygen for up to five hours with air-oxygen (Airmix) or up to three hours on 100 per cent oxygen for a sustained cruise at 35,000 ft.

A contents gauge and a doll's eye flow indicator are provided, as well as a failure warning light to enable the pilot to monitor the system.

If the normal oxygen supply should fail then the crew can change over to the oxygen bottle carried on their ejection seats. Although this will provide oxygen for a limited duration only, it will be sufficient to return to base. A cylinder of about 70 l capacity is connected so that gas flow is routed through a seat-mounted demand regulator. Selection of emergency oxygen automatically ensures a supply of 100 per cent oxygen irrespective of any previous crew selections. The bottle also provides an automatic supply of oxygen to the pilot upon ejection (Fig. 8.8).

Passenger evacuation

Commercial aircraft and military transports must provide a means of allowing all passengers to evacuate the aircraft in a certain time. Emergency exit doors are provided at strategic locations in the aircraft, and the doors are fitted with escape chutes so that passengers can slide to the ground. The chutes are designed to operate automatically or manually, and to inflate rapidly on command (Fig. 8.9). Doors are designed to open outwards and are of sufficient width to allow passengers to exit rapidly. All doors and exits are identified with illuminated signs.

Life vests are carried beneath the passenger seats, and the aircraft is equipped with life rafts and with locator beacons.

Crew escape

The crew of a commercial aircraft can escape through the passenger emergency exits or by using an escape rope to slide down from the flight-deck through the opening side windows.

Military crews in combat aircraft are provided with ejection seats which allow them to abandon their aircraft at all flight conditions ranging from high speed, high altitude to zero speed and zero height (zero-zero). The seat is provided with a full harness, restraints to pull the legs and arms into the seat to avoid injury, a parachute, dinghy, oxygen and locator beacon. The seat is mounted in the aircraft on a slide rail which permits the seat to travel in a controlled manner upwards and out of the cockpit. The design of the seat and the rail allows the seat and occupant to exit the aircraft with sufficient clearance between the cockpit panels and the pilot to avoid injury. The seat is operated by pulling a handle which initiates a rocket motor to propel the seat up the slide. The ejection system may be synchronized to allow the canopy to be explosively ejected or shattered before the seat reaches it, or the seat top will be designed to shatter

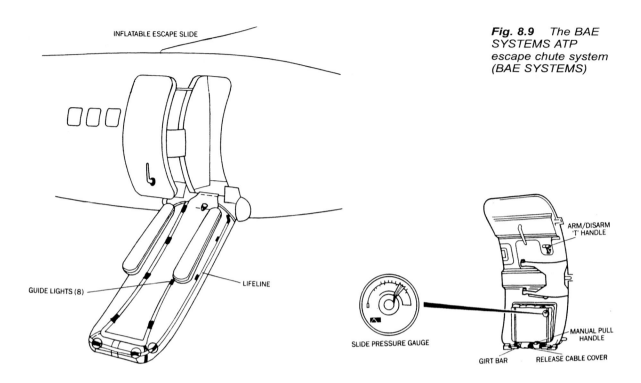

Fig. 8.9 *The BAE SYSTEMS ATP escape chute system (BAE SYSTEMS)*

INFLATABLE ESCAPE SLIDE

GUIDE LIGHTS (8)

LIFELINE

SLIDE PRESSURE GAUGE

ARM/DISARM 'T' HANDLE

MANUAL PULL HANDLE

GIRT BAR   RELEASE CABLE COVER

the canopy. The canopy can be shattered by a pattern of miniature detonating chord embedded in the acrylic (Fig. 8.10). Such cutting systems are essential for polycarbonate transparencies, which cannot easily be shattered mechanically.

The chord is a continuous loop of small-diameter lead tubing filled with explosive material. The loop is bonded to the canopy transparency in a pattern that causes the canopy to fragment before the pilot leaves the aircraft. The fragmentation system can be fired from outside the aircraft to allow rescuers to free the crew of a downed aircraft.

The seat leaves the aircraft in a controlled manner to reduce the effect of acceleration on the crew member and a parachute is deployed to decelerate the seat and to stabilize its position. After an interval the seat detaches from the man and a personal parachute opens.

The Martin Baker Mk 10 ejection seat fitted to the Tornado has a zero-zero capability, enabling safe ejection from zero altitude and zero forward speed. This means that the crew can safely eject from an aircraft on the ground while it is stationary or taxying. A fast, efficient ejection is absolutely essential for an aircraft designed to operate at low level and high speed. Operations in such conditions leave little time for crew escape in the event of a catastrophe. The crew can elect to eject by pulling the seat ejection handle. The escape sequence is then fully automatic, and takes about 2.5 sec for the parachute to be fully deployed. A baro-static mechanism ensures that the seat detaches from the pilot automatically below 10,000 ft.

The Tornado is typical of present-day, in-service systems. It is a two-crew aircraft with a fully automatic escape system, which needs a single input from either crew member in order to initiate it. Each cockpit is provided with a Martin Baker Mk 10A ejection seat and both cockpits are covered by a single transparent canopy. Its escape system provides all of the following functions automatically:

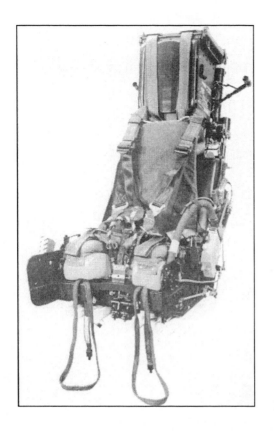

Fig. 8.10 Crew
Escape (BAE
SYSTEMS/Martin
Baker Engineering)

- Primary canopy removal by jettison
- Secondary, back-up canopy removal
- Jettison of night vision devices
- Ejection of the rear seat
- Ejection of the front seat
- Seat/occupant separation
- Personal locator beacon switch on
- Parachute deployment
- Lowering of survival aids container
- Inflation of lifejacket and liferaft on water entry

Computer-controlled seats

At the forefront of current developments is the escape system for the Eurofighter 2000/Typhoon. This is based around the Martin Baker Mk 16 ejection seat, which is controlled by a microprocessor-based electronic sequencer. It has two configurations; one single-seat and one twin-seat and both variants have a single canopy. The twin-seat variant provides the same fully automatic sequence as for the Tornado, with a similar sequence for the single-seat version. This system is now flying on prototype aircraft, so the age of the computerized escape system has well and truly begun.

Ejection system timing

Current escape system sequences have fixed time delays built into them, to ensure safe separation between the individual elements that are launched from the aircraft. For example, the Tornado has a fully automatic sequence to manage:

- Jettison of the canopy
- Ejection of the rear seat
- Ejection of the front seat

Two fixed timers are used to sequence these three elements such that there is a nominal delay of 0.30 sec between the canopy and the rear seat and a nominal delay of 0.34 between the front and rear seats. These delays are set to give safe separation across the whole escape envelope. They are also subject to production tolerances. The total delay deliberately introduced into the sequence is 0.79 sec or approximately 80 per cent of the overall time taken for the canopy and both seats to separate from the aircraft.

Future improvements in low-level escape capability will come from the introduction of variable time delays, based on actual conditions rather than a single worst design case. An aircraft travelling at 450 kts in a 60-degree dive will descend 520 ft during the 0.79 sec delay of the Tornado system. In a 30-degree dive, it will descend 300 ft. As fast jet aircraft routinely operate at or below such altitudes, any reduction in sequence delays will reduce the height required at system initiation for a safe escape and increase the probability of a survivable ejection. At 450 kts, the total time delay could be reduced considerably, by 50 per cent or more, reducing the safe ejection height for the Tornado front seat by some 200 ft or so. For higher sink rates at ejection, the gain would be even greater.

In order to achieve variable sequence timings, technologies that allow position-sensing and algorithms that can establish the appropriate timing for the prevailing conditions will have to be developed. The introduction of computer-controlled sequencers on to the ejection seats will facilitate the development and integration of these more intelligent overall system sequences.

High-speed escape

Over the years, esScape speeds have been slowly increasing. The table below (Table 8.1) shows the percentage of ejections occurring at or above given speeds. The Hunter is fitted with a Martin Baker MK 4 ejection seat and is typical of the aircraft in service from the mid-1950s to the mid-1970s. The Tornado is fitted with a Martin Baker Mk 10 ejection seat and is typical of aircraft in service from the mid-1970s to today. These are predominantly peacetime ejections, in wartime ejection speeds tend to increase overall.

Table 8.1   *Comparison of Ejection Speeds*

Aircraft	350kts	400kts	450kts
Hunter	19%	8%	4%
Tornado	36%	16%	8%

As ejection speeds increase, the potential for injury from air blast increases. The face and limbs are particularly vulnerable to air blast damage.

In some multi-crew combat aircraft such as the F-111, B-58 or XB-70, the crew escape in a module or capsule, the entire crew compartment being designed to be jettisoned and parachuted to the ground.

Crash recorder

It is a mandatory requirement to carry a recorder in commercial aircraft and in military aircraft operating in civilian airspace so that certain critical parameters are continuously recorded for analysis after an accident. The recorder, variously known as crash recorder, accident data recorder, flight recorder or – as it is referred to in the press – black box recorder, is a crash survivable machine which may be ejected from the aircraft after a crash and contains a radio and sonar locator to guide rescue crews to its location.

The recorder is connected to the aircraft systems so that flight-critical parameters are continuously recorded together with information about the aircraft's flight conditions. For example, control column and throttle positions, flight control surface positions, engine speed, pressure and temperature will be recorded together with altitude, air speed, attitude, position and time. Analysis of this data after an accident will be used to determine the cause of the incident. Recording all crew conversations and communications with the outside world is also carried out, either on the same recorder or on a separate cockpit voice recorder (Fig. 8.11).

Crash switch

On many military aircraft it is accepted that an aircraft may have to be landed in a dangerous condition, either wheels up or wheels down. The crew will have to exit the aircraft quickly and safely in these circumstances and the risk of fire must be reduced as far as possible. A crash switch is designed to do this by providing a single means of

Fig. 8.11 *Examples of crash recorders (BAE SYSTEMS)*

shutting down engines, closing fuel cocks, disconnecting the aircraft battery from the busbars and discharging the fire extinguishers into the engine bays.

These precautions can be provided manually or automatically. The manual method provides a number of switches in the cockpit which are linked by a bar so that a single action will operate all the switches. The pilot will do this immediately before or as soon as the aircraft hits the ground. The automatic method is provided by inertia switches that operate under crash conditions.

Testing

The emergency systems described in this chapter are crucial to the safety of the aircraft, crew and passengers. For this reason they must work when required to do so. Wherever possible a means of testing the systems prior to flight is made available so that the crew can have confidence in the ability of the system to provide its correct function.

Some systems, however, cannot be tested – it would obviously be impractical to test an ejection seat. There are other examples where the crew must depend on periodic testing or have confidence in the correct assembly of the system. This is a dilemma for designers and users – to establish a balance between confidence in design, and proof of design, and practical pre-flight testing.

Reference

(1) **Institution of Mechanical Engineers** (1991) Seminar S969 on the Philosophy of Warning Systems, March.

CHAPTER 9

Helicopter Systems

Introduction

The helicopter was a late arrival on the aviation scene compared to more conventional fixed-wing aircraft. A number of designers experimented with autogiros in the late 1920s and 1930s but it was not until the 1940s that serious helicopter developments began. The Royal Air Force used an autogiro which was a Cierva design, licence-built by Avro, and some Sikorsky Hoverfly I and II examples were used for limited squadron service and evaluation purposes. In general, the helicopter was regarded at the time as something of an anachronism and it was not until the post-war years that its serious development began, most of it being undertaken in the US.

In the UK, Bristol produced the Sycamore Type 171, which entered service with the Royal Air Force in 1953. Bristol also produced the twin-rotor Type 173 which was developed for the military as the Type 192 and subsequently named Belvedere, entering service in the 1960s. The development of helicopters in the UK was in the main based upon UK derivatives of US designs of which the Dragonfly, Whirlwind, Wessex and Sea King have been notable examples.

In the late 1960s and early 1970s Westland Helicopters became involved with the joint design of a family of helicopters together with Aerospatiale of France. This led to the development of the Gazelle, Lynx and Puma helicopters all of which have served with various branches of the UK Armed Forces.

The helicopter came of age as a fighting vehicle in the late 1960s and the US involvement in the Vietnam War was probably the first large-scale conflict in which it played a major part in a variety of roles. This pattern has been followed by the British involvement in the Falklands Campaign where the shortage of helicopters imposed severe operational limitations upon the ground troops. More recently the role of the helicopter in the Gulf War has emphasized its place in the order of battle – in particular the heavy battlefield attack machine (Apache) and the missile-equipped helicopter (Lynx).

Fig. 9.1 *European
Helicopter Industries
EH 101 Merlin
(GKN/Westland
Helicopters)*

As their roles became more demanding so the helicopters became more sophisticated and complex. As the number of systems fitted increased to satisfy greater and more difficult tasks, so too has the amount of propulsive power required and both the power and the number of engines fitted have increased to accommodate these needs. The Dragonfly of the 1950s required a single 550 hp engine to power the 5,500 lb fully loaded helicopter. The EH101 Merlin of the 1990s (see Fig. 9.1) has three T700 engines, each rated at 1,437 shp to lift the helicopter with an all-up-weight of around 30,000 lb. As the size of the helicopter and engines has increased so has the complexity of the various systems. The amount of electrical power required by a large helicopter of this type equates to that needed for most jet fighters a few years ago. The EH101 Merlin also requires a complex autopilot and flight control system to provide the necessary handling characteristics so that the crew can devote their attention to the demands of the mission. Electrical and hydraulic systems also require higher levels of redundancy to support the mission requirements. Finally, the avionic equipment required to undertake a range of missions also places additional demands upon the baseline helicopter systems.

Special requirements of helicopters

The unique nature of the helicopter compared to conventional fixed-wing aircraft deserves special consideration in relation to aircraft systems. Despite the fact that many of the same principles apply, the vertical take-off and landing features of the helicopter place a different emphasis upon their embodiment. Vertical take-off imposes a requirement for a high power to weight ratio. It is generally reckoned that for an aircraft to take off vertically with an adequate control margin, a thrust to weight ratio of 1.25:1 is required. This ratio applies after various transmission losses have been taken into account.

The means of controlling a helicopter is by its very nature totally different to the methods used by fixed-wing aircraft. Also, due to unique properties such as hovering flight, and the ability to land vertically in confined areas, some system requirements are unusual. These lead to the adoption of autopilot control modes such as auto-hover, which are not possible on fixed-wing aircraft. The ability to hover also dictates the need for winch systems and has led to the development of specialized autopilot modes. The need to land and remain tethered on ship decks in high seas has resulted in the introduction and use of deck-locking systems.

Principles of helicopter flight

Whereas the lift force for a fixed-wing system is produced by the passage of air over the wing aerofoil, the helicopter rotor blades are aerofoils which generate the lift force to counteract the vehicle weight. See Fig. 9.2. While it is more usual to have one rotor, there are a number of twin-rotor helicopters where the rotors may be located fore-and-aft in tandem, while others may have the rotors located side-by-side on either side of the fuselage. The rotors may comprise a number of blades which may vary between two and six.

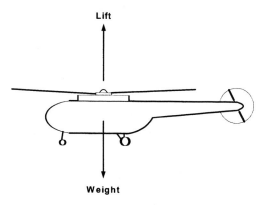

Lift

Weight

Fig. 9.2 Helicopter lift forces

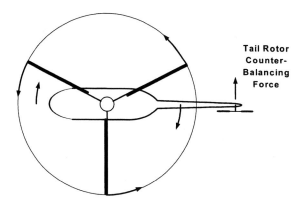

Tail Rotor Counter-Balancing Force

Fig. 9.3 Rotor torque effects and the need for a tail rotor

The fact that the helicopter lift force is generated by rotation of the rotor causes additional complication for the helicopter. As the helicopter propulsion system drives the rotor head in one direction, a Newtonian equal and opposite reaction tends to rotate the fuselage in the other direction and clearly this would be unacceptable for normal controlled flight. Refer to Fig. 9.3. This problem is overcome by using a tail rotor which applies a counter-acting force (effectively a horizontal 'lift' force) which prevents the helicopter fuselage from rotating. The tail rotor is driven by an extension of the gearbox transmission system which couples the rotor head to the propulsion. An alternative method, called NOTAR™, or no tail rotor has been developed recently and this is described later in this chapter. Twin-rotor helicopters do not suffer from this torque problem as the two rotor heads effectively cancel each other out.

Basic helicopter control

Tilting the rotor head provides the longitudinal (fore-and-aft) and lateral (side-to-side) forces necessary to give the helicopter horizontal movement. This is achieved by varying the cyclic pitch of the rotor head. Moving the pilot's stick forward alters the cyclic pitch such that the rotor tilts forward, thereby adding a forward component to the lift force and enabling the helicopter to move forwards. Moving the pilot's stick back causes the rotor to tilt backwards and the resulting aft component of the lift force makes the helicopter fly backwards. Figure 9.4 shows the effect of the pilot's controls on the rotor head and the subsequent helicopter motion.

Movement of the pilot's control column from side to side tilts the rotor accordingly and causes the helicopter to move laterally from left to right. Yaw control is by means of rudder pedals as for a fixed-wing aircraft. In the case of the helicopter, movement of the rudder pedals modifies the pitch of the tail rotor blades and therefore the thrust force generated by the tail rotor. Moving the rudder pedals to the left to initiate a yaw movement to the left increases the thrust of the tail rotor and causes the helicopter to rotate (yaw) to the left. Moving the rudder pedals to the right causes a corresponding reduction in the tail rotor thrust and the helicopter yaws right.

Vertical movement of the helicopter is initiated by varying the pitch of all of the rotor blades and thereby increasing or decreasing total rotor lift. Reducing rotor lift results in a downward force which causes the helicopter to descend. Increasing rotor lift generates a resulting upward force which causes it to ascend. The pitch of the main rotor blades is varied by means of a collective pitch lever. The engine power, or torque, is controlled by a throttle twist-grip located at the end of the collective lever and is usually operated in conjunction with the collective pitch lever to cause the helicopter to climb smoothly or descend as required. Flying the helicopter is therefore achieved by a smooth co-ordination of pitch and lateral cyclic control, together with rudder pedals, power and collective pitch controls.

In general the helicopter is more unstable than its conventional fixed-wing counterpart. Furthermore, the secondary effects of some of the helicopter controls are more pronounced, thus requiring greater compensatory control corrections by the pilot. It follows that flying a helicopter is generally much more difficult than flying a fixed-wing aircraft, particularly when an attempt is made to execute precise tracking or positional tasks in gusty or turbulent conditions. For this reason, some sophisticated helicopters possess auto-stabilization and multimode autopilot systems to minimize

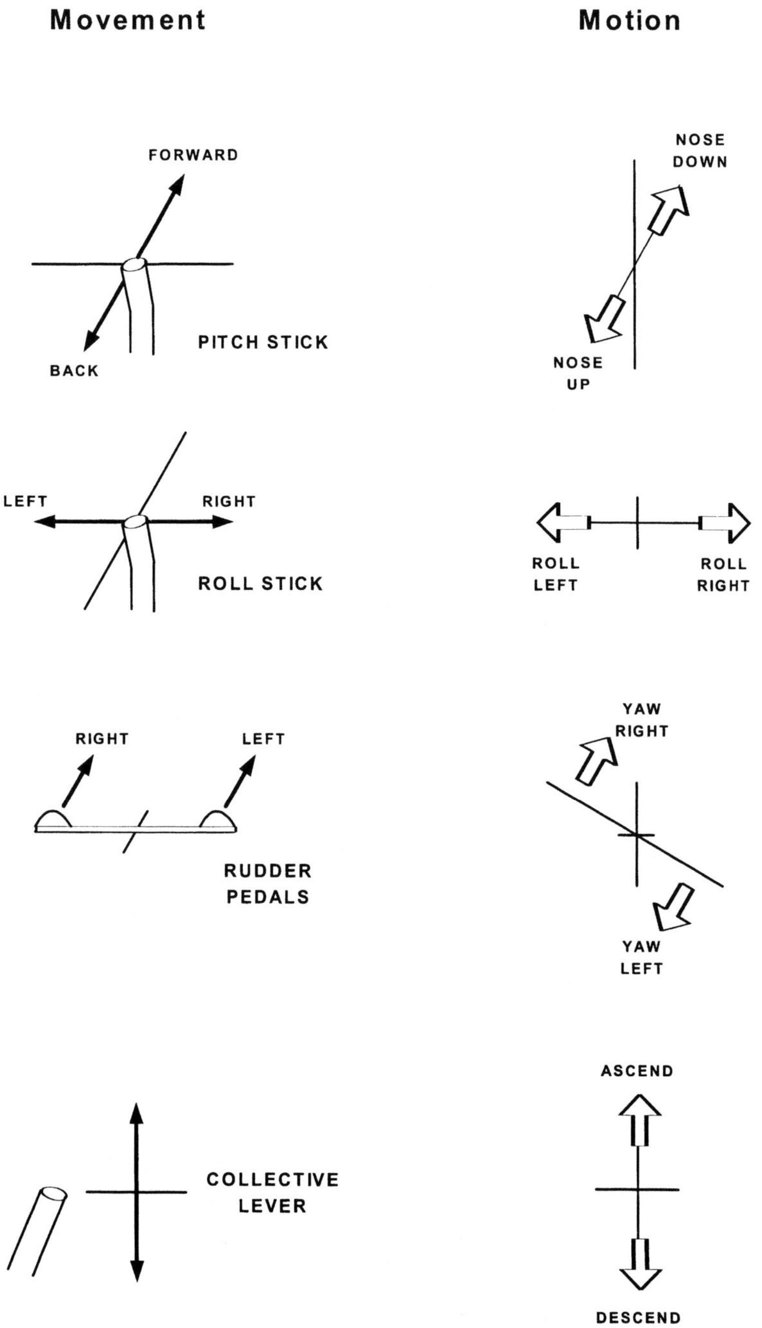

Movement

Motion

Fig. 9.4 *The effect of pilot's controls on helicopter motion*

FORWARD

PITCH STICK

BACK

NOSE
DOWN

NOSE
UP

LEFT   RIGHT

ROLL STICK

ROLL
LEFT

ROLL
RIGHT

RIGHT   LEFT

RUDDER
PEDALS

YAW
RIGHT

YAW
LEFT

COLLECTIVE
LEVER

ASCEND

DESCEND

the effects of inter-reactions , thereby reducing pilot workload and thus enabling the pilot to concentrate on crucial aspects of the flight or mission. For an easily digestible, but nonetheless comprehensive description of the helicopter and how it flies, see reference (1).

Key helicopter systems

The basic principles of many helicopter systems are identical to similar systems in fixed-wing aircraft. However, the unique nature of the helicopter places a different emphasis upon how these systems are implemented and also introduces a requirement for some totally new systems. A range of these systems is described so that a comparison might be made with the fixed-wing aircraft equivalent. They are:

- Engine and transmission system.
- Hydraulic system.
- Electrical system.
- Health monitoring system.
- Specialized helicopter systems.

A separate section will address the flight control system.

Engine and transmission system

Many helicopters today have a number of engines to supply motive power to the rotor and transmission system. In fact, all but the smallest helicopters usually have two engines, and some larger ones have three. The need for multiple engines is obvious; helicopter lift is wholly dependent upon rotor speed, which in turn depends upon the power provided by the engines. In the event of engine failure it is still necessary to have power available to drive the rotor, therefore multiple engines are needed so that the remaining engine(s) can satisfy this requirement. Although it is possible to land a single-engined helicopter following engine failure, using a technique called auto-rotation, this mode of unpowered flight takes time to establish. If the helicopter is flying at around 500 ft or less then it is unlikely that safe auto-rotation recovery can be carried out. Engines are usually sized so that the aircraft can fly for a period of time with one engine failed, except in the most extreme flight conditions: when the helicopter is flying heavily loaded or 'hot and high'. Reference (2) and Figs 20 and 21 are indicative of the performance of the European Helicopter Industries EH 101 Merlin helicopter operating in single and two engine-out cases. The EH 101 Merlin is fitted with a variant of the General Electric T700-GE-401 turbo-shaft engines in the naval version while civil and military versions are powered by the General Electric CT7-6, a variant of the T700 developed specifically for the EH 101 Merlin. See Reference (3) for more detail regarding the development of the T700 family of engines.

A more recently developed engine available for this class of helicopter is the Rolls/Turbomeca RTM 322 which is designed to operate at 2,100 shp (shaft horse power) and weighs around 530 lb. This engine is of a suitable size to power up-rated versions of the EH 101 Merlin. It is being produced with a 50/50 work share by Rolls-Royce and Turbomeca and an indication of the engine configuration and work share is given in Fig. 9.5. A description of the development programme of the RTM 322 is given in references (4) and (5).

The majority of new helicopters use gas turbines rather than internal combustion engines, for a variety of reasons. Most engines are electronically controlled using computers and over recent years control has become digital in nature, using Digital Engine Control Units (DECUs). These units are usually configured with two lanes or channels of control, though, for a single-engined helicopter, a dual channel and a hydromechanical stand-by channel may be provided. Typical control laws which would be embodied are:

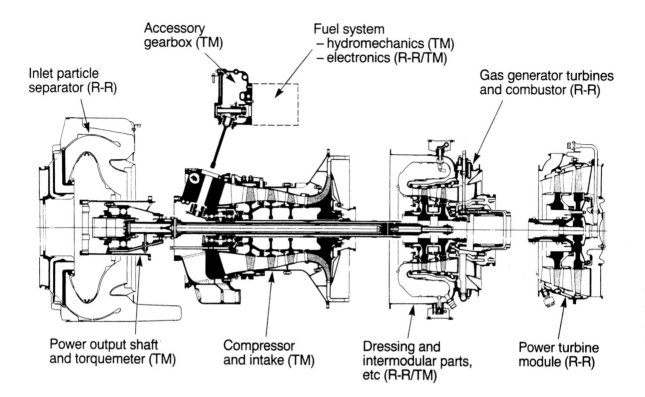

Accessory gearbox (TM)

Fuel system
– hydromechanics (TM)
– electronics (R-R/TM)

Inlet particle separator (R-R)

Gas generator turbines and combustor (R-R)

Power output shaft and torquemeter (TM)

Compressor and intake (TM)

Dressing and intermodular parts, etc (R-R/TM)

Power turbine module (R-R)

Fig. 9.5 RTM 322 configuration and work share (Rolls-Royce Turbomeca)

- Acceleration control.
- NH control.
- Surge prevention.
- Fuel flow max/min.
- Torque limiter.
- Torque/free turbine droop governor.

A fuller description of these control laws and their implementation is given in reference (6). A system comprising more than one engine/DECU may also incorporate features whereby one will be accelerated to maximum power if one of the other engines fails or the thrust drops below a predetermined level. Such a system is likely to apply power more quickly than the pilot when operating in a critical flight mode such as the hover.

An idea of the complexity of the transmission system needed for a three-engined helicopter may be gained from Fig. 9.6. This depicts how each of the three engines drive though a series of reduction gears to the third-stage collector gear. The collector gear drives the rotor at 210 rpm through a sun and planet gear. The tail rotor shaft is driven off the collector gear at 3,312 rpm. The accessory gearbox is also driven off the collector gear, however when the rotor is stationary it is possible to drive the accessory gearbox by the APU or from No. 1 engine by pilot selection. The accessory gearbox drives two of the three hydraulic pumps and the two AC generators. The third hydraulic pump is driven directly off the main gearbox. The main gearbox lubrication system comprises two independent lubrication circuits, each with its own oil pump filter and cooler.

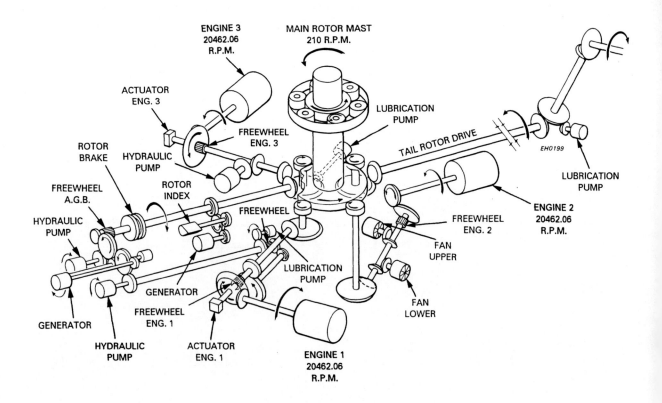

ENGINE 3
20462.06
R.P.M.

MAIN ROTOR MAST
210 R.P.M.

ACTUATOR
ENG. 3

LUBRICATION
PUMP

FREEWHEEL
ENG. 3

ROTOR
BRAKE

HYDRAULIC
PUMP

TAIL ROTOR DRIVE

EH0199

ROTOR
INDEX

FREEWHEEL
A.G.B.

FREEWHEEL

LUBRICATION
PUMP

HYDRAULIC
PUMP

ENGINE 2
20462.06
R.P.M.

FREEWHEEL
ENG. 2

FAN
UPPER

GENERATOR

LUBRICATION
PUMP

FREEWHEEL
ENG. 1

FAN
LOWER

GENERATOR

HYDRAULIC
PUMP

ACTUATOR
ENG. 1

ENGINE 1
20462.06
R.P.M.

***Fig. 9.6** EH 101 Merlin main transmission system (GKN Westland Helicopters)*

The EH 101 Merlin main gearbox and engine installation are shown in Fig. 9.7. The nose of the helicopter is to the left of this diagram. The three engines can be clearly seen as can the APU which is to the rear of the main gearbox and just above the tail rotor drive shaft. The accessory gearbox is located on the front of the main gearbox and the main rotor drive rises vertically from the main gearbox. Due to the obvious importance of the transmission system a considerable degree of monitoring is in-built to detect failures at an incipient stage. Typical parameters which are monitored are oil pressures and temperatures, bearing temperatures, wear, and in some cases accelerations. The role of the health and usage monitoring system on board helicopters is assuming paramount importance and will be discussed later in this chapter.

Hydraulic systems

For helicopters, the hydraulic systems are a major source of power for the flying controls as for various other ancillary services. A typical large helicopter, such as the EH 101 Merlin, has three hydraulic systems, though smaller vehicles may not be so well-endowed. The number of hydraulic systems will depend upon integrity requirements and helicopter handling following loss of hydraulic power.

The main hydraulic loads supplied are:

- Powered flying controls
  - 3 dual main rotor jacks
  - 1 dual tail rotor jack

Fig. 9.7 *EH 101 Merlin main gearbox and engine installation (GKN Westland Helicopters)*

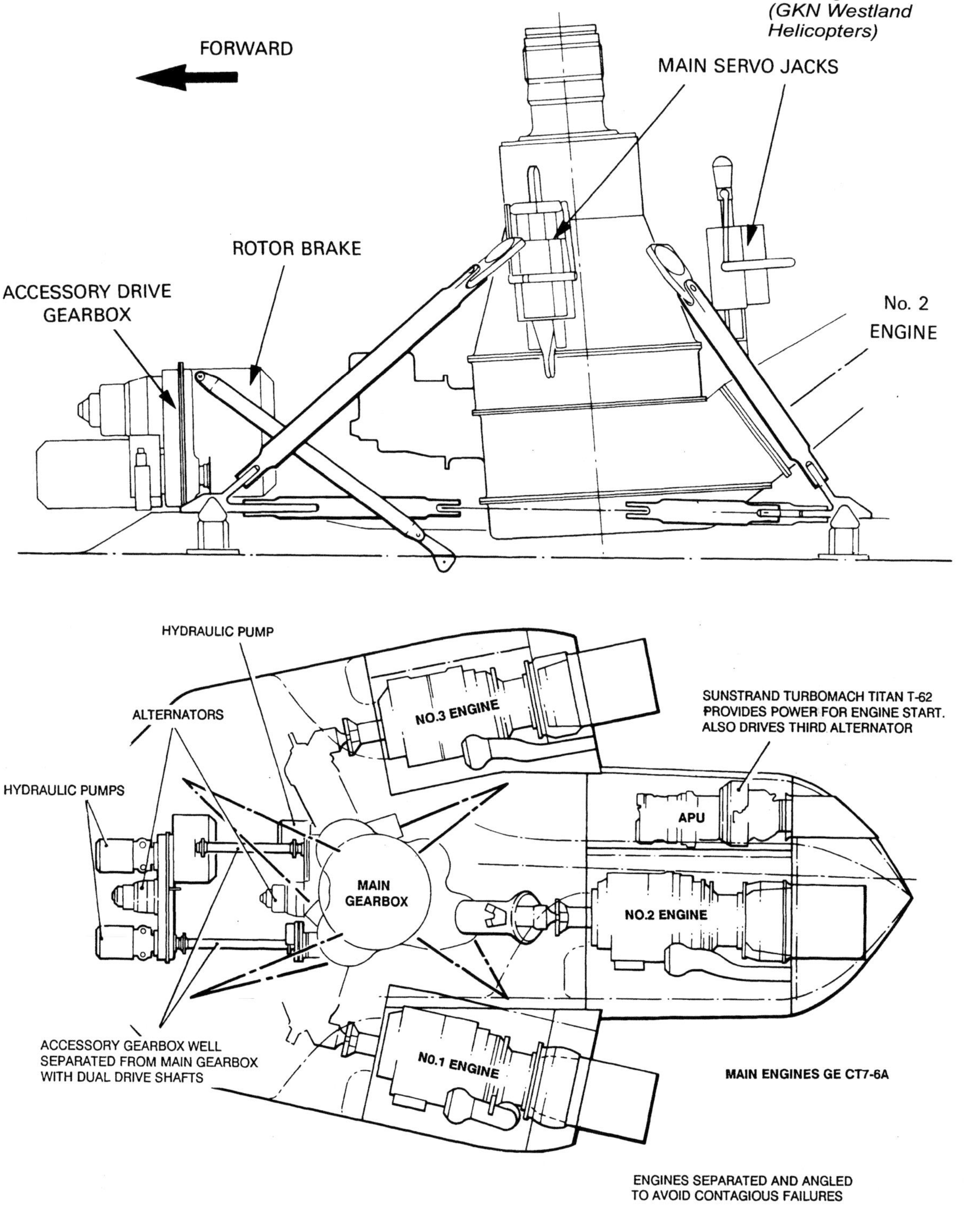

FORWARD

MAIN SERVO JACKS

ROTOR BRAKE

ACCESSORY DRIVE
GEARBOX

No. 2
ENGINE

HYDRAULIC PUMP

ALTERNATORS

NO.3 ENGINE

SUNSTRAND TURBOMACH TITAN T-62
PROVIDES POWER FOR ENGINE START.
ALSO DRIVES THIRD ALTERNATOR

HYDRAULIC PUMPS

APU

MAIN
GEARBOX

NO.2 ENGINE

NO.1 ENGINE

ACCESSORY GEARBOX WELL
SEPARATED FROM MAIN GEARBOX
WITH DUAL DRIVE SHAFTS

MAIN ENGINES GE CT7-6A

ENGINES SEPARATED AND ANGLED
TO AVOID CONTAGIOUS FAILURES

Fig. 9.8 EH 101
Merlin simplified
electrical system
(GKN Westland
Helicopters)

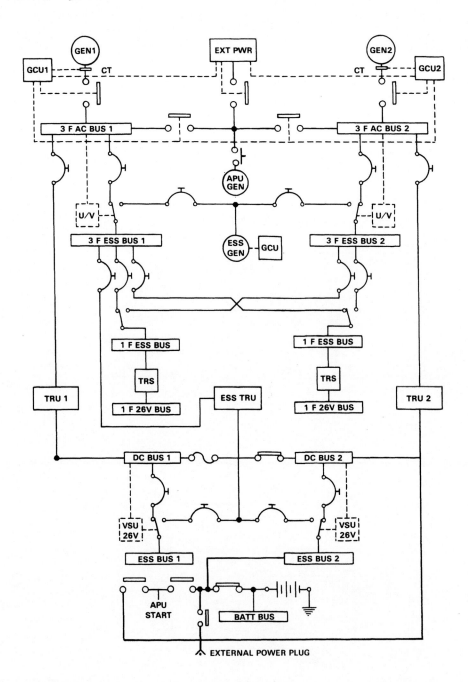

- Ancillary services
  - Landing-gear
  - Steering
  - Wheel brakes
  - Rotor brake
  - Winch (if needed)

Electrical system

The EH 101 Merlin electrical system shown in the simplified block schematic in Fig. 9.8 is typical of the electrical system of this size of helicopter. AC generation is supplied by two main generators each of 45 kVA capacity driven by the accessory gearbox. An emergency AC generator is driven directly off the main gearbox. The arrangement of the main generator tie contactors and the bus tie contactors, controlled by the two generator control units, is typical of a system of this configuration. In the event of an under-voltage condition being sensed, bus transfer relays switch the output of the essential AC generator on to No. 1 and No. 2 essential buses as appropriate. These feed, in turn, single-phase essential buses and the essential TRU. In normal conditions No. 1 and No. 2 TRUs feed DC buses No. 1 and No. 2 respectively. A battery is provided, mainly to start the APU; however this can provide short duration emergency power in the event of a triple electrical systems failure.

Health monitoring system

The importance of the health monitoring system has already been briefly mentioned in the section on the engine and transmission system. The health and usage monitoring systems, or HUMS are now considered to be so important that the UK Civil Aviation Authority (CAA) now specifies the equipment as mandatory for all helicopters certified in the UK.

There are two notable aspects to the use of HUMS. The first relates to criticality and flight safety, the second to cost savings. If the correct critical parameters in an engine and transmission system are monitored then it is possible to identify deterioration of components before a critical failure occurs. This is done by establishing a time-history of the parameter during normal operation of the aircraft, and carrying out trend analysis using computers and data reduction techniques. The tendency for a parameter to exceed set thresholds on either an occasional or regular basis can be readily identified as may a steadily rising trend in a component vibration measurement. Such trends may be identified as heralding a gearbox failure – possibly an impending gear tooth failure – or increasing torque levels in a transmission shaft which might indicate that the component is being overstressed and may fail in a catastrophic fashion. Many such failures in a helicopter gearbox and transmission system could cause the loss of the helicopter and occupants.

With regard to cost savings HUMS helps to avoid the expense of a major failure and the significant engine damage and expense which this entails. As has been shown, a multi-engined helicopter is well capable of flying and landing with two or even one remaining engine, so the flight hazard is of a lower order. However, the expense of overhauling an engine after a major failure is considerable. It therefore makes sound economic sense to monitor key engine parameters and forestall the problem by removing the engine for overhaul when certain critical exceedances have been attained. The ability to monitor the consumption of component life may be used to modify the way in which the helicopter is operated or maintained. If it is apparent that operating the aircraft in a certain way consumes component life in an excessive manner then the pilots may be instructed to modify the flight envelope to avoid the flight condition responsible. From a maintenance standpoint it may be possible to extend the life of certain components if more information is available regarding true degradation or wear. In some cases it may be possible to dispense with a rigid component lifing policy and replace units in a more intelligent way based upon component condition.

The parameters which may be monitored are extensive and may depend to some degree upon the precise engine/gearbox/rotor configuration. Listed below is a range of typical parameters together with the reason for their use.

- Speed probes and tachometer generators: the measurement of speed is of importance to ensure that a rotating component does not exceed limits with the risk of being overstressed.
- Temperature measurement: the exceedance of temperature limits or a tendency to run hot is often a prelude to a major component or system failure.
- Pressure measurement: a tendency to over-pressure or low pressure may be an indication of impending failure or a loss of vital system fluids.
- Acceleration: higher acceleration readings than normal may indicate that a component has been overstressed or that abnormal wear is occurring. The use of low-cycle fatigue algorithms may indicate blade fatigue which could result in blade failure.
- Particle detection: metal particle detection may indicate higher than normal metal composition in an engine or gearbox oil system resulting from abnormal or excessive wear of a bearing which could fail if left unchecked.

Most HUMS systems continuously monitor and log the above-mentioned parameters and would only indicate to the pilot when an exceedance had occurred. The data accumulated is regularly downloaded from the aircraft using a data transfer unit. The data is then transferred to a ground-based computer and replay facility which performs the necessary data reduction and performance/trend algorithms, as well as providing a means of displaying the data. In this way it is possible to maintain a record of every helicopter in the fleet and to take the necessary actions when any exceedances or unhealthy trends have been identified. Reference (7) describes the health and usage monitoring system of the Westland 30 helicopter.

Specialized helicopter systems

A number of systems to be found on a helicopter are specific to the nature of its mode of operation and would find no equivalent application on a fixed-wing aircraft. Two such systems are the winch and the deck-locking system.

Winch system

The helicopter's ability to hover, when coupled with the provision of a winch system, clearly enhances its flexibility in a range of roles such as the lifting and handling of loads or the recovery of personnel from the ground in an emergency situation. The winch may be either electrically or hydraulically operated and some aircraft may offer both. The winch operates by using either source of power to drive a reversible motor which pays out or retrieves the winch cable. The winch control system has the ability to lock the cable at any position while under load. Winch power may be controlled by the pilots using a control unit in the cockpit. However, it is usual to provide a control station adjacent to the cargo door where the winch may be controlled by a dedicated cable operator. The system may include a guillotine arrangement whereby it may be severed should the winch operation endanger the helicopter. This could occur if the winch hook became entangled with an object on the ground or if the helicopter suffered an engine failure or power loss while lifting a heavy load.

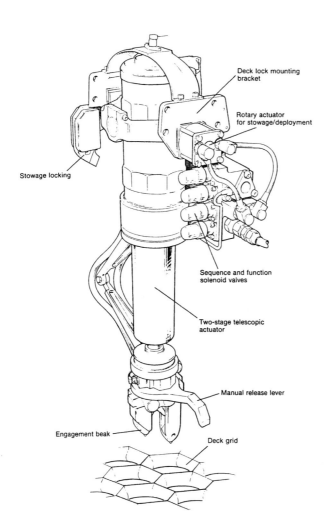

Fig. 9.9 *EH 101 Merlin deck-lock system (Claverham/FHL)*

Deck lock mounting bracket

Rotary actuator for stowage/deployment

Stowage locking

Sequence and function solenoid valves

Two-stage telescopic actuator

Manual release lever

Engagement beak

Deck grid

Deck-locking system

The deck-locking system enables a helicopter to land on, and remain secured to, the deck of a heaving ship in gale force winds up to 50 kts. The principle has been in use since the early 1960s when a rudimentary system was tested by the Royal Aircraft Establishment (now the Defence Research Agency) using a Dragonfly helicopter. The system allows the pilot to 'capture' the deck, either for a final recovery landing or to re-arm or refuel prior to an additional sortie.

The deck-lock system was developed and produced by Claverham/FHL and is in use for the recovery of helicopters up to 20,000 lb. Later systems under development for use with the EH 101 Merlin will enable operation with helicopters up to 30,000 lb. The ship deck has a grid into which a helicopter-mounted harpoon arrangement may engage. The helicopter hovers above the deck as the pilot 'arms' the system. This causes the deck-lock to be lowered from the stowage bay into an extended position. By judging the movement of the ship, the pilot elects to touch down and activates the system by pressing a switch located on his collective lever. This enables the engagement beak and jaws to engage the deck grid and secure the helicopter to the deck. If for any reason the beak misses the grid, or encounters solid deck, the system automatically recycles and the pilot may re-attempt engagement. The engagement sequence is complete within 1.5 sec. The deck-lock system for the EH 101 Merlin is shown in Fig. 9.9.

Helicopter flight control

EH 101 Merlin flight control system

Most helicopter flight control systems use conventional rod and lever mechanical control runs with mechanical mixer units. Figure 9.10 shows the flight control scheme for the EH 101 Merlin which has many similarities to the Hawk 200 system described in Chapter 3, Flight Controls. In the cockpit, dual controls for cyclic (pitch and roll), collective and yaw are provided. Behind the cockpit the Automatic Flight Control System (AFCS) parallel actuators are located for pitch, roll, yaw and collective channels. The same location houses the trim actuators and the artificial spring feel units for pitch, roll and yaw. The AFCS series actuators for pitch, roll, and collective are located in the cabin roof, forward of the primary and secondary mechanical mixing units. Rotor actuation is undertaken by three duplex parallel actuators powered by the three hydraulic systems as described earlier in this chapter. The yaw control run has a quick-disconnect coupling at the rear fuselage break. The AFCS yaw series actuator is located between the quick-disconnect coupling and duplex tandem tail rotor actuator.

Handling a large helicopter such as the EH 101 Merlin requires a great deal of effort and concentration by pilots who have other considerable demands placed upon them, for instance by Air Traffic Control or mission requirements. The need for an advanced AFCS is paramount and the system developed by Smiths Industries and OMI Agusta will provide the necessary automatic flight control. The AFCS functions may be split into two main areas:

- Autostabilization functions
  - Pitch, roll and yaw autostabilization
  - Pitch and roll attitude hold
  - Heading hold
  - Turn co-ordination
  - Autotrim
- Autopilot functions
  - Barometric altitude hold
  - Radar altitude hold
  - Air speed hold
  - Heading acquire
  - Vertical speed acquire
  - Navigation mode
  - Approach mode
  - Back course
  - Go-around
  - Hover hold
  - Hover trim
  - Cable hover
  - Transition up/down

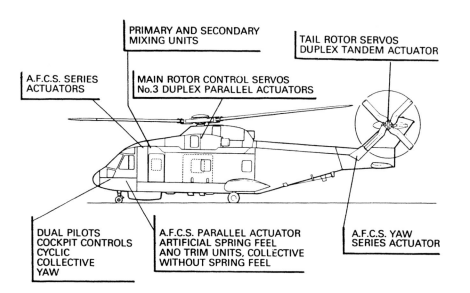

PRIMARY AND SECONDARY
MIXING UNITS

TAIL ROTOR SERVOS
DUPLEX TANDEM ACTUATOR

A.F.C.S. SERIES
ACTUATORS

MAIN ROTOR CONTROL SERVOS
No.3 DUPLEX PARALLEL ACTUATORS

DUAL PILOTS
COCKPIT CONTROLS
CYCLIC
COLLECTIVE
YAW

A.F.C.S. PARALLEL ACTUATOR
ARTIFICIAL SPRING FEEL
ANO TRIM UNITS, COLLECTIVE
WITHOUT SPRING FEEL

A.F.C.S. YAW
SERIES ACTUATOR

Fig. 9.10 *EH 101
Merlin flight control
schematic (GKN
Westland Helicopters)*

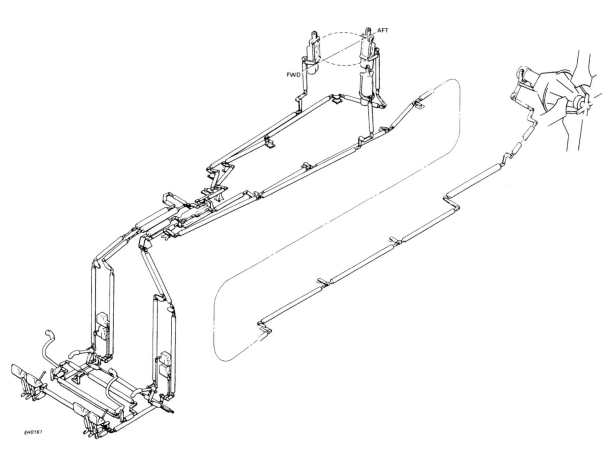

EH0161

Fig. 9.11 Simplified
EH 101 Merlin AFCS
architecture (Smiths
Industries)

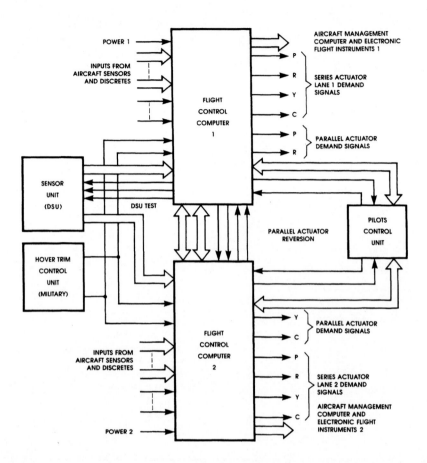

The AFCS developed by Smiths Industries/OMI Agusta is based upon a dual duplex architecture. Dissimilar microprocessors and software are utilized to meet the high integrity requirements. The simplified AFCS architecture is shown in Fig. 9.11. At the heart of the system are the two Flight Control Computers (FCCs). Each FCC receives sensor information from the sensor unit as well as discrete and digital information from the aircraft sensors and systems. Both FCCs communicate with the other via digital and hardwired links. Both FCCs also communicate with the pilot's control unit shown in Fig. 9.12. The control unit conveys to the pilot the status of the system and enables the pilot to monitor hover and radar altimeter altitude in feet and helicopter air speed in knots.

Both FCCs output information to the aircraft management computer and to the Electronic Flight Instruments System (EFIS) displays. FCC 1 feeds lane 1 commands to the pitch, roll, yaw and collective series actuators. FCC 1 also supplies the parallel actuator pitch and roll commands. FCC 2 supplies lane 2 commands to the pitch, roll, yaw and collective series actuators. FCC 2 also supplies the yaw and collective parallel actuator commands.

NOTAR™ *method of yaw control*

The helicopter systems described so far have been controlled in yaw by means of conventional use of the tail rotor. Boeing (formerly the McDonnell Douglas Helicopter

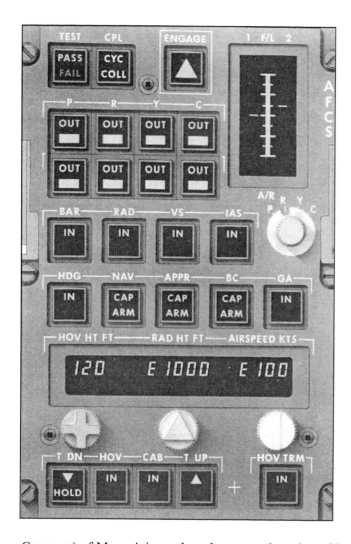

Fig. 9.12 *EH 101 Merlin AFCS pilot's control unit (Smiths Industries)*

Company) of Mesa, Arizona, have been experimenting with an alternative method of yaw control called NOTAR™ (short for NO TAil Rotor). This method replaces the variable pitch tail rotor and the rotating drive shaft which has to pass the length of the tail boom to drive the tail rotor gearbox. It should be noted that the first application of NOTAR™ was the Cierva Weir WQ flown in 1944.

The NOTAR™ principle uses blown air to counteract the main rotor torque effect and it does this by employing two different means. Instead of a conventional tail boom structure the NOTAR™ tail comprises a hollow tube down which air is blown by a variable 13-blade 22 in-diameter fan. At the end of the boom, air is vented through direct jets which counteract the rotor torque. In addition downwash from the rotor passes externally over the boom causing a sideways anti-torque force very similar to the way in which an aircraft wing works. The air flow down the right-hand side of the boom is encouraged to adhere to the boom by means of air bled out of thin longitudinal slots in the boom. The resulting forces induce a counter torque moment due to the Coanda effect. Measurements have indicated that approximately two-thirds of the counter torque force of the NOTAR™ concept is produced by the Coanda effect; the

Fig. 9.13   Boeing helicopter NOTAR™ – concept of operation (Boeing)

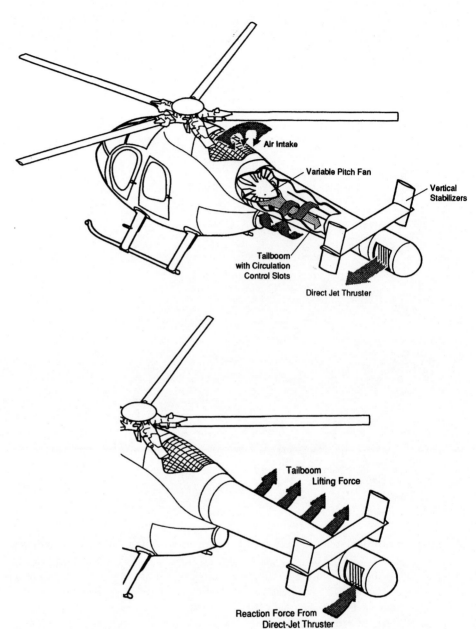

Air Intake

Variable Pitch Fan

Vertical Stabilizers

Tailboom with Circulation Control Slots

Direct Jet Thruster

Tailboom Lifting Force

Reaction Force From Direct-Jet Thruster

remaining third is generated by the low pressure air exhausting from the rear of the boom. See Fig. 9.13.

The advantage of NOTAR™ is that it is relatively simple compared to the conventional tail rotor. The only moving parts are the fan, and significant weight savings are achieved. The McDonnell Douglas demonstrator, MD 530 helicopter achieved up to 40 kts sideways motion using this principle and it is claimed that turns are much easier to co-ordinate, particularly in gusty conditions. Another advantage is that the concept is largely self-correcting with increases in power; as power is increased so does the rotor torque effect; however, so too does the rotor downwash and the Coanda effect and the counteracting force. A further benefit is the absence of rotating parts at the end of the tail boom which reduces the hazard to personnel on the ground and to the aircraft while manoeuvring close to trees in a combat situation.

McDonnell Douglas and their Superteam partner Bell included the NOTAR™ design in their submission for the US Army light helicopter (LH) proposal. This is the next generation of lightweight helicopters for the US Army. The contract award was however given to a Boeing/Sikorsky grouping. The US Army has recently announced that it is modifying 36 H-6-530 helicopters (US Army version of the MD530) to the NOTAR™ configuration. The modification is said to save 20 per cent of the airframe weight and it is expected that handling will be improved, noise reduced and power savings made.

Active control technology

Active Control Technology (ACT) is the term used in the UK to describe full-authority, manoeuvre-demand flight control systems. Such a system would be fly-by-wire using electrical, or possibly fibre-optic signalling, instead of the conventional rod and lever flight control runs of the type already described for the EH 101 Merlin helicopter. The most obvious advantages of ACT as applied to a helicopter are weight savings due to the removal of the mechanical control runs and pilot workload improvement due to enhanced handling characteristics. Future battlefield helicopters will need to be extremely agile compared to those of today and ACT is seen as vital in providing the necessary carefree manoeuvring capabilities. References (8) and (9) are two papers addressing ACT and the helicopter.

The key issues relating to ACT and the helicopter are:

(1) The level of redundancy, i.e. triplex versus quadruplex lane architecture required to meet the integrity levels specified. This decision depends upon whether the helicopter requirements are military or civil and upon the effectiveness of BIT coverage and in-lane monitoring. Present thinking appears to favour a triplex implementation provided the monitoring and dissimilarity issues are addressed in a satisfactory manner.

(2) The degree of dissimilarity between the processing 'strings' is important for high integrity. There is a general fear of the probability of a single catastrophic failure in the electronic computing elements or associated input/output which could cause a common mode failure of all lanes. This concern has become more prevalent due to the proliferation of commercial VLSI microelectronic chips, where it is almost impossible to conduct a Failure Modes and Effects Analysis (FMEA) with a high level of confidence. Equally there is concern regarding latent common mode software failures. The main way of solving these problems is by introducing hardware and software dissimilarity. It will be recalled that such a scheme is utilized in the EH 101 Merlin AFCS which is not fly-by-wire.

(3) The signalling and transmission medium is a further consideration. The use of serial data buses offers great attractions: the main area of debate is whether signalling should be by electrical or fibre-optic transmission. The appeal of the fibre-optic medium is an improved resistance to ElectroMagnetic Interference (EMI).

At present the application of ACT to helicopter flight control systems appears a long way off. The EH 101 Merlin helicopter has suffered significant programme delays and overspend and has recently entered service with the Royal Navy. The next battlefield helicopter is an off-the-shelf purchase of the GKN Westland/Boeing WAH-64-C/D Longbow Apache. It may be that a technology demonstrator for ACT will be forthcoming at some point in the future as has been the case in other countries.

Advanced battlefield helicopter

The most capable battlefield helicopter – arguably in the world – has recently had a chance to prove its combat effectiveness. The Boeing AH-64A Apache was used to great effect during 'Desert Storm', in the Middle East, when its night capability and fearsome firepower was amply demonstrated. Due to the success of the helicopter during that conflict it is considered topical to outline some of the key characteristics in this chapter.

The Apache helicopter was originally designed by the Hughes Helicopter Company which was later acquired by the McDonnell Douglas Corporation, the company being renamed McDonnell Douglas Helicopter Company (MDHC). MDHC later became part of Boeing as has already been mentioned The first Apache prototype flew in September 1975 and the first production aircraft was delivered in January 1984. By the end of 1987 some 300 aircraft had been delivered. The US Army has a total requirement for around 800 helicopters, the last of which was delivered in the 1992/93 timeframe. At that stage some aircraft will be updated or retro-fitted with improved systems including an improved electrical power system and eventually some aircraft will receive an advanced mast-mounted millimetric fire control radar called Longbow. The basic AH-64A Apache configuration is shown in Fig. 9.14.

The helicopter has a four-blade articulated main rotor and a four-blade tail rotor. The tail rotor blades are skewed at 55/125 degrees as this is apparently the optimum position for noise reduction. The helicopter is powered by two 1,696 shp General Electric T700/701 turbo-shaft engines which have an engine-out rating of 1,723 shp. The hydraulic system comprises dual 3,000 psi systems which power dual actuators for main and tail rotors. In the event that both systems fail a reversionary electrical link provides back-up control. The AH-64A has two 35 kVA AC generators. A Garrett APU is provided for engine starting and to provide electrical power for maintenance. The helicopter is operated by two crew: a pilot sitting aft and a Co-Pilot Gunner (CPG) in the front cockpit.

Perhaps the two most striking features of the AH-64A are the night vision system and the extensive range of armaments which may be carried.

Target Acquisition and Designator System(TADS)/Pilots Night Vision System (PNVS)

The night vision system is called the Target Acquisition and Designator System/Pilots Night Vision System or TADS/PNVS for short. The systems, comprising two separate elements, are located in the bulbous protrusions on the aircraft nose. Figure 9.15 shows the TADS/PNVS installation with some of the relevant fields of view.

The target acquisition and designator system (TADS) comprises the following facilities:

Fig. 9.14 *AH-64A Apache (Boeing)*

- Direct vision optics to enhance daylight long-range target recognition. An optical relay tube transmits the direct view optics to the CPG.
- Forward-Looking Infra-Red (FLIR).
- TV.
- Laser designator/range finder.
- Laser tracker.

The Pilot Night Vision System (PNVS) gives the pilot a FLIR image over a 30 degree × 40 degree field of view which can be slaved to the direction the pilot is looking by means of an integrated helmet display.

The combination of TADS and PNVS capabilities gives the Apache a potent system which has demonstrated maturity and durability during the Gulf War. For more

Fig. 9.15 Apache
TADS/PNVS
installation (Boeing)

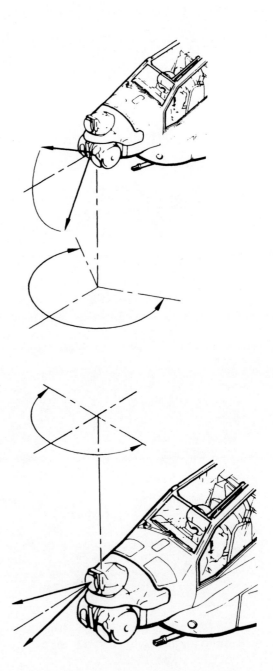

information relating to the Apache helicopter and its capabilities see references (10) and (11).

The aircraft can carry a variety of weapons and missiles. In addition a 30 mm chain cannon is fitted as standard. The weapons which can be carried are:

- 70 mm rockets.
- Hellfire anti-tank missiles.
- Stinger air-to-air missiles.

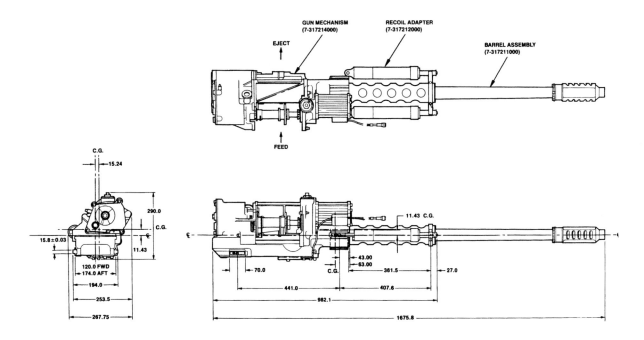

The M230 chain gun is fitted under the forward fuselage as shown in Fig. 9.16. The gun is a 30 mm cannon with a firing rate of 10 rounds/second. The gun may either fire a single shot or 10, 20, or 50 round bursts; it can be slaved to the pilot's helmet system and can therefore traverse over the field of view of the pilot to engage the target. A typical load of ammunition would be 1,200 rounds.

Fig. 9.16 Apache M230 chain gun installation (Boeing)

AH-64 C/D Longbow Apache

In 1989 McDonnell Douglas embarked upon a development programme to upgrade the Apache helicopter to a configuration called Longbow Apache. The Longbow prefix related mainly to the addition of a millimetric, mast-mounted, fire control radar previously called the Airborne Adverse-Weather Weapon System (AAWWS). This advanced radar system allows the Apache to hover in a screened position behind a tree or ridge line with the radar illuminating and identifying targets out to several kilometres range. The helicopter system may then rise above the ridge line, launch missiles at several pre-designated targets and then drop out of sight of the defending forces. This greatly improves the capability of the helicopter while reducing its vulnerability. The radar also operates well in conditions not suited to the use of the electro-optic TADS/PNVS. Refer to Fig. 9.17.

One of the improved system features of the Longbow Apache is a new electrical system called the Electrical Power Management System (EPMS). During the development of the Longbow configuration it was found necessary to increase the rating of the main AC generators from 35 kVA to 45 kVA to provide more power to feed the new systems being fitted to the aircraft. Improvements in system architecture were also made to eradicate certain single point failures in the system as well as to provide better reliability and maintainability. A circuit-breaker panel was removed from the left upper part of the pilot's cockpit thereby improving the field of view for air-to-air operations. Most circuit-breakers were removed from the cockpit and a significant

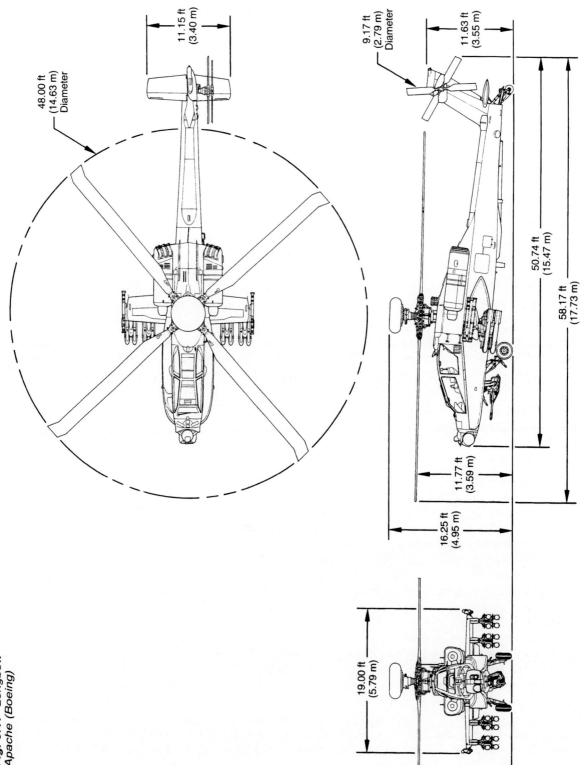

48.00 ft (14.63 m) Diameter

11.15 ft (3.40 m)

9.17 ft (2.79 m) Diameter

11.63 ft (3.55 m)

50.74 ft (15.47 m)

58.17 ft (17.73 m)

11.77 ft (3.59 m)

16.25 ft (4.95 m)

19.00 ft (5.79 m)

Fig. 9.17 Longbow Apache (Boeing)

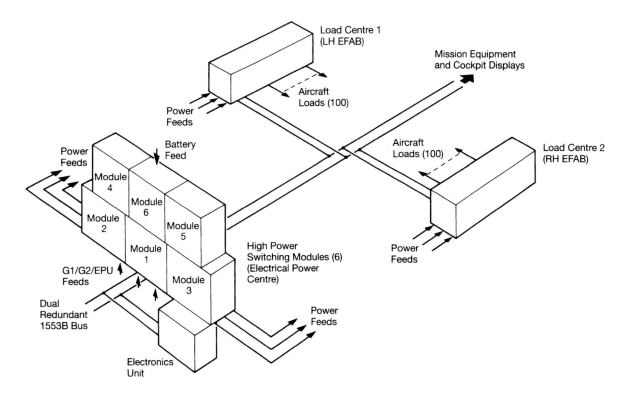

proportion of the aircraft electrical loads are now remotely switched from the cockpit using the touch-sensitive screens.

An early version of the EPMS developed and manufactured by Smiths Industries is shown in Fig. 9.18. The system has progressively undergone a series of upgrades, improving packaging and increasing system functionality.

The prototype system comprised a total of nine line replaceable units (LRUs). Six LRUs are high-power switching modules which contain the primary power switching contactors and the aircraft primary 115 V three-phase 400 Hz VAC, 28 VDC and battery bus bars. These LRUs may be quickly replaced following failure and are partitioned and monitored using built-in test (BIT) such that electrical faults may be quickly traced to the correct module. Certain contactors have I^2t trip characteristics which enable fault conditions to be identified and removed within tighter tolerances than previously possible, thereby enabling reduction in the size of the busbars with consequent weight savings. The high-power modules receive 115 VAC power from either main generator or the external power unit and DC power from the transformer rectifiers (TRs) or the battery. High-power 115 VAC, 28 VDC, and battery feeds run forward to two load centres which control the aircraft secondary loads; those loads which are less than 20 amps. See reference (12).

The electronic unit has hardwired connections to the six high-power modules enabling the primary contactors to be switched and status-monitored as necessary. A large amount of system monitoring of current and voltage is possible. The electronic unit is connected to the aircraft 1553B dual redundant avionics bus. The six high-power switching modules and the electronics unit are mounted on a bulkhead in the transmission bay, just behind the pilot's seat.

Fig. 9.18 Longbow Apache prototype EPMS (Smiths Industries)

The two load centres are mounted in the forward avionics bays. The primary power fed from the high-power switching modules is distributed to all the secondary loads and protected within the load centres. Each load centre protects and feeds around 100 secondary loads of which approximately 30–35 are remotely switched from the cockpit via the touch-sensitive displays and the 1553B avionics data bus. The load centres are supplied with conditioned air to remove excess heat.

The repackaging exercise for initial production aircraft rationalized the system to a total of four LRUs and introduced more processing capability. Subsequent modifications have included greater functionality to include the control of utilities systems.

References

(1)   **Fay, J.** (1987) *The Helicopter, History, Piloting and How it Flies*, Fourth edition, David and Charles.
(2)   **Hague, C.W.** (1984) EH 101, *Aerospace*, July/August.
(3)   **Martin, E.E.** (1984) T-700 – A program designed for early maturity and growth potential, Tenth European Rotorcraft Forum, The Hague, Netherlands.
(4)   **Bryanton, R.** (1985) RTM 322 – Europe's newest helicopter engine, *Aerospace*, May.
(5)   **Buller, M.J.** and **Lewis, D.** (1985) The conception and development of a family of small engines for the 1990s, Eleventh European Rotorcraft Forum, London, Great Britain.
(6)   **Saunders, A.F.** (1983) An advanced helicopter engine control system, *Aircraft Engineering*, March.
(7)   **Astridge, D.G.** and **Roe, J.D.** (1984) The health and usage monitoring system of the Westland 30 Series 300 helicopter, Tenth European Rotorcraft Forum, The Hague, Netherlands.
(8)   **Richards, W.R.** ACT applied to helicopter flight control, AGARD Conference Proceedings, 384.
(9)   **Wyatt, G.C.F.** The evolution of active control technology for the 1990s helicopter, AGARD Conference Proceedings, 384.
(10) **Rorke, J.B.** Apache for the battlefield of today and the 21st century, AGARD Conference Proceedings, 423.
(11) **Green, D.L.** (1985) Flying in the Army's latest warrior – the Hughes AH-64A Apache, *Rotor & Wing International*, April.
(12) **Moir, I.** and **Filler, T.** (1992) The Longbow Apache electrical power management system, *Aerotech '92*, January.

CHAPTER 10

Advanced Systems

Introduction

The advanced systems chapter addresses some of those systems which broach new areas, having been either recently developed or under development. In many cases the concepts may have been under study for a number of years and recent developments in technology may have given the impetus and the means of implementation.

Some of these developments relate to the improved integration of aircraft systems to achieve hitherto unattainable benefits. Others embrace low-observability or 'stealth' technology. The following range of developments are addressed in this chapter.

STOL manoeuvre technology demonstrator (SMTD)

The US Air Force SMTD F-15 upon which Integrated Flight and Propulsion Control (IFPC) allows closer integration of the aircraft flight control and engine control systems. Flight control systems are virtually all fly-by-wire in the modern fighter aircraft of today; the benefits being weight reduction and improved handling characteristics. New engines are likewise adopting Full-Authority Digital Engine Control (FADEC) for the benefits offered by digital control. On aircraft such as the US Air Force SMTD F-15 these systems are being integrated to evaluate new control techniques applied to a modified F-15. This type of system could find application on the new generation of V/STOL aircraft to replace the Harrier in the early twenty-first century.

Vehicle management systems (VMS)

Vehicle Management Systems (VMS) carry this integration still further, combining flight control and propulsion control with the control of utility and power management. This further improves the control of the aircraft systems and permits the integration of functions such as thermal management which will be vital to the performance of fighter aircraft cruising for extended periods at Mach 1.6 which is a requirement for the US Air Force F-

22 Raptor. Thermal management is presently spread across several aircraft subsystems and these boundaries will need to be revised if the problem is to be properly tackled.

All-electric aircraft

There is considerable interest on both sides of the Atlantic in addressing the move toward the use of more electrical power on aircraft – both in the civil and military fields. These issues are studied under the mantle of the more-electric aircraft, the all-electric aircraft and the more-electric or all-electric engine. In general, aircraft power levels are increasing for a number of reasons. On civil aircraft, galley loads and advanced In-Flight Entertainment (IFE) systems are requiring higher levels of electrical power. Aircraft system loads are also increasing. Although technology breakthroughs are being made in a number of key areas, and many of the technologies have or will be demonstrated, it is a major step to embrace all the changes on one programme.

The Bell-Boeing V-22 Osprey tilt rotor aircraft

The novel Bell-Boeing V-22 Osprey tilt rotor development has survived despite earlier attempts by the Pentagon to have it cancelled. Congress continued to fund the programme without the existence of a production order as it perceived that maturity in the necessary technologies was imminent. The US Marines are now beginning to deploy the MV-22 marine version. The Japanese are also attracted to the use of tilt rotor because of its ability to operate into and out of dense environments in a manner which can only be achieved at present by the helicopter. In the civil field the tilt rotor/Bell Augusta 609 is taking shape as a definite project.

Stealth

The development of 'low-observable' aircraft has been given a high priority by the US Air Force in particular in the last decade as a way of improving the combat effectiveness of the combat vehicle. The Lockheed F-117A 'stealth fighter', Northrop B-2 'stealth bomber' and the former Advanced Tactical Fighter (ATF) Dem/Val YF-22A and YF-23A projects were designed with this feature in mind. The selected F-22 Raptor is now at an advanced stage of development. The F-117A in particular graphically indicated the benefits of this technology during the Gulf War; both the F-117A and the B-2 bomber were deployed during the 1999 Kosovo conflict. The F-117A has recently begun a standard configuration fleet modification to standardize the low-observable coatings used across the fleet – at present a number of different techniques are utilized which evolved during the development and production phases. Recent reports of modifications to the B-2 bomber fleet have suggested that the stealth technology, while operationally highly effective, does have a maintenance penalty.

Joint Strike Fighter (JSF)

The Joint Strike Fighter (JSF) is fielding competing teams from Boeing (incorporating the former McDonnell Douglas) with the X-32, and Lockheed Martin with the X-35. Both these aircraft first flew in the autumn of 2000 with a down-selection to an overall winner in 2001. These designs also embody stealth technology. The aircraft are designed to meet the requirements of four Services: the US Air Force; US Navy; US Marines and British Royal Navy. Three main vehicle configurations are being developed: conventional take-off and landing (CTOL) for the US air force; carrier vehicle (CV) for the US Navy and Short Take-Off Vertical Landing (STOVL) for the US Marines and Royal Navy.

Fig. 10.1(a) *F-15 SMTD (Boeing)*

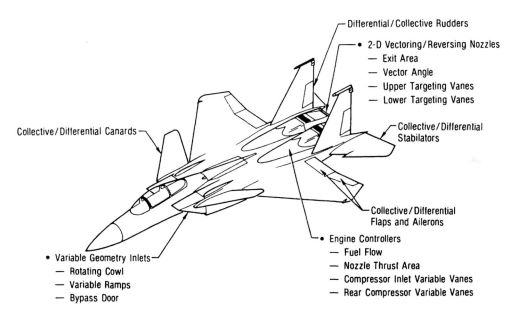

Differential/Collective Rudders

• 2-D Vectoring/Reversing Nozzles
— Exit Area
— Vector Angle
— Upper Targeting Vanes
— Lower Targeting Vanes

Collective/Differential Canards

Collective/Differential Stabilators

Collective/Differential Flaps and Ailerons

• Variable Geometry Inlets
— Rotating Cowl
— Variable Ramps
— Bypass Door

• Engine Controllers
— Fuel Flow
— Nozzle Thrust Area
— Compressor Inlet Variable Vanes
— Rear Compressor Variable Vanes

Fig. 10.1(b) *F-15 SMTD flight control configuration*

Integrated flight and propulsion control

As avionics technologies have developed in the last decade, it has become commonplace for the control of major systems to be vested in electronic implementations; such systems may have previously been solely mechanically or electromechanically controlled. Moreover, the availability and maturity of the technologies required to satisfy avionics system integration have proved equally appealing in satisfying the requirements of more basic aircraft systems. The benefits of digital electronic control of mechanical systems are evident in greater precision and an ability to measure or predict performance degradation and incipient failure. Typical examples of this are digital implementations of flight control or fly-by-wire and digital engine control, or Full-Authority Digital Engine Control (FADEC). As substantial benefits of improved performance and reliability are realized, e.g. weight reduction and other improvements in system integration and data flow, so the level of systems integration becomes correspondingly more ambitious.

It is therefore a logical progression that the demonstrated benefits of digital flight control and engine control systems has instigated development programmes which are examining the next level of integration – that of Integrated Flight and Propulsion Control (IFPC). IFPC is actively being developed in the US. The vehicle for this US Air Force funded programme is the F-15 STOL/Manoeuvre Technology Demonstrator (SMTD), a highly modified F-15B which has been flying for some years from Edwards Air Force Base. Other aims of the technology demonstrators were to show that a high performance fighter could land upon a roughly constructed (or repaired) concrete strip 1,500 × 50 ft. This requires a sophisticated guidance system and an IFPC system to improve the aircraft response and therefore the precision with which the pilot can fly the aircraft during the approach. The configuration of the F-15 SMTD aircraft is shown in Figs 10.1(a) and 10.1(b).

Of particular interest are the multiple effectors utilized on the SMTD aircraft which may be summarized as follows:

- Collective/differential canards.
- Collective/differential flaps and ailerons.
- Collective/differential stabilators.
- Collective/differential rudders.
- Variable-geometry inlets.
- Engine control.
- Two-dimensional (2-D) vectoring/reversing nozzles.

The collective/differential flight control surfaces allow a significant enhancement of the aircraft performance over and above that normally possible in an F-15 in the approach configuration. In addition normal control modes and the use of collective flight control surfaces should offer direct translational flight; that is, operation of those control surfaces should allow the aircraft to move, say vertically, without altering the pitch vector or attitude. The thrust vectoring control adds an additional facility and the aircraft has been flying with 2-D nozzles operational since May 1989. These may be operated in a thrust reverser mode. The F-15 SMTD has been under test since 1988 and has demonstrated operation of the thrust reversers in flight.

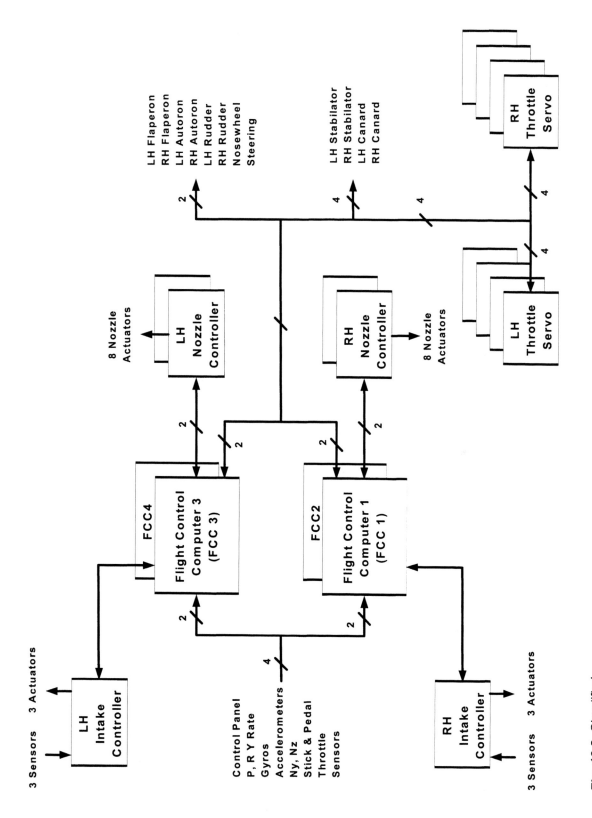

Fig. 10.2 Simplified
F-15 SMTD IFPC
architecture

In order to gain some idea of the complexity of the IFPC, the following summarizes the number of sensors and effectors associated with the system:

- Flight control
    - 11 quadruplex sensors
    - 6 quadruplex actuators
    - 7 dual-redundant actuators
- Intake control (per engine)
    - 3 sensors
    - 3 actuators
- Engine control (per engine)
    - 8 sensors (4 dual-redundant)
    - 6 actuators
- Nozzle control (per nozzle)
    - 8 actuators

See Fig. 10.2. For a detailed paper describing the IFPC fault-tolerant design see reference (1).

Vehicle management systems

The Integrated Flight and Propulsion Control (IFPC) described above is an integration of two main aircraft control systems into one. Vehicle Management Systems (VMS) relate to a higher level of system integration, that is the combination of flight control, propulsion control, and utilities/power management. One reason for combining these systems into a VMS is that the aircraft performance demands an improvement in the integration of these major systems. For example, twenty years ago no fighter aircraft would have been fitted with a fly-by-wire system. Stability augmentation systems were used as a matter of course but the flight control system was implemented using the push–pull rod systems of the type outlined in Chapter 1, Flight Control Systems. Nowadays, virtually all front line fighters routinely employ fly-by-wire systems: they offer artificial stability if the aircraft is unstable, or may merely improve aircraft handling. In either case, they improve handling and performance from the pilot's point of view. Fly-by-wire systems also save weight and can greatly ease or limit structural loading by curtailing demands where necessary. Of course this has all been made possible by advances in microelectronics and actuation techniques. The point is that these techniques have become the stock-in-trade of implementing flight control, as is shown by the extensive use of such systems in the new generation of stealth aircraft described later in this chapter.

In recent years engine control has moved toward Full-Authority Digital Engine Control (FADEC) solutions and the F-15 SMTD programme, already covered, shows how intake and nozzle control may need to be more closely integrated with digital engine control to satisfy some requirements. The air intake or inlet must be correctly matched to the engine or optimum performance will never be achieved, especially for supersonic aircraft. The F-15 SMTD two-dimensional nozzles require a total of six actuators to control the thrust vectoring in the vertical plane and reverse thrust modes for each engine. Whereas pure raw performance may be the objective for some applications, others may seek to improve performance in more subtle ways. An F-117A stealth fighter seeks low observability as a primary mission goal, not the utmost in speed or excess thrust. The technical solutions adopted to achieve the primary goal of stealth most probably directly detract from performance; the means used to reduce the temperature and size of the exhaust plume reducing propulsive power. In this situation more elegant control methods may be required to ensure that these losses are not prohibitive.

Many aircraft systems, such as utilities management and electrical power management, require better control to meet more demanding problem statements. Systems such as fuel, hydraulics, secondary power, environmental control and electrical power systems are being improved by the use of digital control techniques. The UK Experimental Aircraft Programme (EAP) employed a Utilities Management System (UMS) which fully integrated many of these control functions into four dedicated control units as shown in Fig. 10.3. This system first flew August 1986 and a similar system – Utility Control System (UCS) is fitted to Eurofighter. For more detail on the EAP system see references (2) and (3). The Boeing AH-64C/D Longbow Apache employs an integrated Electrical Power Management System (EPMS) to improve the control and distribution of the primary electrical system on this advanced battlefield attack helicopter. See Chapter 9, Helicopter Systems.

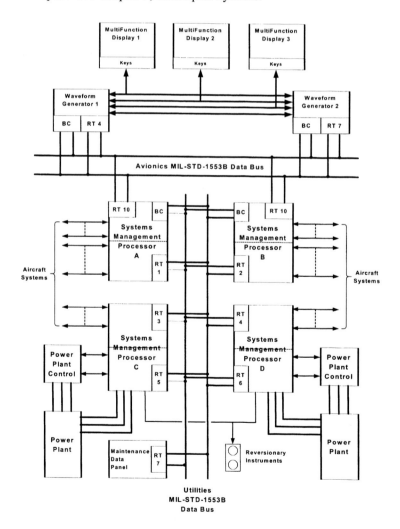

Fig. 10.3 *EAP utilities system management control units (Smiths Industries)*

The VMS concept seeks to integrate all these major systems into one system responsible for controlling the air vehicle or aircraft. All of the systems utilize digital computer control and data buses which allow them to communicate with each other and with the remaining aircraft systems. This leads to the possibility of integrating the VMS using a series of data buses and one such architecture is shown in Fig. 10.4. A major difference between the EAP and Eurofighter USM/UCS and the VMS proposed for future aircraft is that high-rate, closed-loop servo systems have been included in the control concept.

This generic architecture shows a number of control units associated with flight control, engine control and utilities/power management. This allows the units to be closely tied to each other and to the sensors and actuators associated with the control task. In this scheme certain computers have responsibility for interfacing the VMS as a whole to the avionics system and to the pilot. This type of closely coupled control permits modes of operation that would be much more difficult to control if the systems were not integrated into a VMS. For example the fuel management system on a fighter can be used to control the aircraft CG. The position of the CG in relation to the centre of lift determines the aircraft stability and trim drag. For optimum cruise the CG could be positioned at or near the neutral point to minimize trim drag. For combat the CG could be moved aft to make the aircraft more manoeuvrable. Therefore in this example there is an inter-reaction between flight control and utility control which allows optimum modes to be selected for various phases of flight.

Thermal management is an area which is becoming more important in combat aircraft such as the F-22 Raptor which is designed for 'persistent supersonic cruise' operation. That is, the aircraft is designed to cruise for long periods at speeds of

Fig. 10.4 *Generic VMS architecture*

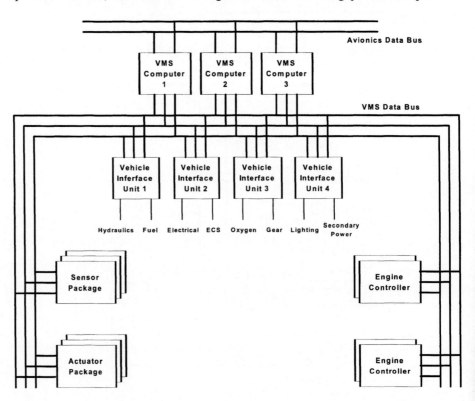

Mach 1.6 whereas previous fighters could only operate at such speeds during a short 'supersonic dash'. This leads to the problem of where to sink all the thermal energy generated during high-speed cruise. The inter-reaction of the fuel system (fuel being used as a heat sink) and the environmental control system, is of great importance in solving the problem. More energy-efficient methods of extracting and utilizing power from the engines can also help and is one of the reasons for studying the all-electric aircraft concept which is described in detail elsewhere in this chapter. Technology demonstration programmes associated with the Joint Strike Fighter (JSF) are also making major advances in this area as will be described later in this chapter.

The US Air Force has embraced the VMS on recent programmes in order that these improvements may be realized. Though the precise architectures may vary by programme depending upon the maturity of the various technologies, it is clear that many of the necessary technologies and building blocks are available and that such systems may be embodied without significant risk.

All-electric aircraft concept

For a number of years the concept of the 'all-electric aircraft' has been espoused. The Bristol Brabazon utilized a great number of electrical systems and the Vickers Valiant V-Bomber was also highly electrical in nature. At the time – mid-1950s – the concept did not fully catch on, though over the years there has been a great deal of debate relating to the advantages of electrical versus other forms of secondary power, such as hydraulics or high-pressure bleed air systems. The dialogue may be summarized by referring to the papers produced by Mike Cronin, formerly the Chief Electrical Engineer on the Brabazon Project and an employee of the Lockheed Aeronautical Systems Company for many years prior to his retirement in 1990. See references (4)–(10). These merely represent a summary of Cronin's work on the subject; more of his and other authors' papers on the subject are noted if more exhaustive research is required.

Over the past decade, examination of the benefits of the all-electric aircraft has been promoted by a number of aeronautical agencies in the US. In the early 1980s NASA funded a number of studies addressing the Integrated Digital Electrical Airplane (IDEA). The IDEA concept studies embraced a range of technologies which could improve the efficiency of a 250–300 seater replacement for an aircraft such as the Lockheed L1011 (Tristar). The areas covered were:

(1) Flight control technology – relaxed stability augmentation leading to a reduction in trim drag with consequent down-sizing of the tailplane and fuel savings.
(2) Wing technology – use of efficient high-aspect ratio wings using gust alleviation modes of the FCS to improve range and fuel consumption and reduce wing bending moments.
(3) Engine power extraction – the reduction of engine power extraction losses by minimizing the use of high-pressure bleed air and hydraulic power and maximizing the use of more efficient electrical power extraction techniques.
(4) Flight control actuation – the use of electromechanical actuation in lieu of hydromechanical actuation systems.
(5) Advanced electrical power systems – the development of new systems to generate and distribute electrical power as an adjunct to more efficient engine power extraction.

Flight control system and flight control actuation developments are already under way or are embodied in major civil programmes as evidenced by systems on the Airbus A320/A330/A340 and Boeing 777 aircraft. Improved wing technology is already being implemented on modern commercial airliners. The more revolutionary techniques of modifying engine power extraction and embodying advanced electrical power systems have yet to be employed to any significant degree on a major civil aircraft programme.

More-electric aircraft generation options

The US Air Force Aeronautical Laboratories have also given considerable consideration to the use of all-electric system technologies for military aircraft; both for fighter and large transport aircraft. While electrical power generation is addressed fully in Chapter 5, Electrical Systems, it is worth briefly contrasting some of the existing techniques with the newer candidates. The main options for power generation are as follows:

(1)   Integrated drive generators (IDGs) using constant speed drives (CSDs).
(2)   Variable speed constant frequency (VSCF) cycloconverter.
(3)   VSCF DC link.
(4)   Variable frequency (VF).
(5)   230 VAC.
(6)   270 VDC.
(7)   Switched reluctance.

The principles of operation and examples of options 1–4 have been described in Chapter 5. In this section they will be briefly reviewed in terms of the maximum power levels they produce in existing or proposed systems.

Integrated drive generator

The constant speed drive or CSD is the most widely used method of providing constant frequency AC supplies (400 Hz) for airborne systems. CSDs are produced exclusively by the Hamilton/Sundstrand Corporation with licences to other companies such as TRW Lucas Aerospace in the UK. CSDs are fairly complex and are therefore generally mechanically unreliable compared with some other methods of power generation. The generation unit is now an Integrated Drive Generator (IDG) encompassing both CSD and generator elements. At present IDGs are still competitive in terms of weight and cost compared with other forms of generation. The Boeing 777 aircraft uses two 120 kVA IDGs, one powered by each engine, as the source of primary power. Specialized military applications are believed to have elevated the power to 150 kVA per channel.

Variable Speed Constant Frequency (VSCF) cycloconverter

The VSCF cycloconverter uses a high frequency generator to generate electrical power at much higher frequencies than the desired constant frequency (400 Hz). General Electric (GE) was generally associated with the lead in this technology. The VSCF cycloconverter system was used on the former McDonnell Douglas F/A-18 aircraft among others and with over 4,000 systems in service it is generally considered to be a mature technology. The 60/65 kVA systems developed for the Boeing F/A-18E/F presently represents the highest level of power achieved using this technique in an airborne application.

VSCF DC link

The VSCF DC link converts the variable frequency generator output to DC using a phase delay rectifier before converting the DC power to AC using an inverter. The former Westinghouse Company (now part of Hamilton/Sundstrand) was the lead company in this technology which does not require the high generator frequencies of the cycloconverter system. Originally the DC link systems were limited to around 20 to 40 kVA though the availability of improved high-voltage/high-current power switching devices has elevated the capacity to around 75kVA in a civil application (MD-90)

Variable Frequency (VF)

Variable frequency electrical power is the cheapest and most reliable contender. As yet extremely high levels of power greater than 40–50 kVA per channel have not been used. However the Airbus A380 large aircraft development has reportedly adopted a baseline of 150 kVA VF power per channel. This four-engined aircraft will therefore have a total of 600 kVA of electrical power on-board, not including ancillary sources such as the APU. As has already been mentioned, VF potentially passes some problems down to the aircraft subsystems which hitherto have been optimized to operate with constant frequency 400Hz power.

270 VDC

The use of 270 VDC systems has been primarily sponsored by the US Navy/Naval Air Development Centre (NADC). One of the reasons is the ease of deriving 270 VDC from the 200 VAC 60 Hz power readily available on US Navy aircraft carriers. 270 VDC systems are easier to parallel than AC systems; a characteristic which it shares with 28 VDC systems (Chapter 5 refers). DC generation systems offer an effective 'constant power' generation system from a 'variable speed' prime mover, i.e. the engine which may have 2:1 variation in speed range. An original disadvantage of 270 VDC is that most constant speed motors used on aircraft are AC synchronous motors and therefore inverters are required to supply constant frequency AC power to these motors. A further significant disadvantage is that switching devices, such as contactors and relays, must withstand higher voltage, although the development of 270 VDC Solid State Power Controllers (SSPCs) or switching devices in recent years has offered some alleviation of this problem.

Notwithstanding these difficulties, the US Air Force Wright Research and Development Centre (WRDC) funded development activities in relation to 270 VDC systems which have been directed as the 'More-Electric Aircraft', suggesting perhaps some compromise as to the degree to which all-electric aircraft principles may be applied. Contracts were awarded for the management and distribution of the more-electric aircraft (MADMEL) programme to study and fabricate a 270 VDC aircraft electrical systems. 270 VDC technology has moved ahead in recent years and DC high voltage systems are now baseline on the following aircraft:

- Lockheed/Martin F-22 Raptor
- Sikorsky RAH-22 Comanche
- Lockheed/Martin X-35 JSF Contender

An advantage of 270 VDC is that it lends itself readily to power-by-wire technologies such as Electro-Hydrostatic Actuators (EHAs). The Joint Strike Fighter/Integrated Subsystems Technology (J/IST) is demonstrating a 270 VDC power-by-wire Fly-By-Wire (FBW) flight control on a modified F-16, completely replacing the existing conventional hydromechanical flight control system. This system flew in Autumn 2000 and will be described under the JSF section. Present 270 VDC generation systems appear to be sized at around 60 to 65 kVA per channel.

230 VAC

230 VAC has been suggested as a possible way of increasing power on three-phase AC systems. By elevating the supply voltage, feeder sizes and losses may be reduced and higher power levels switched with existing switchgear. A disadvantage, similar to 270 VDC and VF power, is that many of the aircraft systems and equipment are designed to operate at 115 VAC, constant frequency 400 Hz. One way round this problem is to use a 230/115 VAC step-down transformer within the aircraft primary electrical power distribution system to accommodate normal 115 VAC loads.

Switched reluctance machines

Another possibility to satisfy the aircraft electrical power generation demands is the Switched Reluctance (SR) machine. This has the significant advantage that it may be designed to operate as a motor as well as a generator allowing the machine to start the engine, and thereafter to act as a generator.

For an aircraft application this opens up the possibility of using a SR machine as a starter/generator, effectively solving two problems in one. The US Air Force Laboratories, Power Division have sponsored a SR development called an integrated power unit (IPU) for a 250 kW (peak), 125 kW (continuous) integrated unit to take the place of an APU. This development is designed to demonstrate an alternative to the conventional APU with an oil-less bearing system, the bearings being electromagnetic.

The J/IST programme also uses a SR machine as will be described later. Switched reluctance machines are therefore on the point of progressing beyond the demonstrator stage.

MEA aircraft subsystem implications

The implications of adopting all/more-electric aircraft designs have a significant effect upon the way some of the aircraft subsystems are implemented today and this fact obviously has a bearing upon acceptance or otherwise of the concept.

Present power off-takes from the engine are:
- Hydraulic power
- Bleed air
- Electrical power

In the case of a civil engine the power is extracted by means of Engine-Driven Pumps (EDPs) and electrical generators shaft-driven by an engine-mounted accessory gearbox. Military fighter aircraft similarly extract hydraulic and electrical power by using an Airframe-Mounted Accessory Gearbox (AMAD). Bleed air is extracted from the engine by means of a bleed air control system which is pneumatically powered and controlled but electrically initiated; refer to Chapter 6, Pneumatic Systems.

The removal of the AMAD implied by driving electrical power generators direct from the engine may save weight but would clearly not find favour with the companies who supply these gearboxes. Similarly, reducing the bleed air off-take from the engine may be welcomed by the engine manufacturer but not by the companies who supply bleed air valves and ducting. The deletion of many hydraulic services would clearly run counter to the interests of the manufacturers of hydraulic equipment. The removal of bleed air also raises the need to examine alternative methods of starting the engine and of providing the aircraft anti-icing services, cabin conditioning and cabin pressurization services which engine bleed air presently supplies. Furthermore, there are some who argue that the flight control actuation requirements cannot be met other than by the use of hydraulically activated flight control actuators: the development of EHAs would now appear to counter this argument.

In many subsystem areas, technology is moving into a phase where many of the methods of implementing aircraft subsystems may be engineered in differing, more energy-efficient ways than those that have prevailed for the past 40 or 50 years. It will be commercial and competitive pressures which will ultimately dictate which changes are adopted and when. A brief summary of some of the subsystem development efforts to support the MEA concept follows; many of these are approaching demonstration status.

More-electric engine/all-electric engine

Oil-less engine

The engine oil system is complex on many engines, usually comprising a number of oil pumps, filter assemblies, coolers etc. The generation/conversion losses from the aircraft electrical generators reject heat into the engine. Great savings could be made if the oil system could be replaced with an alternative form of supporting the rotating engine assemblies. Electromagnetic bearing technology has been demonstrated on both sides of the Atlantic. However in order to be totally practicable, additional technologies have to be developed which permit the removal of the accessory gearbox and its associated power off-takes from the engine.

IGV/VSV control

Many engines use Variable Inlet-Guide Vanes (VIGVs), and Variable Stator Vanes (VSVs), to control the air flow into the engine central core. These may be variously powered by hydraulics, pneumatics (bleed air) or by fueldraulic means where pressurized fuel is used as a source of hydraulic power. Programmes are under way to examine the feasibility of using electrical actuation techniques to replace the fluidic power media.

Distributed engine control

Present primary engine control is by means of a Full-Authority Digital Engine Control (FADEC) which is normally located on the engine fan casing. However there are many features of engine control which are distributed around the engine – such as reverse thrust, presently pneumatically actuated – which would need to be actuated by alternative means in a more-electric engine. This leads to the possibility of using distributed engine control.

Electrically driven fuel pump

Engine fuel is pressurized by means of a shaft-driven High-Pressure (HP) pump. The pump control is not refined and much energy is wasted in the pumping process. In an all-electric engine this shaft would be removed and the engine HP pump electrically driven.

Electric brakes

The aircraft braking system is a key system that is hydraulically actuated in present aircraft. The hydraulic fluid can cause a problem when it is in the proximity of hot brake pads as fires can result. The electric brake replaces the hydraulic brake actuators with Electro Mechanical Actuators (EMAs) controlled by dual channel computation and actuator drive electronics. Such systems have been demonstrated on a rolling dynamometer but not yet on an aircraft.

Power generation and conversion

Although the primary power generation methods have been described, other generator types and associated power conversion may be required in future for novel solutions. Rolls-Royce have proposed shaft-mounted starter generators as means of generating aircraft level electrical power, starting the engine and interchanging energy between shafts on a triple-shaft civil engine. Other concepts include driving PM machines off the LP fan where emergency power could be extracted when the engine is windmilling. These techniques will need new power electronics solutions to convert the electrical power into a useable form, particularly if the electronics is to be engine-mounted in a harsh environment.

FBW/power-by-wire

A number of flight control actuator possibilities are being pursued:
- Switched reluctance and permanent magnet actuators for civil aircraft spoiler systems.
- AFTI F-16 J/IST 270 VDC dual tandem primary flight control actuators as will be described later.

Power distribution and load management

The increasing levels of power which the more-electric aircraft will require will place an increasing demand on the power distribution and load management systems. Furthermore, as more of the flight critical systems such as the flight control system and the engine become more-electric, so the levels of integrity required of the electrical power system will also increase. Levels of equipment reliability and system availability will be increased to meet the higher goals.

The UK Department of Trade & Industry (DTI) has been part-funding many of these initiatives as part of a programme called MEA challenge. Many of the topics summarized above are addressed by papers in references (11), (12), and (13).

V-22 tilt rotor system

The tilt rotor concept as demonstrated by the Bell/Boeing V-22 is a concept that has been under development for a number of years with a view to combining the most advantageous characteristics of both fixed- and rotary-wing aircraft. Fixed-wing aircraft are efficient in the cruise configuration when the propulsion and aerodynamic configurations are operating somewhere near to optimum conditions. The disadvantage of fixed-wing aircraft is that their flight characteristics are far from optimum during the low-speed take-off and landing. Special provisions such as flaps, slats and other similar high lift devices are required to reduce aircraft speed to acceptable levels during these critical stages of flight.

Helicopters suffer from the reverse problem, being extremely effective for low- or zero-speed vertical landing and take-off but very limited for high-speed cruise, due to rotor tip stall and other features of the rotary wing. Helicopters and other vertical take-off aircraft require a very high ratio of thrust to weight; generally reckoned to exceed 1.25:1 for effective vertical take-off and landing. It therefore follows that a vehicle capable of exploiting both the characteristics of vertical take-off and landing and conventional flight should have a lot of advantages to offer even though it might not compete with the optimum machine designed for either regime. While aircraft such as the BAE SYSTEMS Harrier and the Soviet Yakovlev YAK-36 (Forger) aircraft represent solutions to the problem when approached from the 'conventional' viewpoint, they both have limitations in that they are highly specialized military aircraft which can offer little possibility of adoption in the commercial aircraft arena.

A potential solution to this problem – starting from a helicopter baseline – was initially explored by the Bell Helicopter Company as long ago as 1944. It resulted in the development of the Bell XV-3 which first flew in 1953. The principle employed was that of the tilt rotor with twin engines located at the extremities of a conventional wing. During take-off the rotors were positioned with axes such that the aircraft operated as a conventional helicopter, albeit a twin-rotor helicopter. As the aircraft transitioned into the cruise the rotor tilted forward until eventually both rotors acted as conventional propellers or airscrews pulling the aircraft forward in the normal way. For landing the situation was reversed with the rotors being tilted aft until the aircraft was flying in the helicopter mode once again. The XV-3 was powered by a single 450 hp piston engine which transmitted power to the rotors via a complex mechanical arrangement. In 1956 the aircraft suffered a serious crash which halted the development. It appears that the fundamental problem was a lack of structural rigidity due to rotor pylon coupling which led to a catastrophic failure while in the hover. Nevertheless the Bell XV-3 flew and demonstrated the concept of the tilt rotor with the transition to and from the contrasting conventional and helicopter modes of flight. Bell was followed by the Vought Corporation (later Ling Temco Vought or LTV) who produced and flew several prototypes of the XC-142. This aircraft was a multiple-engined machine which was not fail-safe in the hover mode and was also very complex mechanically – this programme was eventually terminated.

However, interest persisted in the NASA organization and eventually the XV-15 programme was initiated in 1971 with Bell selected in preference to Boeing Vertol; the twin-engined Bell XV-15 first flew in 1977 and executed a successful flight test programme including the demonstration of an engine-out capability. This aircraft benefited from turbine engines rated as 1,550 shp with a higher thrust-to-weight ratio than had been possible on the XV-3.

Eventually, after many problems of funding, the concept was revived and the programme got fully under way in 1985 with US Navy and US Marine sponsorship as the V-22 Osprey. The US Army and US Air Force also showed an interest in limited quantities of the aircraft for specialized 'special forces' roles. At the time of the programme launch in 1985 the joint Bell/Boeing Vertol team saw prospects for the production of over 1,000 aircraft for all four major US Services with the Marines being by far the largest customer with the MV-22 variant.

The V-22 configuration is shown in Fig. 10.5. The aircraft is powered by two 6,000 shp Allison T-406 turbine engines each of which is contained within the tilting nacelles. It is interesting to note that each engine/nacelle combination weighs about 5,000 lb which

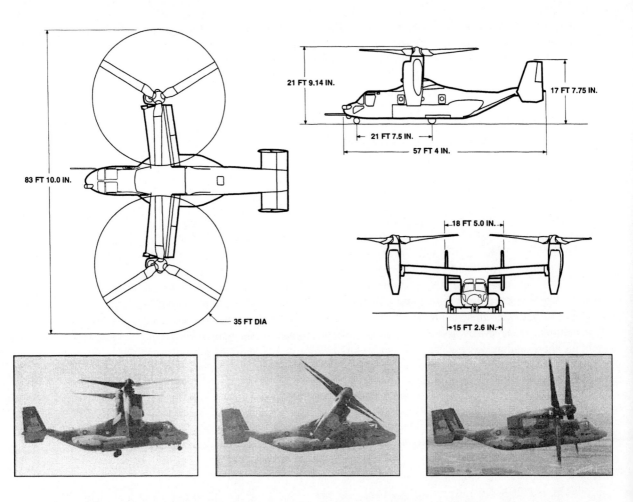

Fig. 10.5 *V-22 Osprey configuration (Boeing Vertol)*

is almost the same as the total weight of the original Bell XV-3. The total production aircraft weight is in the region of 32,000 lb empty. To minimize structural weight and maximize the payload the aircraft makes extensive use of composite materials.

Of particular interest on the V-22 are the propulsion drive and fuel and flight control systems. Figure 10.6 shows the mechanical drive system interconnection between the nacelles. Each engine drives a prop rotor gearbox located in the nacelles from which each rotor is driven in the opposing direction to the other thereby counter balancing torque effects. Each prop rotor gearbox also drives through a tilt axis gearbox and mechanical linkage running through the wing to a midwing gearbox. This effectively interconnects the two systems and also acts as the main aircraft accessory gearbox driving the constant frequency AC generators (two per aircraft, each rated at 40 kVA) and variable frequency AC generators (two per aircraft, each rated at 50/80 kVA) and 5,000 psi hydraulic systems pumps. A recent innovation is the adoption of two 40 kVA VSCF cycloconverters in place of the present IDGs for the constant frequency generation system. The APU also has the capability of driving the accessory gearbox which also drives the environmental system compressor.

The V-22 fuel system is much more complex than most helicopter systems reflecting the various ambitious mission scenarios. A number of possible configurations exist.

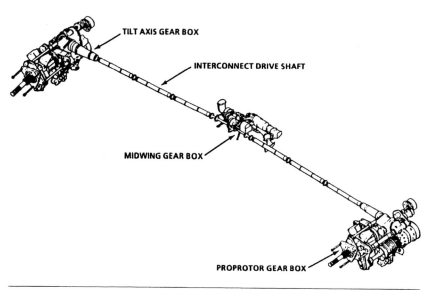

Fig. 10.6 *V-22 drive system (Bell Boeing V-22 Tiltrotor Team)*

TILT AXIS GEAR BOX

INTERCONNECT DRIVE SHAFT

MIDWING GEAR BOX

PROPROTOR GEAR BOX

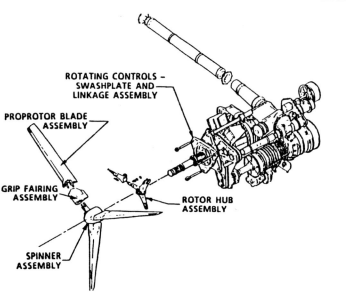

ROTATING CONTROLS –
SWASHPLATE AND
LINKAGE ASSEMBLY

PROPROTOR BLADE
ASSEMBLY

GRIP FAIRING
ASSEMBLY

ROTOR HUB
ASSEMBLY

SPINNER
ASSEMBLY

Fuel is carried in tanks in the forward sections of the left and right sponsons (lower fuselage fairings) and in feed tanks just inboard of the nacelles. The capacity of the sponson tanks and feed tanks is 3,155 lb and 675 lb respectively, giving a total of 7,660 lb fuel for the basic version. Additional fuel may be carried in a right aft sponson tank for some variants; this tank contains a further 2,040 lb. For specific variants up to four internal wing tanks or auxiliary tanks may be fitted adding a further 2,000 lb per wing. Finally, for long range ferry flights cabin mounted rigid ferry tanks may be fitted which add an additional tankage of around 16,370 lb. If all the internal tanks are fitted, the total fuel capacity is in the region of 30,000 lb. See Fig. 10.7 for an outline of the tank configurations. To control this system (excluding the ferry tanks) requires a total of 17 motorized valves and 9 fuel pumps and a miscellany of other valves. This is more in line with the complexity of a high-performance aircraft than a normal fixed-rotor helicopter.

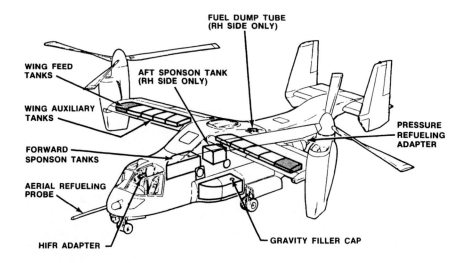

Fig. 10.7 *V-22 Osprey fuel system (Bell Boeing V-22 Tiltrotor Team)*

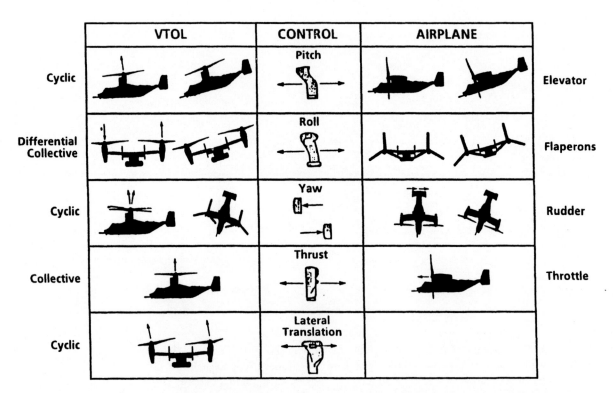

Fig. 10.8 *V-22 flight control modes (Bell Boeing V-22 Tiltrotor Team)*

The flight control system has effectively to control two different modes of flight and the transition between them. Figure 10.8 shows the different flight control modes for the V-22 in the VTOL and aeroplane modes. VTOL control modes are shown in the left column and aeroplane control modes in the right column. In the VTOL condition, power and cyclic and collective pitch are used as for a conventional helicopter except that differential cyclic pitch provides aircraft roll and differential longitudinal cyclic pitch provides aircraft yaw. In the aeroplane mode of flight-pitch, roll and yaw are provided by elevators, flaperons, and rudders respectively. At an appropriate point in the nacelle tilt operation the vertical flight

Fig. 10.9 *V-22 rotor and wing stowage sequence (Bell Boeing V-22 Tiltrotor Team)*

control functions are 'washed out' and the aircraft is established in the aeroplane mode. The flight control computations are provided by a triple-redundant all digital fly-by-wire system developed by General Electric. See reference (14).

A further interesting and probably unique feature of the V-22 is the rotor and wing stowage facility. The need to stow the aircraft onboard aircraft carriers and amphibious assault ships dictates severe stowage constraints. The rotor/wing stowage occurs in the sequence shown in Fig. 10.9. First the rotor blades are folded inboard to align with the wing. Then the nacelles are tilted forwards to place the rotor blades parallel with the wing leading edges. Finally, the whole wing is rotated 90 degrees clockwise to be positioned along the top of the fuselage. For articles which further detail the V-22 Osprey see references (15) and (16).

Impact of stealth design

Over the past ten years or so the term 'stealth' has become a common expression in relation to new combat aircraft programmes, particularly recent developments in the US. The term 'stealth' relates to the ability of an aircraft to remain undetected and hence deny an adversary the opportunity to engage in combat. The main aircraft detection techniques involve the use of radar or infra-red thermal detection principles. It follows that stealth techniques aim to reduce radar and infra-red 'signature' emissions from the aircraft; this being what the use of stealth, or 'low observability' is all about. Though not totally new in principle, a range of new military aircraft developments by the US has, in recent years, given further impetus to the application of stealth techniques, to the point where military aircraft design, construction and manufacture, and operations are ruled by the stealth or low-observability requirements.

This principle is perhaps best illustrated by a simple example. The radar range equation governs the parameters which dictate the distance at which an aircraft will be detected. One of the key factors is the reflecting area of the target or aircraft. Typically for a fighter aircraft a radar reflecting area may be of the order of 10 m². For a stealth aircraft it may be assumed that this is reduced to 0.1 m² – that is reduced by a factor of 100. The range at which an aircraft may be detected is proportional to the fourth root of the radar reflecting area. The fourth root of 100 is 3.16 and therefore the maximum detection range would have been reduced by this value. A radar previously able to detect a conventional target at 158 miles would now only be able to detect a stealthy target at 158/3.16 or 50 mi. Detail of precisely how small the radar signature can be made is highly classified and it is likely to be much smaller than that given in the example. If the equivalent radar area were reduced by 10,000 rather than the factor of 100 used above, then the radar range would be reduced by a factor of 10 rather than 3.16 and the detection range would be reduced to 15.8 miles which would mean that the aircraft would be detected almost too late to engage successfully. The difficulty in rear aspect radar detection is almost certainly linked to the reduction in infra-red or IR signature upon which many missile terminal guidance systems are based. The combination of significant reductions of both radar and IR signatures must make a stealthy aircraft very difficult to detect and engage by conventional means, herein lies the attraction.

The suppression of these two signatures has an impact upon aircraft design in the following areas:

- Most aircraft reflections are from the engine intake and exhausts and therefore considerable efforts may be expended to avoid these orifices acting as radar reflectors.

Fig. 10.10(a) *F-117A configuration (US Air Force)*

- Intakes and jet pipes apart, angular corners or large plane reflecting surfaces should be avoided. Even straight edges such as wing leading or trailing edges may increase the reflecting area for some aircraft aspects.
- Aircraft metal skins offer a good reflecting surface for radar emissions and the use of radar absorbent materials may also be considered.
- To suppress the aircraft IR signature, efforts may be made to reduce the temperature of the jet plume issuing from the jet pipes by shielding the emissions or by diffusing cooler air into the jet exhaust to reduce the temperature.

None of the techniques outlined above may be applied without accompanying penalties and it is interesting to contrast the differing stealth designs flying today as solutions to the problem, though the relative performance gains or losses must be purely a matter for speculation. The aircraft currently known today are:

- Lockheed/Martin F-117A stealth fighter.
- Northrop B-2 stealth bomber.
- Lockheed/Martin F-22 Raptor.
- Other aircraft embodying stealth characteristics are the RAH-66 Comanche and competing X-32 and X-35 JSF designs. The Lockheed SR-71 Blackbird made considerable use of stealth techniques.

Fig. 10.10(b) *F-117A engine exhaust ducts (US Air Force)*

Fig. 10.10(c) *F-117A in-flight refuelling (US Air Force)*

Lockheed F-117A stealth fighter

The F-117A programme was commenced in 1978 and the aircraft first flew in 1981, though the US Air Force did not admit to its existence until November 1988 when the aircraft had already entered service. The general planform of the aircraft is depicted in Fig. 10.10(a) from which it can be seen that it has a highly angular almost prismatic construction comprising relatively few facets; the wings and fins are highly swept such that any incident radar energy which is reflected does not scatter in an organized fashion. The relatively simple polyhedron approach of the F-117A was presumably easier to model during early assessment of the low-observability features of the design. It is also believed that the planar facets would have facilitated aircraft manufacture using radar absorbent material.

The aircraft is of subsonic performance, powered by the same General Electric 404 engines used on the McDonnell Douglas F/A-18 Hornet though no reheat is provided for the F-117A. The aircraft uses the same 40/45 kVA VSCF cycloconverter that was used on the F/A-18E/F. The engine air inlets are covered with grilles, supposedly using composite materials for a 0.75 x 1.25 in mesh which prevents any reflections from the engine inlet turbine blades. The engine exhaust is diffused with cool air after exiting the engine and is spread by vanes to exhaust through wide shallow apertures across the entire inboard trailing edges of both wings – Fig. 10.10(b). The aircraft has a fly-by-wire control system though it is not known whether the aircraft is dynamically unstable. It is more likely that the fly-by-wire system is employed primarily to reduce weight and improve handling qualities.

Weapons are carried internally to preserve the low radar signature as is the case on all other stealth aircraft. Otherwise the aircraft systems are believed to be relatively conventional, some being purloined from other aircraft. The fuel system is certainly conventional if the in-flight refuelling photographs are anything to judge by – see Fig. 10.10(c). The aircraft was used operationally during the US intervention in Panama in 1990 and a number of aircraft were deployed to the Gulf in 1990 as part of the US response to that crisis. All the reports of the performance of the F-117A during Desert Storm suggest that the aircraft was extremely effective in terms of stealth and as a weapon delivery platform. A total of 59 aircraft were built under the US Force F-117A procurement contract.

Northrop B-2 stealth bomber

The B-2 stealth bomber programme was publicly acknowledged before the US Air Force finally lifted the security veil in November 1988 at the aircraft roll-out. It is produced by Northrop with the Boeing Company as a major subcontractor. The flying wing design had been anticipated; however, what was unexpected was the angular wing platform with totally straight leading edge and the now customary zig-zag trailing edge. The aircraft also differed considerably from the previously unveiled F-117A in the degree of smooth fuselage/wing blended contours that are in stark contrast to the stealth fighter's polyhedral, planar faceted features. See Fig. 10.11.

The aircraft owes its pedigree to the Northrop flying wing designs of the immediate post-war era. One of them, the Northrop YB-49, was developed to the stage of having two flying prototypes. One crashed and the other was destroyed on take-off; the main difficulty being that of maintaining longitudinal stability. It is virtually certain that the B-2 uses a quadruplex computer-controlled fly-by-wire flight control system to provide stability. Unlike the F-117A the B-2 bomber is smoothly contoured with blended wing fuselage so that there are no abrupt changes of form. This probably offers a better or lower radar signature than the F-117A though it is probably correspondingly more difficult to manufacture. It has been reported in the aviation press that the prototypes have been manufactured to very precise production tooling standards and this may be a prerequisite to the smooth contouring of the aircraft.

Fig. 10.11 B-2 stealth bomber planform (Northrop Corporation)

The aircraft is controlled entirely by flying control surfaces along the wing trailing edge. Yaw is controlled by means of split ailerons on the outboard section of each wing. These have upper and lower surfaces that may be opened independently like air brakes. Differential operation of the split ailerons allows differential drag to be applied to the aircraft allowing control in yaw. See Fig. 10.12. The centre rear portion of the fuselage, called the 'beaver's tail', is also believed to move vertically in a limited fashion and may permit trimming of the aircraft in pitch. The engine intakes and exhausts are situated on the upper surface of the wing where they are shielded from ground-based radars. Most of the fuel is believed to be carried in the outboard sections of the wing. The aircraft indicated a conventional in-flight refuelling capability at an early stage in the flight test programme as shown in Fig. 10.12. The centre and inboard wing sections house the engines, intakes and exhausts and the internal weapon bay as the B-2 carries its weapons internally in common with the other stealth aircraft. During a much-publicized fault at an early stage in the flight test programme it was revealed that the aircraft was experiencing oil leaks from the AMAD – aircraft-mounted accessory drives or gearboxes. This suggests that the aircraft is fairly conventional in terms of hydraulic and electrical systems.

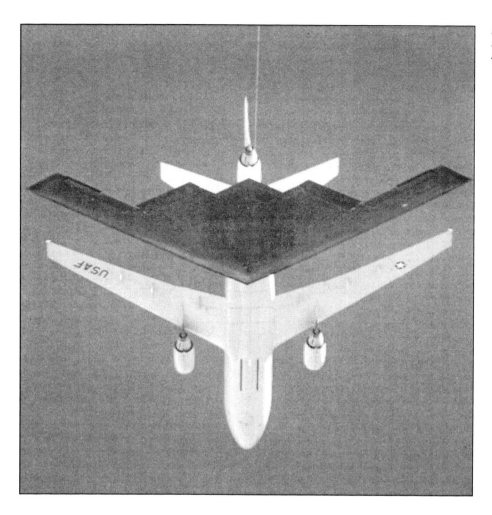

Fig. 10.12 *B-2 refuelling in-flight (US Air Force)*

The B-2 has been the subject of intense political debate due to the high programme costs and extremely high unit production costs of several hundred million dollars per aircraft. Congress finally permitted production of 15 aircraft as opposed to the 132 that the US Air Force originally wished to purchase. Therefore the B-2 is unlikely to make the intended major contribution to the air-launched portion of the deterrent. In recent years the aircraft has been demonstrated or has over-flown a number of international Air Shows in Singapore and the UK. In both cases the aircraft flew directly to the location from its operating base in the central US. In 1999 B-2 bombers were deployed directly from the US to bomb Yugoslavia using precision-guided munitions during the Kosovo crisis.

Joint Strike Fighter (JSF)

The latest fighter aircraft development programme is the Joint Strike Fighter (JSF) in which two competing teams are developing flying demonstrators to prove the respective technologies and operating concepts. This phase is termed the Concept Demonstration Aircraft (CDA). The competing aircraft are:

Fig. 10.13 *Boeing
JSF (X-32)
(JSF Program Office)*

- Boeing JSF Team – X-32
- Lockheed/Martin JSF Team – X-35

Both aircraft are intended to address the requirements of the following customers:

- US Air Force. The US Air Force or Conventional Take-Off and Landing (CTOL) has a fairly conventional set of requirements which include internal and external weapons carriage and a multi-role supersonic capability.
- US Navy. The Carrier Vehicle (CV) has similar characteristics to the Air Force version but requires additional structural strength to accommodate the additional stresses associated with deck landings. Other key requirements are identical to those of the Air Force.
- US Marines. The Marines version has similar requirements to the Air Force and Navy variants but mandates a Short Take-Off and Vertical Landing (STOVL) capability. This leads to the need for a direct lift propulsion system.
- UK Royal Navy. The Royal Navy requirement is directly equivalent to that for the US Marines.

In addition, as the JSF is intended as a possible replacement for F-16, AV-8B, Sea Harrier/Harrier and other present front-line fighter aircraft, other nations have observer status in the programme, being granted access to briefings and status updates as the programme develops. The competing designs both flew during the autumn of 2000 and flight demonstration will allow the two configurations to be evaluated as has been the case previously with the YF-16/YF-17 and the YF-22A/YF-23A on previous fighter selection or Demonstration/Evaluation (Dem/Val) programmes.

The engine system selected for the JSF is a derivative of the Pratt & Whitney F119 which is the engine well into development for the F-22 Raptor and has several thousand hours of ground test experience plus the flying experience gathered so far in the F-22 flight test programme. Demonstration engines for both teams were successfully run in

Fig. 10.14(a)
*Lockheed Martin JSF
configuration (X-35)
(JSF Program Office)*

the middle of 1998.

As far as the independent observer is concerned the requirements of the four sponsoring Services appear to be diametrically opposing in terms of achieving a final solution. Nevertheless, if a high degree of commonality can be maintained between the competing variants and/or requirements then the US military authorities will have achieved a degree of standardization which will doubtless yield significant benefits: both to the operational Services and the taxpayer on both sides of the Atlantic.

Boeing X-32 configuration

The Boeing team, comprising the former McDonnell Douglas company based primarily at St Louis, have a wealth of carrier-based fighter aircraft experience based upon relatively recent F/A-18 variants and AV-8B – not to mention the long pedigree of previous fighter aircraft. The X-32 configuration is shown in Fig. 10.13.

From an aircraft system viewpoint the X-32 is believed to be relatively conservative using variable frequency three-phase 115 VAC electrical power with conventionally powered hydromechanical flight controls.

Lockheed Martin X-35 configuration

The pedigree of the Lockheed team is based upon the highly successful Lockheed F-117A stealth fighter and (General Dynamics) F-16, of which over 4,000 aircraft have been produced. The X-35 is very similar to its F-22 stable-mate in technology terms. The X-35 is shown in Fig. 10.14.

The X-35, judging from the available literature is designed to a more ambitious technology baseline, as will be deduced from the thrust of some of the associated J/IST technology programmes. The X-35 would appear to be following the 270 VDC

Fig. 10.14(b) X-35
concept demonstrator
aircraft after assembly
of major structural
components (BAE
SYSTEMS)

electrical system lineage established by the stable-mate F-22 and building upon that experience. From the FBW/power-by-wire viewpoint the approach appears to be inclining toward a 270 VDC/EHA architecture as envisaged within the J/IST programme and being flight demonstrated on the AFTI F-16.

Technology developments/demonstrators

Supporting the JSF flight demonstration programme is the JSF Integrated Subsystems Technology (J/IST) demonstrator program. Key among the aircraft systems related demonstrations are:

- Fault-tolerant 270 VDC electrical power generation system.
- Thermal and Energy Management Module (T/EMM).
- AFTI F-16 EHA demonstration.

Fault-tolerant 270 VDC electrical power generation system

The J/IST electrical power generation and distribution system as fitted to the NASA Dryden Advanced Fighter Technology Integration (AFTI) F-16 is based upon a 270 VDC 80 kW switched reluctance starter/generator incorporating a dual channel

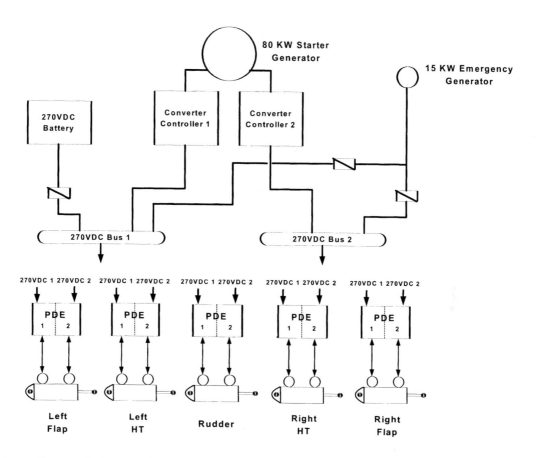

converter/controller supplied by Sundstrand. The aircraft also has a 270 VDC 15 kW emergency generator. This system provides flight critical power by means of two independent 270 VDC aircraft buses as shown in Fig. 10.15. Each 270 VDC bus feeds one-half of a power drive electronics unit (PDE) of which there is one per primary flight control surface. The PDE in turn controls one-half of the Parker Aerospace dual-tandem 270 VDC EHA.

Fig. 10.15 AFTI F-16 simplified 270VDC system

Five main flight control actuators so powered are:

- Left flaperon
- Right flaperon
- Left horizontal tail
- Right horizontal tail
- Rudder

Thermal and Energy Management Module

The Thermal and Energy Management Module (T/EMM) combines the function of a traditional APU, emergency power unit and environmental control system. This allows the conventional AMAD to be removed as is the aircraft central hydraulics system. The engine fan duct air is used as the heat sink thereby removing the usual heat exchangers and associated ducting. Extensive of the Honeywell (Allied Signal) supplied T/EMM has been undergoing rig testing prior to engine and T/EMM integration in early 2000.

Fig. 10.16 Simplified
schematic of J/IST
dual-tandem actuator

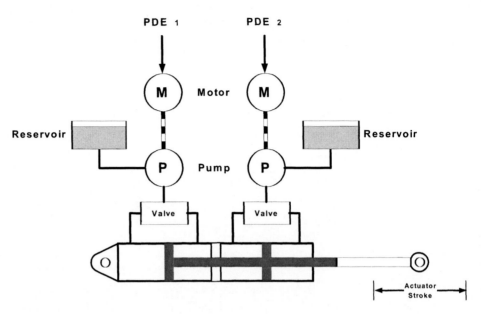

Fig. 10.16 Simplified schematic of J/IST dual-tandem actuator

AFTI F-16 flight demonstration

The AFTI F-16 is the flight test bed for the flying elements of the J/IST demonstration programme. The aircraft has been modified to accommodate the 270 VDC architecture shown in Fig. 10.15. The five PDEs each drive a dual-tandem actuator supplied by Parker Aerospace; one for each flight control surface as already mentioned. PDE channels 1 and 2 each drive a brushless DC motor which in turn powers half of the actuator package; a PDE channel performs loop closure around its respective components. Each half of the actuator comprises a motor, pump, local fluid reservoir and a valve assembly. As the name suggests, normally both channels operate in tandem. The valve assemblies ensure that each channel can drive the actuator ram if the other channel fails. In the event of both channels failing, aerodynamic pressure drives the control surface to central position where it becomes hydraulically locked. A simplified schematic of the dual-tandem actuator is at Fig. 10.16.

The control side of the implementation posed the problem of interfacing the existing quadruple-redundant Digital Flight Control Computer (DFCC), with the five new PDE actuator drive packages. This was achieved by the introduction of a new triple-redundant control electronics unit which interfaces the 'old' digital flight control system with the 'new' PDEs and actuators. For a comprehensive overview of the AFTI F-16 system see reference (17). A three-dimensional diagram of the AFTI F-16 is at Fig. 10.17, though the aircraft external appearance yields no clue as to the major systems modifications which are contained therein.

Prognostics

An important system that is emerging from future system studies is that of prognostics. For some while it has been the practice to log failures as they occur in flight to aid rapid detection and repair on the ground. However, increasing demands to reduce support

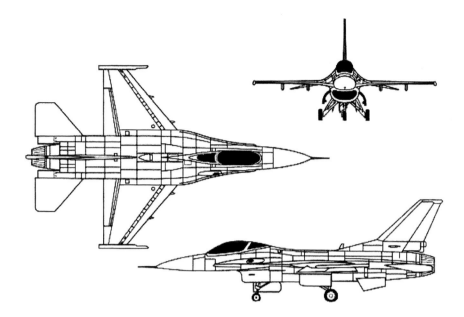

Fig. 10.17 *NASA Dryden AFTI F-16 (NASA)*

costs and improve turn-around times have led to a demand for something more sophisticated – the ability to predict and plan for failures.

The modern aircraft computing architecture contains a wealth of information that characterizes the normal and potentially degrading performance of a system and its components. Knowledge of such information as flow rates, pressures, loss rates and actuator positions and states, number of excursions, and elapsed operating time can be compared with input data of known wear characteristics to form the basis of an analysis of degrading performance.

The introduction of knowledge-based systems and the application of Bayesian statistics allows models to be constructed that can draw inferences from measured performance. This inferential data can be used to predict the time at which system or component performance becomes unacceptable or to estimate time to failure.

This is of importance to operators who can support the aircraft by arranging for maintenance to be performed at preferred maintenance centres before failure occurs; this allows them to plan for parts, tools and staff to be available for a rapid repair. This leads on to a concept of Maintenance-Free Operating Periods (MFOP) as a contractual requirement rather than working to a scheduled maintenance period.

References

(1) **Tuttle, F.L., Kisslinger, R.L.,** and **Ritzema, D.F.** (1990) F-15 S/MTD IFPC *Fault Tolerant Design*, IEE.

(2) **Moir, I.** and **Seabridge, A.G.** (1986) Management of utilities systems in the Experimental Aircraft Programme, *Aerospace*, September.

(3) **Lowry, A., Moir, I.** and **Seabridge, A.G.** (1987) Integration of secondary power systems on the EAP. SEA Aerotech 87, Long Beach, California.

(4) **Cronin, M.J.** (1951) The Development of Electrical System in the Bristol Brabazon Mk1 Institution of Electrical Engineers, London, Great Britain, 5 April.

(5)   **Cronin, M.J.** All Electric Technologies in Future Advanced Aircraft.

(6)   **Cronin, M.J.** The role of Avionics in the All Electric Airplane, *American Institution of Aeronautics and Astronautics*.

(7)   **Cronin, M.J.** (1982) All-Electric vs Conventional Aircraft: The Production/Operational Aspects. *American Institution of Aeronautics and Astronautics*, Long Beach, 12-14 May.

(8)   **Cronin, M.J.** (1983) Advanced Electric Power Systems for All-Electric Aircraft, IEEE.

(9)   **Cronin, M.J.** The Déjà Vu of All Electric/All Digital Aircraft, *AIAA/IEEE Sixth Digital Avionics Systems Conference*.

(10)   **Cronin, M.J.** The All Electric Airplane Revisited. *Society of Automative Engineers*.

(11)   All Electric Aircraft, IEE Colloquium, London, June 1998.

(12)   Electrical Machines and Systems for the More-Electric Aircraft, IEE Colloquium, London, November 1999.

(13)   The More Electric Aircraft and Beyond, I Mech E Conference, May 2000.

(14)   **McManus, B.L.** (1987) V-22 tilt rotor fly-by-wire flight control system. *NAECON*.

(15)   **Mark, H.** (1986) Aircraft without airports. The tilt-rotor concept and VTOL aviation, *75th Wilbur and Orville Wright Lecture, Royal Aeronautical Society*, December.

(16)   **Moxon, J.** (1988) V-22 Osprey changing the way man flies, *Flight International*, 14th May.

(17)   **Schley, W.R.** and **Kotalik, R.J.** Implementation of flightworthy electrical actuators for the F-16, I Mech E Conference. May 2000.

CHAPTER 11

Systems Design and Development

Introduction

As the reader will judge from the contents of this book, aircraft systems are becoming more complex and more sophisticated for a number of technology and performance reasons. In addition, avionics technology, while bringing the benefits of improved control by using digital computing and greatly increased integration by the adoption of digital data buses, is also bringing greater levels of complexity to the development process. The disciplines of avionics system development – including hardware and software integration – are now being applied to virtually every aircraft system.

The increasing level of system sophistication and the increased interrelation of systems is also making the development process more difficult. The ability to capture all of the system requirements and interdependencies between systems has to be established at an early stage in the programme. Safety and integrity analyses have to be undertaken to ensure that the system meets the necessary safety goals, and a variety of other trades studies and analytical activities have to be carried out.

These increasing strictures need to be met by following a set of rules and this chapter gives a brief overview of the regulations, development processes and analyses which are employed in the development of modern aircraft systems; particularly where avionics technology is also extensively employed.

The design of an aircraft system is subject to many rigours and has to satisfy a multitude of requirements derived from specifications and regulations. There are also many development processes to be embraced. The purpose of this chapter is not to document these ad nauseam but to give the reader an appreciation of the depth and breadth of the issues which need to be addressed.

Systems Design

There are references to some of the better known specifications and requirements, but the chapter also attempts to act as a tutorial in terms of giving examples of how the various design techniques and methods are applied. As the complexity and increasing interrelationship and reliance between aircraft systems has progressed it has become necessary to provide a framework of documents for the designer of complex aircraft systems.

Development processes

An overview of a typical life cycle for an aircraft or equipment is given and the various activities described. Further, some of the programme management disciplines are briefly visited.

System design

Key documentation is applied under the auspices of a number of agencies. A list of the major documents which apply are included in the reference section of this chapter and it is not intended to dwell on chapter and verse of those documents in this brief overview. There are several agencies who provide material in the form of regulations, advisory information and design guidelines whereby aircraft and system designers may satisfy mandatory requirements.

Key agencies and documentation

These agencies include:

- Society of Automobile Engineers (SAE):
    ARP 4761 (1)
    ARP 4754 (2)
- Federal Aviation Authority (FAA):
    AC 25.1309-1A (3)
- Joint Airworthiness Authority (JAA):
    AMJ 25.1309 (4)
- Air Transport Association (ATA):
    ATA-100 (5)
- Radio Technical Committee Association (RTCA):
    DO-178b (6)
    DO-xxx (7)

This list should not be regarded as exhaustive but merely indicative of the range of documentation which exists.

Design guidelines and certification techniques

References (1) and (2) offer a useful starting point in understanding the interrelationships of the design and development process:

- Reference (1) – ARP 4761: Safety Assessment Process Guidelines and Methods
- Reference (2) – ARP 4754: System Development Processes
- Def Stan 00-970 for military aircraft

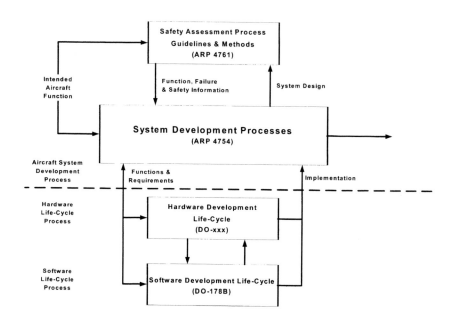

Fig. 11.1 *ARP 4754 system development process*

Figure 11.1 shows the interplay between the major techniques and processes associated with the design and development process.

This figure which is presented as part of the SAE ARP 4761 document gives an overview of the interplay between some of the major references/working documents which apply to the design and development process. In summary:

- ARP 4761 represents a set of tools and techniques
- ARP 4754 is a set of design processes
- DO-xxx will offer guidance for hardware design and development
- DO-178b offers advice for the design and certification of software

Serious students or potential users of this process are advised to procure an updated set of these documents from the appropriate authorities.

Major safety processes

There are a number of interrelated processes that are applied during the safety assessment of an aircraft system. These are:

- Functional Hazard Analysis (FHA)
- Preliminary System Safety Analysis (PSSA)
- System Safety Analysis (SSA)
- Common Cause Analysis (CCA)

Figure 11.2 shows a simplified version of the interplay between these processes as the system design evolves and eventually the system achieves certification.

The diagram effectively splits into two sections: design activities on the left and analysis on the right. As the system evolves from aircraft level requirements, aircraft functions are evolved. These lead in turn to system architectures which in turn define software requirements and the eventual system implementation. At corresponding stages of the design, various analyses are conducted which examine the design in the

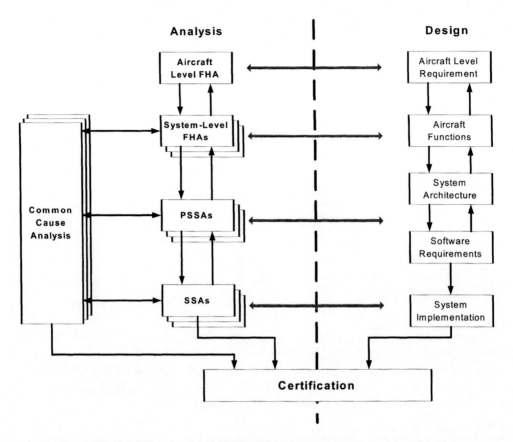

Fig. 11.2 *Simplified portrayal of safety processes*

light of the mandated and recommended practices. At every stage the analyses and the design interact in an evolutionary manner as the design converges upon a solution which is both cost-effective and which meets all the safety requirements.

Functional Hazard Analysis (FHA)

A FHA is carried out at both aircraft and system levels; one flows down from the other. The FHA identifies system failures and identifies the effects of these failures. Failures are tabulated and classified according the effects which that failure may cause and the safety objectives assigned according to the criteria briefly listed in Table 11.1.

The FHA identifies the data in the first two columns of the table: the failure condition classification and the development assurance level. These allow the safety objectives to be assigned for that particular condition and a quantitative probability requirement derived. For a failure which is identified as having a catastrophic effect, the highest assurance level A will be assigned. The system designer will be required to implement fail-safe features in his design and will have to demonstrate by appropriate analysis that the design is capable of meeting or exceeding the probability of failure less than 1×10^{-9} per flight hour. In other words, the particular failure should occur less than once per 1,000,000,000 flight hours or once per 1,000 million flight hours. This category of failure is assigned to systems such as flight controls, structure etc. where a failure could lead to the loss of the aircraft. The vast majority of aircraft systems are categorized at much lower levels where little or no safety concerns apply.

Table 11.1 *Overview of failure classification and safety objectives*

Failure condition classification	Development assurance level	Safety objectives	Safety objectives quantitative requirement (Probability per flight hour)
Catastrophic	A	Required	$< 1 \times 10^{-9}$
Hazardous/ severe major	B	May be required	$< 1 \times 10^{-7}$
Major	C	May be required	$< 1 \times 10^{-5}$
Minor	D	Not required	None
No safety effect	E	Not required	None

A more user-friendly definition quoted in words as used by the Civil Airworthiness Authority (CAA) may be:

Catastrophic: less than 1×10^{-9}; extremely improbable

Hazardous: between 1×10^{-9} and 1×10^{-7}; extremely remote

Major: between 1×10^{-7} and 1×10^{-5}; remote

Minor: between 1×10^{-5} and 1×10^{-3}; reasonably probable
    greater than 1×10^{-3}; frequent

Preliminary System Safety Analysis (PSSA)

The PSSA examines the failure conditions established by the FHA(s) and demonstrates how the system design will meet the specified requirements. Various techniques such as fault tree analysis (FTA), Markov diagrams etc. may be used to identify how the design counters the effects of various failures and may point toward design strategies which need to be incorporated in the system design to meet the safety requirements. Typical analyses may include the identification of system redundancy requirements (how many channels?), what control strategies could be employed and the need for dissimilarity of control; e.g. dissimilar hardware and/or dissimilar software implementation. The PSSA is therefore part of an iterative process which scrutinizes the system design and assists the system designers in ascribing and meeting risk budgets across one or a number of systems. Increasingly, given the high degree of integration and interrelationship between major aircraft systems this is likely to be a multi-system, multi-disciplinary exercise co-ordinating the input of many systems specialists.

System Safety Analysis (SSA)

The SSA is a systematic and comprehensive evaluation of the system design using similar techniques to those employed during the PSSA activities. However whereas the PSSA identifies the requirements, the SSA is intended to verify that the proposed design does in fact meet the specified requirements as identified during the FHA and PSSA analyses conducted previously. As may be seen in the early Fig. 11.2, the SSA occurs at the point in the design cycle where the system implementation is concluded or finalized and prior to system certification.

Common Cause Analysis (CCA)

The CCA begins concurrently with the system FHA and is interactive with this activity and subsequent PSSA and SSA analyses. The purpose of the CCA is – as the name suggests – to identify common cause or common mode failures in the proposed design and assist in directing the designers toward strategies which will obviate the possibility of such failures. Such common cause failures may include:

- Failure to correctly identify the requirement
- Failure to correctly specify the system
- Hardware design errors
- Component failures
- Software design and implementation errors
- Software tool deficiencies
- Maintenance errors
- Operational errors

The CCA is therefore intended to scrutinize a far wider range of issues than the system hardware or software process. Rather it is meant to embrace the whole process of developing, certifying, operating and maintaining the system throughout the life cycle.

Requirements capture

It can be seen from the foregoing that requirements capture is a key activity in identifying and quantifying all the necessary strands of information which contribute to a complete and coherent system design. There are a number of ways in which the requirements capture may be addressed. Two main methods are commonly used:

- Top-down approach
- Bottom-up approach

Top-down approach

The top-down approach is shown in Fig. 11.3. This represents a classical way of tackling the requirements capture by decomposing the system requirements into smaller functional modules. These functional modules may be further decomposed into functional sub-modules. This approach tends to be suited to the decomposition of large software tasks where overall requirements may be flowed down into smaller functional software tasks or modules. This would apply to a task where the hardware boundaries are fairly well understood or inferred by the overall system requirement. An example might be the definition of the requirements for an avionics system such as a Flight Management System (FMS). In such a system basic requirement – the need to improve the navigation function is well understood – and the means by which the various navigation modes are implemented: INS, GPS, VOR, etc. are well defined.

Bottom-up approach

The bottom-up method is shown in Fig. 11.4. The bottom-up approach is best applied to systems where some of the lower level functions may be well understood and documented and represented by a number of sub-modules. However, the process of

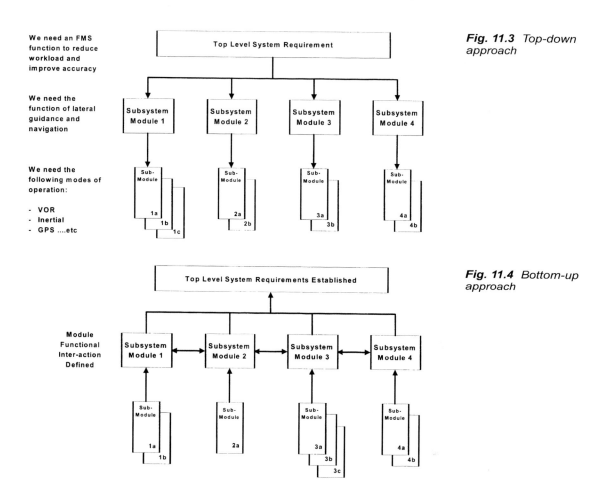

Fig. 11.3 *Top-down approach*

Fig. 11.4 *Bottom-up approach*

integrating these modules into a higher subset presents difficulties as the interaction between the individual subsystems is not fully understood. In this case building up the top level requirements from the bottom may well enable the requirements to be fully captured. An example of this type might be the integration of aircraft systems into an integrated utilities management system. In this case the individual requirements of the fuel system, hydraulic system, environmental control system etc. may be well understood. However, the interrelationships between the candidate systems and the implications of adopting integration may be better understood and documented by working bottom up.

In fact most development projects may use a combination of both of these approaches to best capture the requirements.

Requirements capture example

The example given in Fig. 11.5 shows a functional mapping process which identifies the elements or threads necessary to implement a fuel jettison function. Two main functional subsystems are involved: the fuel quantity measurement function and the fuel management function. Note that this technique merely identifies the data threads which are necessary to perform the system function. No attempt is made at this stage to ascribe particular functions to particular hardware or software entities. Neither

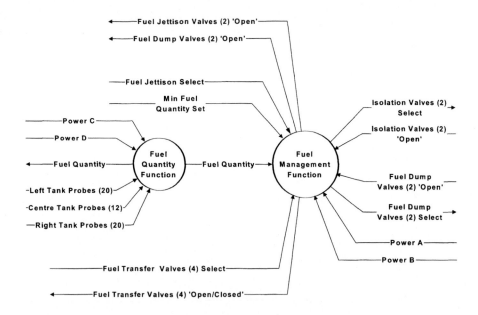

Fig. 11.5 *System requirements capture example*

is any attempt made to determine whether signals are hard-wired or whether they may be transmitted as multiplexed data as part of an aircraft system data bus network. The system requirements from the flight crew perspective are:

- The flight crew need to jettison excess fuel in an emergency situation in order that the aircraft may land under the maximum landing weight.
- The flight crew wish to be able to jettison down to a pre-selected fuel quantity.
- The crew wish to be given indications that fuel jettison is under way.

The information threads associated with the flight crew requirements are shown in the upper centre portion of the diagram. It may be seen that although the system requirements are relatively simple when stated from the flight crew viewpoint, many other subsystem information strands have to be considered to achieve a comprehensive system design.

Fuel quantity function

The fuel quantity function measures the aircraft fuel quantity by sensing fuel in the aircraft fuel tanks; in this example a total of 52 probes are required to sense the fuel held in three tanks. The fuel quantity calculations measure the amount of fuel which the aircraft has on-board taking account of fuel density and temperature. It is usual in this system, as in many others, to have dual power supply inputs to the fuel quantity function to assure availability in the event of an aircraft electrical system busbar failure. Finally, when the calculations have been completed they are passed to the flight deck where the aircraft fuel quantity is available for display to the flight crew. Fuel quantity is also relayed to the fuel management function so that in the event of fuel jettison, the amount of fuel on-board may be compared with the pre-set jettison value. The fuel quantity function interfaces to:

- The fuel quantity system measurement probes and sensors
- The flight deck multi-function displays

- The fuel management system
- The aircraft electrical system

Fuel management function

The fuel management function accepts information regarding the aircraft fuel state from the fuel quantity function. The flight crew inputs a 'fuel jettison select' command and the minimum fuel quantity which the crew wishes to have available at the end of fuel jettison. The fuel management function accepts flight crew commands for the fuel transfer valves (4), fuel dump (jettison) valves (2), and fuel isolation valves (2). It also provides 'open'/'closed' status information on the fuel system valves to the flight crew. As before two separate power inputs are received from the aircraft electrical system. The fuel management function interfaces with:

- The fuel system valves
- The flight deck multi-function displays and overhead panel
- The fuel quantity function
- The aircraft electrical system

This example shows how a relatively simple function interfaces to various aircraft systems and illustrates some of the difficulties which exist in correctly capturing system requirements in a modern integrated aircraft system.

Fault tree analysis

The Fault Tree Analysis (FTA) is one of the tools described in SAE document ARP 4761 (Reference (1)). This analysis technique uses probability to assess whether a particular system configuration or architecture will meet the mandated requirements. For example: assume that the total loss of aircraft electrical power on-board an aircraft has catastrophic failure consequences as identified by the functional hazard analysis – see Fig. 11.2 and Table 1 above. Then the safety objective quantitative requirement established by FAA/JAA 25.1309 and as amplified in ARP 4754 will be such that this event cannot occur with a probability of greater than 1×10^{-9} per flight hour (or once per 1000 million flight hours). The ability of a system design to meet these requirements is established by a FTA which uses the following probability techniques. In the example it is assumed:

- That the aircraft has two independent electrical power generation systems, the main components of which are the generator and the Generator Control Unit (GCU) which governs voltage regulation and system protection.
- The aircraft has an independent emergency system such as a Ram Air Turbine (RAT).
- That the failure rates of these components may be established and agreed due to the availability of in-service component reliability data or sound engineering rationale which will provide a figure acceptable to the certification authorities.

The FTA analysis – very much simplified – for this example is shown in Fig. 11.6. Starting in the bottom left-hand portion of the diagram: the mean time between failure (MTBF) of a generator is 2,000 h – this means that the failure rate of Generator 1 is 1/2000 or 5.0×10^{-4} per flight hour. Similarly if the MTBF of the generator controller GCU 1 is 5,000 h then the failure rate of GCU 1 is 1/5,000 or 2.0×10^{-4} per flight hour.

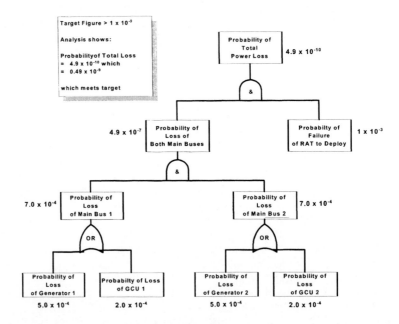

Fig 11.6 *Simplified FTA for an aircraft electrical power system*

The combined failure rate gives the probability of loss of electrical power to Main Bus 1. This is calculated by summing the failure rates of generator and controller as either failing will cause the loss of Main Bus 1:

$$= 5.0 \times 10^{-4} + 2.0 \times 10^{-4} = 7 \times 10^{-4} \text{ per flight hour}$$
$$\quad \text{(Generator 1)} \quad \text{(GCU 1)} \quad \text{(Main Bus 1)}$$

Similarly, assuming generator channels 1 and 2 are identical the failure rate of Main Bus 2 is given by:

$$= 5.0 \times 10^{-4} + 2.0 \times 10^{-4} = 7 \times 10^{-4} \text{ per flight hour}$$
$$\quad \text{(Generator 2)} \quad \text{(GCU 2)} \quad \text{(Main Bus 2)}$$

(Note that at this state the experienced aircraft systems designer would be considering the effect of a common cause or common mode failure.)

The probability of the failure of two independent channels (assuming no common cause failure) is derived by multiplying the respective failure rates. Therefore the probability of both Main Buses failing is:

$$= 7 \times 10^{-4} \quad \times \quad 7 \times 10^{-4} = 49 \times 10^{-8} \text{ or } 4.9 \times 10^{-7} \text{ per flight hour}$$
$$\quad \text{(Main Bus 1)} \quad \text{(Main Bus 2)}$$

Therefore the two independent electrical power channels alone will not meet the requirement. Assuming the addition of the ram air turbine (RAT) emergency channel as shown in the figure the probability of total loss of electrical power:

$$= 4.9 \times 10^{-7} \quad \times \quad 1 \times 10^{-3} = 4.9 \times 10^{-10} \text{ per flight hour which meets the}$$
$$\text{requirements}$$

$$\quad \text{(Main Bus 1} \quad \text{(RAT failure)}$$
$$\quad \text{\& Main Bus 2)}$$

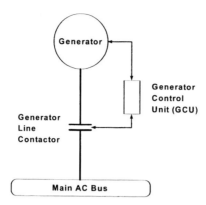

Fig. 11.7 *Main generator, GCU and power contactor relationship*

This very simple example is illustrative of the FTA which is one of the techniques used during the PSSA and SSA processes. However, even this simple example outlines some of the issues and interactions which need to be considered. Real systems are very complex with many more system variables and interlinks between a number of aircraft systems.

Failure Modes and Effects Analysis (FMEA)

The example given is a useful tool to examine total system integrity using a bottom-up approach. Certain parts of systems may be subject to scrutiny as they represent single point failures and as such more detailed analysis is warranted. The analysis used in this situation is the Failure Modes and Effects Analysis (FMEA).

Again, the process used in the FMEA is best illustrated by the use of a simple example. In this case an electrical generator is feeding an aircraft main electrical busbar via an electrical power line contactor. The line contactor is operated under the control of the GCU as shown in Fig. 11.7.

A FMEA on this portion of the aircraft electrical system will examine the possible failures of all the elements:

The generator failures and effects; that is examine in detail all the failures which contribute to the generator failure rate of 5×10^{-4} per flight hour as used in the previous analysis and the effects of those failures.

The GCU failures and effects: examining all the failures which contributed to the overall failure rate of 2×10^{-4} per flight hour as used above and the effects of those failures.

The line contactor failures and effects. If a line contactor has an MTBF of 100,000 hour/failure rate of 1×10^{-5} per flight hour, the ways in which the contactor may fail are ascribed portions of this failure rate for the different failures and effects:

- The contactor may fail open
- The contactor may fail closed
- The contactor may fail with one contact welded shut but the others open and so on until all the failures have been allocated a budget.

This process is conducted in a tabular form such that:

- Failure modes are identified
- Mode failure rates are ascribed

- Failure effects are identified
- The means by which the failure is detected is identified

An FMEA should therefore respond to the questions asked of the system or element under examination in a quantitative manner.

Component reliability

In the analyses described, a great deal of emphasis is placed upon the failure rate of a component or element within the system under review. This clearly calls into question how reliability values for different type of component are established. There are two main methods of determining component reliability:

- Analytical by component count
- Historical by means of accumulated in-service experience

Analytical methods

Mil-Std-781E is a standard developed by the US military over a period of years to use an analytical bottom-up approach to predicting reliability. This method uses a component count to build up and analyse the reliability of a unit. This approach has probably best been applied to electronics over the years as the use of electronic components within a design tends to be replicated within a design and across a family of designs. This method uses type of component, environment and quality factor as major discriminators in predicting the failure of a particular component, module and ultimately subsystem. Component failure rates are extracted from the US military standard and then applied with the appropriate factors to establish the predicted value as shown in the simplified example below:

Failure Rate, $\lambda = \pi Q \times (K1\pi T + K2 \times \pi E) \times \pi L$

Where πQ is a device quality factor
πT is a temperature factor
πE is an environmental factor
πL is a maturity factor
K1 and K2 are constants

There are a number of issues associated with this method:

It is only as good as the data base of components and the factors used.

- Experience has generally shown that – if anything – predicted values are generally pessimistic thereby generating predicted failure rates worse than might be expected in real life.
- The technique has merit in comparing competing design options in a quantitative manner when using a common baseline for each design.
- It is difficult to continue to update the data base; particularly with the growing levels of integration with Integrated Circuits (ICs) which makes device failure rates difficult to establish.
- The increasing number of Commercial Off-The-Shelf (COTS) components also confuses the comparison.

- The technique is particularly valuable when it can be compared with in-service experience and appropriate correction factors applied.

Reference (8) is a paper presented at a recent international aerospace conference which gives a very good overview of this technique when applied to power electronics.

In-service data

The use of in-service data is the best way of approaching the assessment of mechanical components used in the same environment. It does depend upon correspondence between the components which the design is contemplating with the in-service data base being used. Any significant variation in component usage, technology baseline or location in the aircraft/environment may nullify the comparison. Nevertheless, when used in conjunction with other methods this is a valid method. The manufacturers of civil, fighter aircraft and helicopters and their associated suppliers will generally be able to make 'industry standard' estimates using this technique.

Dispatch reliability

Dispatch availability is key to an aircraft fulfilling its mission, whether a military or civil aircraft. The ability to be able to continue to dispatch an aircraft with given faults has been given impetus by the commercial pressures of the air transport environment where the use of dual-redundancy for integrity reasons has been also used to aid aircraft dispatch. On the Boeing 777 the need for high rates of dispatch availability was specified in many systems and in some systems this leads to the adoption of dual-redundancy for dispatch availability reasons rather than for reasons of integrity. A simplified version of the dispatch requirements is shown in Fig. 11.8.

This means of specifying the dispatch requirement of part of an aircraft system leads to an operational philosophy far beyond a 'get-you-home' mode of operation. In fact it is the first step towards a philosophy of no unscheduled maintenance. For an aircraft flying 12 hours per day – a typical utilization for a wide-bodied civil transport – this

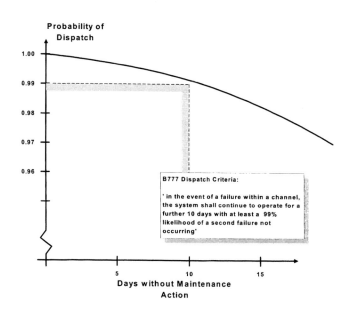

Fig. 11.8 *Simplified dispatch criteria*

Fig. 11.9 Simplified
FADEC architecture

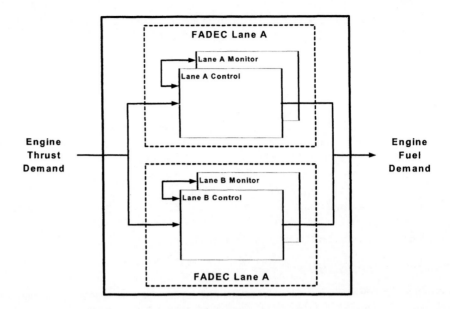

Fig. 11.9 Simplified
FADEC architecture

definition dictates a high level of availability for up to a 120 hour flying period. The
ability to stretch this period in the future – perhaps to a 500 hour operating period – as
more reliable systems become available, could lead to a true system of unscheduled
maintenance. A 500 hour operating period roughly equates to 8–9 weeks of flying, at
which time the aircraft will probably be entering the hangar for other specified
maintenance checks and inspections.

This leads to a more subtle requirement to examine the system's ability to meet
integrity requirements when several failures have already occurred and this requires
different techniques to be utilized.

Markov Analysis

Another technique used to assist in system analysis is the Markov Analysis (MA). This
approach is useful when investigating systems where a number of states may be valid
and also are interrelated. This could be the case in a multi-channel system where
certain failures may be tolerated but not in conjunction with some failure conditions.
The question of whether a system is airworthy is not a simple mathematical calculation
as in previous analyses but depends upon the relative states of parts of the system. The
simple methods used are insufficient in this case and another approach is required. The
Markov Analysis is the technique to be applied in these circumstances.

As before a simple example will be used to illustrate the MA technique: in this case
the dual-channel Full-Authority Digital Engine Control (FADEC) example outlined in
Fig. 11.9.

This simplified architecture is typical of many dual-channel FADECs. There are
two independent lanes: Lane A and Lane B. Each lane comprises a Command and
Monitor portion, which are interconnected for cross-monitoring purposes, and
undertakes the task of metering the fuel flow to the engine in accordance with the
necessary control laws to satisfy the flight crew thrust command. The analysis required
to decide upon the impact of certain failures in conjunction with others, requires a

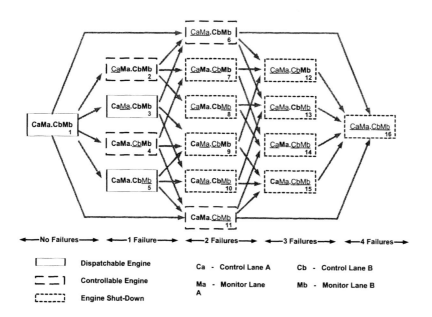

Fig. 11.10 *Use of Markov analysis to examine engine dispatch reliability*

Markov model in order to be able to understand the dependencies.

Figure 11.10 depicts a simple Markov model which equates to this architecture. By using this model the effects of interrelated failures can be examined. The model has a total of 16 states as shown by the number in the bottom right-hand corner of the appropriate box. Each box relates to the serviceability state of the Lane A Command (Ca) and Monitor (Ma) channels and Lane B Command (Cb) and Monitor (Mb) channels. These range from the fully serviceable state in box 1 through a series of failure conditions to the totally failed state in box 16. Clearly most normal operating conditions are going to be in the left-hand region of the model.

Concentrating on the left-hand side of the model it can be seen that the fully serviceable state in box 1 can migrate to any one of six states:

- Failure of Command channel A results in state 2 being reached.
- Failure of Monitor channel A results in state 3 being reached.
- Failure of Command channel B results in state 4 being reached.
- Failure of Monitor channel B results in state 5 being reached.
- Failure of the cross-monitor between Command A and Monitor A results in both being lost simultaneously and reaching state 6.
- Failure of the cross-monitor between Command B and Monitor B results in both being lost simultaneously and reaching state 11.

All of these failure states result in an engine which may still be controlled by the FADEC.

However, further failures beyond this point may result in an engine which may not be controllable either because both control channels are inoperative or because the 'good' control and monitor lanes are in opposing channels or worse. The model shown above is constructed according to the following rules: an engine may be dispatched as a 'get-you-home' measure provided that only one monitor channel has failed. This means that states 3 and 5 are dispatchable: but not states 2, 4, 6, or 11 as subsequent failures could result in engine shut-down.

By knowing the failure rates of the command channels, monitor channels and cross-monitors, quantitative values may be inserted into the model and probabilities assigned to the various states. By summing the probabilities so calculated, numerical values may be derived.

Development processes

The product life cycle

Fig. 11.11 Typical aircraft product life cycle

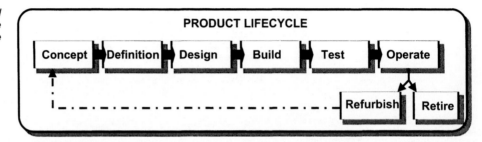

Figure 11.11 shows a typical aircraft product life cycle from concept through to disposal at the end of the product's useful life.

Individual products or equipment may vary from this model, however it is a sufficiently good portrayal to illustrate the role of systems engineering and the equipment life cycle. The major phases of this model are:

- Concept phase
- Definition phase
- Design phase
- Build phase
- Test phase
- Operate phase
- Refurbish or retire

This model closely aligns and closely approximates to the Downey cycle used by the UK Ministry of Defence, for the competitive procurement of defence systems – reference (3). The model is equally applicable for systems used in commercial aircraft as it is for military applications. It is used to describe the role of systems engineering in each phase of the product life cycle.

Concept phase

The concept phase is about understanding the customer's emerging needs and arriving at a conceptual model of a solution to address those needs. The customer continuously assesses his current assets and determines their effectiveness to meet future requirements. The need for a new military system can arise from a change in the local or world political scene that requires a change in defence policy. The need for a new commercial system may be driven by changing national and global travel patterns resulting from business or leisure traveller demands.

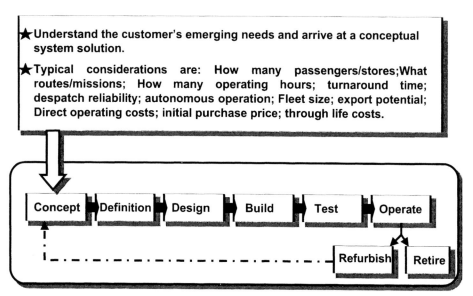

Fig. 11.12 Concept phase

The customer's requirement will be made available to industry so that solutions can be developed specifically for that purpose, or that can be adapted from the current Research and Development (R&D) base. This is an ideal opportunity for industry to discuss and understand the requirements to the mutual benefit of the customer and his industrial suppliers, to understand the implications of providing a fully compliant solution or one which is aggressive and sympathetic to marketplace requirements. Refer to Fig. 11.12.

Typical considerations at this phase are:

- Establishing and understanding the primary role and functions of the required system.
- Establishing and understanding desired performance and market drivers such as:
  - Range
  - Endurance
  - Routes or missions
  - Technology baseline
  - Operational roles
  - Number of passengers
  - Mass, number, and type of weapons
  - Availability and dispatch reliability
  - Fleet size to perform the role or satisfy the routes
  - Purchase budget available
  - Operating or through-life costs
  - Commonality or model range
  - Market size and export potential
  - Customer preference

This phase is focussed on establishing confidence that the requirement can be met within acceptable commercial or technological risk.   The establishment of a baseline of mature technologies may be first solicited by means of a Request For Information (RFI). This process allows possible vendors to establish their technical and other capabilities and represents an opportunity for the platform integrator to assess and quantify the relative strengths of competing vendors and also to capture mature technology of which he was previously unaware for the benefit of the programme.

Definition phase

Fig. 11.13 Definition
Phase

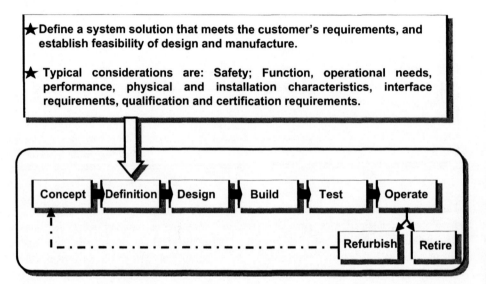

See Fig. 11.13. The customer will usually consolidate all the information gathered during the concept phase to firm up his requirement. A common feature used more frequently by platform integrators is to establish engineering joint concept teams to establish the major system requirements. These teams are sometimes called Integrated Product Teams (IPTs). They may develop a cardinal points specification; perhaps even undertake a preliminary system or baseline design against which all vendors might bid. This results in the issue of a specification or a Request For Proposal (RFP). This allows industry to develop their concepts into a firm definition, to evaluate the technical, technological and commercial risks, and to examine the feasibility of completing the design and moving to a series production solution. Typical considerations at this stage are:

- Developing the concept into a firm definition of a solution.
- Developing system architectures and system configurations.
- Re-evaluating the supplier base to establish what equipment, components and materials are available or may be needed to support the emerging design.
- Defining physical and installation characteristics and interface requirements.
- Developing operational and initial safety models of the individual systems.
- Quantifying key systems performance such as:
  - Mass
  - Volume
  - Growth capability
  - Range/endurance

The output from this phase is usually in the form of feasibility study reports, performance estimates, sets of mathematical models of individual system's behaviour and an operational performance model. This may be complemented by breadboard or experimental models or laboratory or technology demonstrators Preliminary design is also likely to examine installation issues with mock-ups in three-dimensional computer model form (CATIA) which replaces in the main the former need for wooden and metal models.

Design phase

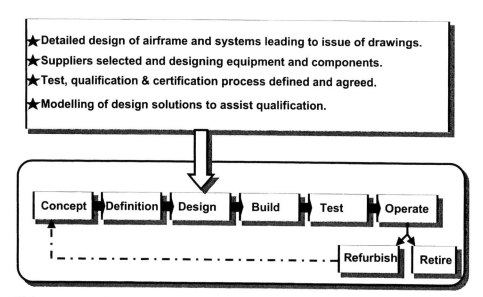

- ★ Detailed design of airframe and systems leading to issue of drawings.
- ★ Suppliers selected and designing equipment and components.
- ★ Test, qualification & certification process defined and agreed.
- ★ Modelling of design solutions to assist qualification.

Concept → Definition → Design → Build → Test → Operate → Refurbish → Retire

Fig. 11.14 Design phase

If the outcome of the definition phase is successful and a decision is made to proceed further, then industry embarks on the design phase within the programme constraints as described later in the chapter. Design takes the definition phase architectures and schemes and refines them to a standard that can be manufactured. Refer to Fig. 11.14.

Detailed design of the airframe ensures that the structure is aerodynamically sound, is of appropriate strength, and is able to carry the crew, passengers, fuel and systems that are required to turn it into a useful product. As part of the detailed design cognisance needs to be made of mandated rules and regulations which apply to the design of an aircraft or airborne equipment. The processes and techniques used to conduct the necessary safety assessments and analyses are described a little later in the chapter.

Three-dimensional solid modelling tools are used to produce the design drawings, in a format that can be used to drive machine tools to manufacture parts for assembly.

Systems are developed beyond the block diagram architectural drawings into detailed wiring diagrams. At the stage suppliers of bought-in equipment are selected and become an inherent part of the process of specifying and designing equipment that can be used in the aircraft or systems. Indeed in order to achieve a fully certifiable design, many of the complex and integrated systems found on aircraft today, an integrated design team comprising platform integrators and supplier(s) is essential. Many of these processes are iterative extending into and even beyond the build and test phases.

Build phase

Fig. 11.15 *Build phase*

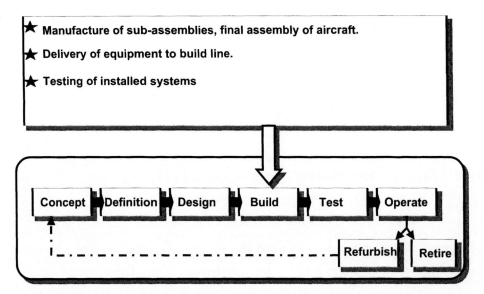

★ **Manufacture of sub-assemblies, final assembly of aircraft.**

★ **Delivery of equipment to build line.**

★ **Testing of installed systems**

The aircraft is manufactured to the drawings and data issued by design as shown in Fig. 11.15. During the early stages of the programme, a delivery schedule would have been established. Some long-lead time items – those which take a long time to build – may need to be ordered well ahead of aircraft build commencing. In the case of some of the more complex, software-driven equipment, design will be overlapping well into the test phase. This is usually accommodated by a phased equipment delivery embracing the following:

- Electrical models – equipment electrically equivalent to the final product but not physically representative.
- Red label hardware – equipment which is physically representative but not cleared for flight.
- Black label hardware – equipment which is physically representative and is cleared for flight either by virtue of the flightworthy testing carried out and/or the standard of the software load incorporated.

These standards are usually accompanied by a staged software release which enables a software load progressively to become more representative of the final functionality.

Test phase

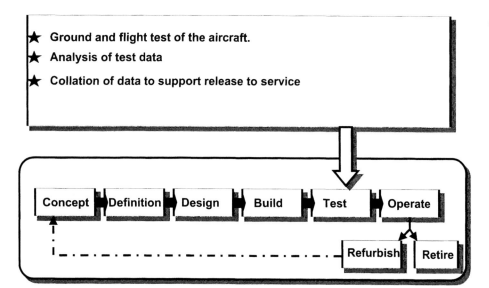

Fig. 11.16 *Test phase*

The aircraft and its components are subject to a rigorous test programme to verify fitness for purpose as shown in Fig. 11.16. This phase includes testing of, and integration of equipment, components, sub-assemblies, and eventually the complete aircraft. Functional testing of equipment and systems on the ground and flight trials verify that the performance and the operation of the equipment is as specified. Conclusion of the test programme and the associated design, analysis and documentation process leads to certification of the aircraft or equipment.

In the event of a new aircraft, responsibility for the certification of the aircraft lies with the aircraft manufacturer. However, where an equipment is to be improved or modified in the civil arena, equipment suppliers or other agencies can certify the equipment by means of the Supplementary Type Certificate (STC) in a process defined by the certification authorities. This permits discrete equipment – for example a more accurate fuel quantity gauging – in a particular aircraft model to be changed without affecting other equipment.

Operate phase

During this phase the customer is operating the aircraft on a routine basis. Its performance will be monitored by means of a formal defect reporting process, so that any defects or faults that arise are analysed by the manufacturer. It is possible to attribute causes to faults such as random component failures, operator mishandling, or design errors. The aircraft manufacturer and his suppliers are expected to participate in the attribution and rectification of problems arising during aircraft operations, as determined by the contract. See Fig 11.17.

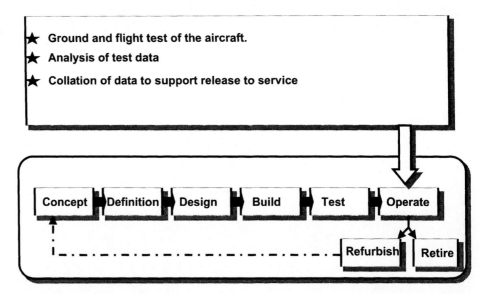

Fig. 11.17 Operate phase

Disposal or refurbish

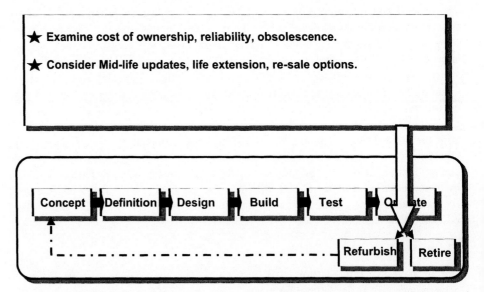

Fig. 11.18 Disposal or refurbishment

At the end of the useful or predicted life of the aircraft, decisions have to be made about its future as depicted in Fig. 11.18. The end of life may be determined by unacceptably high operating costs, unacceptable environmental considerations – noise, pollution etc. – or by predicted failure of mechanical or structural components determined by the supplier's test rigs. If it is not possible to continue to operate the aircraft, then it may be disposed of – sold for scrap or alternative use, such as aircraft enthusiast or gate guardian.

If the aircraft still has some residual and commercially viable life, then it may be refurbished. This latter activity is often known as a mid-life update, or even a conversion to a different role, e.g. VC10 passenger aircraft converted to in-flight refuelling use as has happened with the Royal Air Force. Similarly, in the civil arena, many former passenger aircraft are being converted to the cargo role.

Development programme

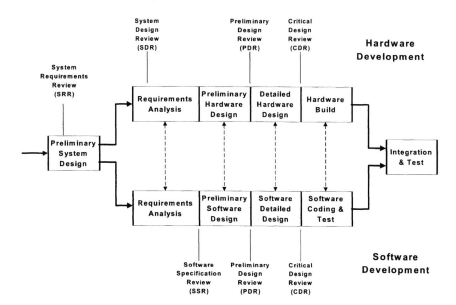

Fig. 11.19 *Typical development programme*

So far, the processes, methods and techniques used during aircraft system design have been described. However these need to applied and controlled within an overall programme management framework. Figure 11.19 shows the major milestones associated with the aircraft systems development process. It is assumed – as is the case for the majority of aircraft systems developed today – that the system has electronics associated with the control function and that the electronics has a software development content.

The main characteristic of the development is the bifurcation of hardware and software development processes into two separate paths though it can be seen that there is considerable interaction between the two. The key steps in the avionics development programme which are primarily designed to contain and mitigate against risk are:

System Requirements Review (SRR)

The SRR is the first top-level, multi-disciplinary review of the perceived system requirements. It is effectively a sanity check upon what the system is required to achieve; a top-level overview of requirements and review against the original objectives. Successful attainment of this milestone leads to a preliminary system design leading in turn to the parallel development of the hardware and software requirements analysis, albeit with significant co-ordination between the two.

System Design Review (SDR)

The hardware SDR immediately follows the preliminary design phase and will encompass a top-level review of the system hardware characteristics such that preliminary design may proceed with confidence. Key hardware characteristics will be reviewed at this stage to ensure that there are no major mismatches between the system requirements and what the hardware is capable of supporting.

Software Specification Review (SSR)

The SSR is essentially a similar process to the hardware SDR but applying to the software when a better appreciation of the software requirements has become apparent and possibly embracing any limitations such as throughput, timing or memory which the adopted hardware solution may impose. Both the SDR and SSR allow the preliminary design to be developed up to the preliminary design review (PDR).

Preliminary Design Review (PDR)

The preliminary design review process is the first detailed review of the initial design (both hardware and software) versus the derived requirements. This is usually the last review before committing major design resource to the detailed design process. This stage in the design process is the last before major commitment to providing the necessary programme resources and investment.

Critical Design Review (CDR)

By the time of the CDR major effort will have been committed to the programme design effort. The CDR offers the possibility of identifying final design flaws or, more likely, trading the risks of one implementation path versus another. The CDR represents the last opportunity to review and alter the direction of the design before very large commitments and final design decisions are taken. Major changes in system design – both hardware and software – after the CDR will be very costly in terms of cost and schedule loss, to the total detriment of the programme.

The final stages following CDR will realize the hardware build and software coding and test processes which bring together the hardware and software into the eventual product realization. Even following system validation and equipment certification it is unusual for there to be a period free of modification either at this stage or later in service when airlines may demand equipment changes for performance, reliability or maintainability reasons.

'V' diagram

The rigours of software development are particularly strict and are dictated by reference (6).
For obvious reasons, the level of criticality of software used in avionics systems determines the rigour applied to the development process. Reference (6) also defines three levels of software:

- Level 1: Used in critical systems application and subject to the greatest levels of control in terms of methodology: quality, design, test, certification, tools and documentation.
- Level 2: Used for essential applications with standards comparable to Level 1 but less stringent in terms of test and documentation.
- Level 3: Used in non-essential applications and with less stringent standards generally equivalent to a good standard of commercial software.

The software development process is generally of the form shown in Fig. 11.20 which shows the development activities evolving down the right of the diagram and the verification activities down the left. This shows how the activities eventually converge in the software validation test at the foot of the diagram that is the confluence of hardware and software design and development activities. Down the centre of the diagram the various development software stages are shown. It can be seen that there is

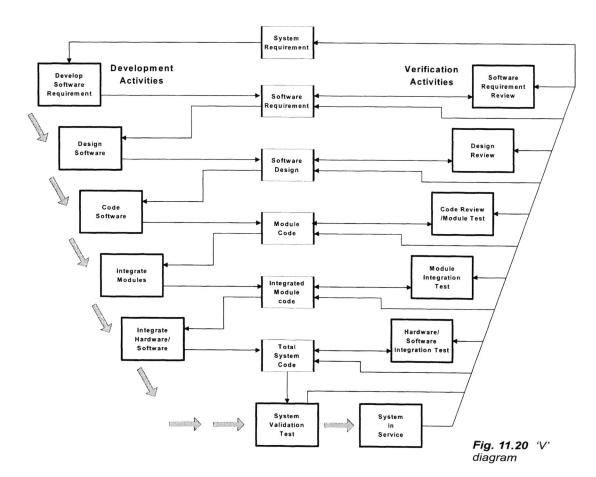

Fig. 11.20 'V' diagram

considerable interaction between all the processes that represent the validation of the requirements and of the hardware and software design at each level. Any problems or issues discovered during the software validation tests are fed back up the chain, if necessary back into the top level. Therefore any minor deviations are reflected back into all the requirements stages to maintain a consistent documentation set and a consistent hardware and software design.

Whereas the earlier stages of software development and test might be hosted in a synthetic software environment it is increasingly important as testing proceeds to undertake testing in a representative hardware environment. This testing represents the culmination of functional testing for the LRU or equipment short of flight test.

References

(1)   ARP 4761   Guidelines and Methods for Conducting the Safety Assessment Process on Civil Airborne Systems.
(2)   ARP 4754   Certification Considerations for Highly-Integrated or Complex Aircraft Systems.
(3)   AC 25.1309-1A   System Design and Analysis, Advisory Circular.
(4)   AMJ 25.1309   System Design and Analysis, Advisory Material Joint.

(5) ATA-100   ATA Specification for Manufacturer's Technical Data.
(6) DO-178b   Software Considerations in Airborne systems and Equipment Certification.
(7) DO-xxx   Document under development by SC-1870 Working title: Design Assurance Guidance for Airborne Electronic Hardware.
(8) Dual use of Variable Speed Constant Frequency (VSCF) cyclo-converter technology, V Bonneau, FITEC 98.

CHAPTER 12

Avionics Technology

Introduction

The first major impetus for use of electronics in aviation occurred during World War II. Communications were maturing and the development of airborne radar using the magnetron and associated technology occurred at a furious pace throughout the conflict. See reference (1).

Fig. 12.1 *Major electronics developments in aviation since 1930*

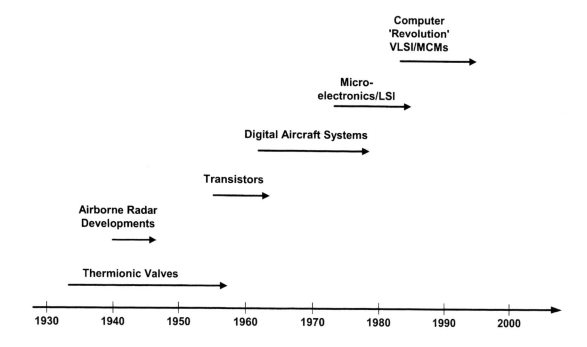

Transistors followed in the late 1950s and 1960s and supplanted thermionic valves for many applications. The improved cost-effectiveness of transistors led to the development of digital aircraft systems throughout the 1960s and 1970s, initially in the military combat aircraft where it was used for Nav/Attack systems. See Fig. 12.1.

For many years the application of electronics to airborne systems was limited to analogue devices and systems with signal levels and voltages generally being related in some linear or predictive way. This type of system was generally prone to heat soak, drift and other non-linearities. The principles of digital computing had been understood for a number of years before the techniques were applied to aircraft. The development of thermionic valves enabled digital computing to be accomplished but at the expense of vast amounts of hardware. During World War II a code-breaking machine called Colossus employed thermionic valves on a large scale. The machine was physically enormous and quite impracticable for use in any airborne application.

The first aircraft to be developed in the US using digital techniques was the North American A-5 Vigilante, a US Navy carrier-borne bomber which became operational in the 1960s. The first aircraft to be developed in the UK intended to use digital techniques on any meaningful scale was the ill-fated TSR 2 which was cancelled by the UK Government in 1965. The technology employed by the TSR 2 was largely based upon solid-state transistors, then in comparative infancy. In the UK, it was not until the development of the Anglo-French Jaguar and the Hawker Siddeley Nimrod in the 1960s that weapon systems began to seriously embody digital computing, albeit on a meagre scale compared to the 1980s.

Since the late 1970s/early 1980s digital technology has become increasingly used in the control of aircraft systems as well as just for mission related systems. A key driver in this application has been the availability of cost-effective digital data buses such as ARINC 429, Mil-Std-1553B and ARINC 629. This technology, coupled with the availability of cheap microprocessors and more advanced software development tools, has led to the widespread application of avionics technology throughout the aircraft. This has advanced to the point that virtually no aircraft system – including the toilet system – has been left untouched.

The evolution and increasing use of avionics technology for civil applications of engine controls and flight controls since the 1950s is shown in Fig. 12.2.

Engine analogue controls were introduced by Ultra in the 1950s which comprised electrical throttle signalling used on aircraft such as the Bristol Britannia. Full-

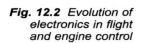

Fig. 12.2 *Evolution of electronics in flight and engine control*

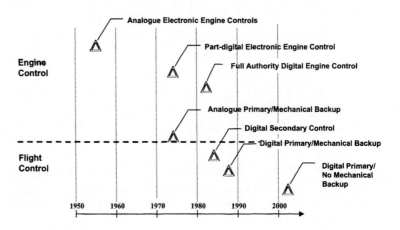

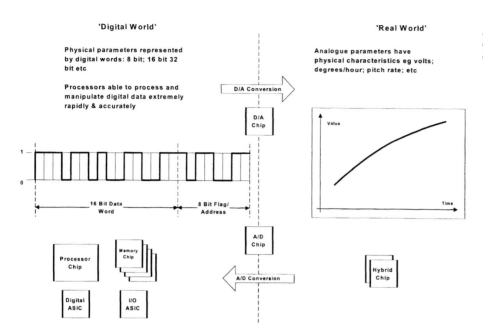

Fig. 12.3 The nature of micro-electronic devices

authority digital engine control became commonly used in the 1980s.

Digital primary flight control with a mechanical backup has been used on the Airbus A320 and A330/A340 families using side-stick controllers and on the Boeing 777 using a conventional control yoke. Aircraft such as the Dornier 728 family and the A380 appear to be adopting flight control without any mechanical backup but with electrically signalled backup.

The application of digital techniques to other aircraft systems – utilities systems – began later as will be described in this chapter. Today, avionics technology is firmly embedded in the control of virtually all aircraft systems. Therefore an understanding of the nature of avionics technology is crucial in understanding how the control of aircraft systems is achieved.

The nature of micro-electronic devices

The development of a wholly digital control system has to accommodate interfaces with the 'real world' which is analogue in nature. The figure shows how the range of micro-electronic devices is used in different applications within a digital system.

Hybrid chips and input/output (i/o) Application-Specific Integrated Circuits (ASICs) are key technologies associated with interfacing to the analogue world. A/D and D/A devices undertake the conversion from analogue to digital and digital to analogue signals respectively. Processor and memory devices, together with digital ASICs perform the digital processing tasks associated with the application. See Fig. 12.3.

Micro-electronic devices are produced from a series of masks that shield various parts of the semiconductor during various processing stages. The resolution of most technology is of the order of $1 - 3$ microns (1 micron is 10^{-6} metres or 1 millionth of a metre, or one thousandth of a millimetre) so the physical attributes are very minute. Thus a device or die about 0.4 in square could have hundreds of thousands of transistors/gates to produce the functionality required of the chip. Devices are produced many at a time on a large circular semiconductor wafer, some devices at the periphery

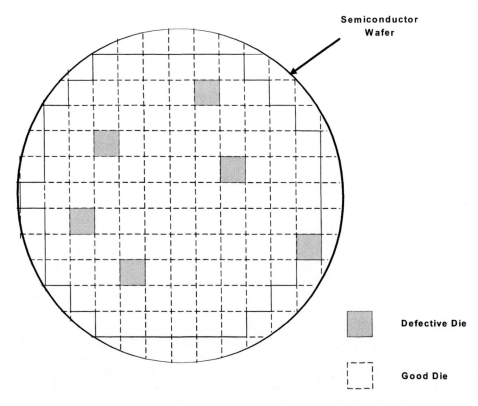

Fig. 12.4
Semiconductor wafer
yield

of the wafer will be incomplete and some of the remaining devices may be flawed and defective. However, the remainder of the good die may be trimmed to size, tested and mounted within the device package. The size of the die, complexity and maturity of the overall semiconductor process and the quality of the material will determine the number of good die yielded by the wafer and this yield will eventually reflect in the cost and availability of the particular device. Note Fig. 12.4.

Micro-electronics devices are environmentally screened according to the severity of the intended application; usually three levels of screening are applied, in increasing levels of test severity:

- Commercial grade
- Industrial grade
- Aerospace military grade – also used in many cases for civil aerospace applications

There is little doubt that this screening technique has helped to improve the maturity of the manufacturing process and quality of the devices in the past. However, as an increasingly small proportion of devices overall are used for aerospace applications, full military screening is difficult to assure for all devices. There is a body of opinion that believes that screening is not beneficial, and adds only to the cost of the device. It is likely that avionics vendors will have to take more responsibility for the quality of devices used in their product in future. There is an increasing and accelerating trend for aerospace micro-electronics to be driven by the computer and telecommunications industries.

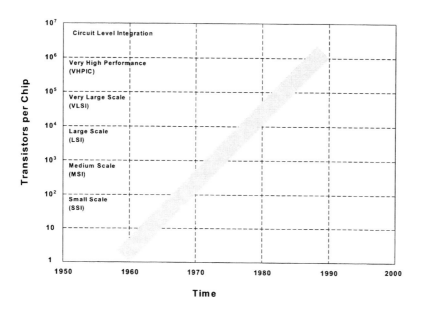

Fig. 12.5 *Trends in integrated circuit development*

The extent of the explosion in IC developments can be judged by reference to Fig. 12.5. This shows a ten-fold increase per decade in the number of transistors per chip. Another factor to consider is the increase in the speed of device switching. The speed of operation is referred to as gate delay; gate delay for a thermionic valve is of the order of 1,000 nanoseconds (1 nanosecond is 10^{-9} or one thousandth of one millionth of a second); transistors are about ten times quicker at 100 nanoseconds. Silicon chips are faster again at approximately 1 nanosecond). This gives an indication of how powerful these devices are and why they have had such an impact upon our daily life.

Another area of major impact for the IC relates to power consumption. ICs consume minuscule amounts. Consumption is related to the technology type and speed of operation. The quicker the speed of operation then the greater the power required and vice versa. The main areas where avionics component technology have developed are:

- Processors
- Memory
- Data buses

Processors

Digital processor devices became available in the early 1970s as 4-bit devices. By the late 1970s 8-bit processors had been superceded by 16-bit devices; these led in turn to 32-bit devices such as the Motorola 68000 which have been widely used on the Eurofighter and Boeing 777. The pace of evolution of processor devices does present a significant concern due to the risk of the chips becoming obsolescent, leading to the prospect of an expensive re-design.

Following adverse experiences with its initial ownership of microprocessor based systems, the US Air Force pressed strong standardization initiatives based upon the MIL-STD-1750A microprocessor with a standardized instruction set architecture (ISA) though this found few applications in aircraft systems computing. For these types of application, starting with the adoption of the Motorola 68020 on Eurofighter, the industry is making extensive use of commercially developed microprocessor or microcontroller products.

Memory devices

Memory devices have experienced a similar explosion in capability. Memory devices comprise two main categories: Read-Only Memory (ROM) represents the memory used to host the application software for a particular function; as the term suggests this type of memory may only be read but not written to. A particular version of ROM used frequently was Electrically Programmable Read-Only Memory (EPROM), however this suffered the disadvantage that memory could only be erased by irradiating the device with ultra-violet (UV) light. For the last few years EPROM has been superseded by the more user-friendly Electrically Erasable Programmable Read-Only Memory (E^{2}PROM). This type of memory may be re-programmed electrically with the memory module still resident within the LRU; using this capability it is now possible to reprogram many units in situ on the aircraft via the aircraft digital data buses.

Random-Access Memory (RAM) is read-write memory that is used as program working memory, storing variable data. Early versions required a power backup in case the aircraft power supply was lost. More recent devices are less demanding in this regard.

Digital data buses

The advent of standard digital data buses began in 1974 with the specification by the US Air Force of MIL-STD-1553. The ARINC 429 data bus became the first standard data bus to be specified and widely used for civil aircraft being widely used on the Boeing 757 and 767 and Airbus A300/A310 in the late 1970s and early 1980s. ARINC 429 (A429) is widely used on a range of civil aircraft today as will become apparent during this chapter. In the early 1980s Boeing developed a more capable digital data bus termed Digital Autonomous Terminal Access Communication (DATAC) which later became an ARINC standard as A629; the Boeing 777 is the first and at present the only aircraft to use this more capable data bus. At the same time, these advances in digital data bus technology were matched by advancements in processor, memory and other micro-electronic devices such as analogue-to-digital and digital-to-analogue devices, logic devices etc. which made the application of digital technology to aircraft systems possible.

The largest single impact of micro-electronics on avionic systems has been the introduction of standardized digital data buses to greatly improve the intercommunication between aircraft systems. Previously, large amounts of aircraft wiring were required to connect each signal with the other equipment. As systems became more complex and more integrated so this problem was aggravated. Digital data transmission techniques use links which send streams of digital data between equipment. These data links comprise only two or four twisted wires and therefore the interconnecting wiring is greatly reduced.

Common types of digital data transmission are:

- Single-source, single-sink This is the earliest application and comprises a dedicated link from one equipment to another. This was developed in the 1970s for use on Tornado and Sea Harrier avionics systems. This technique is not used for the integration of aircraft systems.
- Single-source, multiple-sink This describes a technique where one transmitting equipment can send data to a number of recipient equipment (sinks). ARINC 429 is an example of this data bus which is widely used by civil transports and business jets.

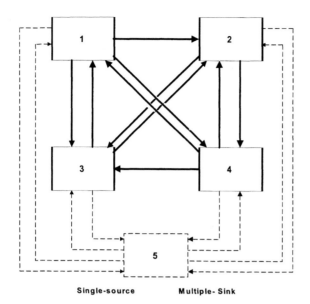

Fig. 12.6 *A429 topology and the effect of adding units*

Single-source   Multiple-Sink

- Multiple-source, multiple-sink In this system multiple transmitting sources may transmit data to multiple receivers. This is known as a full-duplex system and is widely employed by military users (MIL-STD-1553B) and by the Boeing 777 (ARINC 629).

The major digital data buses in use today are:

- ARINC 429 (A429)
- MIL-STD-1553B also covered by Def Stan 00-18 Part 2 and STANAG 3838
- ARINC 629 (A629)

Of these A429 and A629 are commonly in use on civil aircraft. MIL-STD-1553B is a military standard somewhat similar in bus topology, encoding and data encoding to A629 though the command protocol is different.

ARINC 429 data bus

The characteristics of ARINC 429 were agreed among the airlines in 1977/78 and the data bus first used on the Boeing 757/Boeing 767 and Airbus A300 and A310 aircraft. ARINC, short for Aeronautical Radio Inc, is a corporation in the US whose stockholders comprise US and foreign airlines, and aircraft manufacturers. As such it is a powerful organization central to the specification of equipment standards for known and perceived technical requirements.

The ARINC 429 (A429) bus operates in a single-source, multiple-sink mode such that it is a source which may transmit to a number of different terminals or sinks, each of which may receive the data message. However, if any of the sink equipment need to reply then they will each require their own transmitter to do so; and cannot reply down the same wire pair. This half-duplex mode of operation has certain disadvantages. If it is desired to add additional equipment as shown in Fig. 12.6, a new set of buses may be required – up to a maximum of eight new buses in this example if each new link has to possess bi-directional qualities.

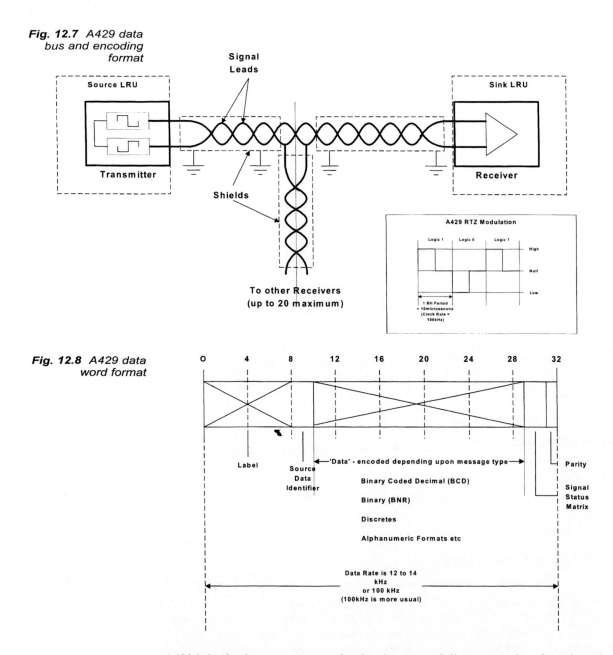

Fig. 12.7 *A429 data bus and encoding format*

Fig. 12.8 *A429 data word format*

A429 is by far the most common data bus in use on civil transport aircraft, regional jets and executive business jets today. Since introduction on the Boeing 757/767 and Airbus aircraft in the early 1980s hardly an aircraft has been produced which does not utilize this data bus.

The physical implementation of the A429 data bus is a screened, twisted wire pair with the screen earthed at both ends and at all intermediate breaks. The transmitting element shown on the left in Fig. 12.7, is embedded in the source equipment and may interface with up to 20 receiving terminals in the sink equipment. Information may be transmitted at a low rate of 12–14 kbits per second or a higher rate of 100 kbits per

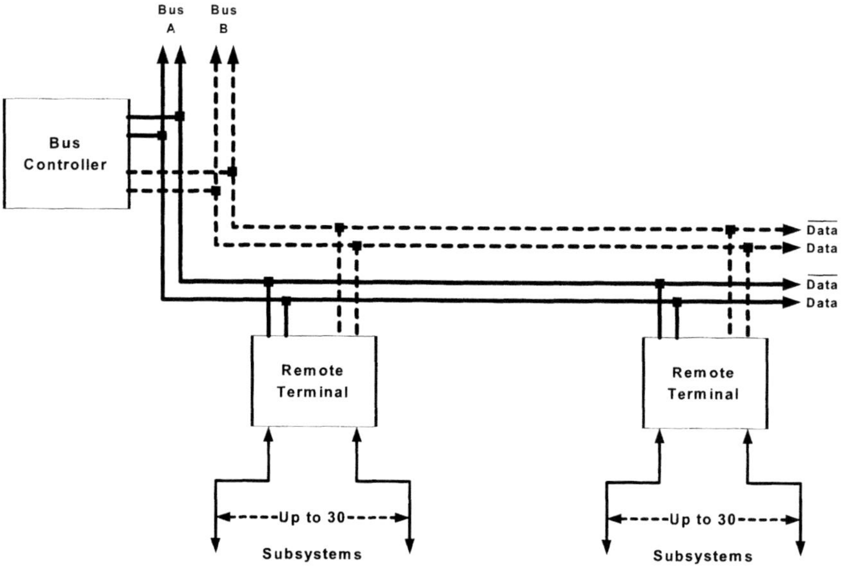

Fig. 12.9 *MIL-STD-1553B data bus*

second; the higher rate is by far the most commonly used. The modulation technique is bipolar Return To Zero (RTZ) as shown in the box in Fig. 12.7. The RTZ modulation technique has three signal levels: high, null and low. A logic state 1 is represented by a high state returning to zero; a logic state 0 is represented by a low state returning to null. Information is transmitted down the bus as 32 bit words as shown in Fig. 12.8.

The standard embraces many fixed labels and formats so that a particular type of equipment always transmits data in a particular way. This standardization has the advantage that all manufacturers of particular equipment know what data to expect. Where necessary, additions to the standard may also be implemented. Further reading for A429 may be found at references (2), (3) and (4).

MIL-STD-1553B

MIL-STD-1553B has evolved since the original publication of MIL-STD-1553 in 1973. The standard has developed through 1553A standard issued in 1975 to the present 1553B standard issued in September 1978. The basic layout of a MIL-STD-1553B data bus is shown in Fig. 12.9. The data bus comprises a twin twisted wire pair along which DATA and $\overline{\text{DATA}}$ are passed. The standard generally allows for dual-redundant operation.

Control of the bus is effected by a Bus Controller (BC) which is connected to a number of Remote Terminals (RTs) (up to a maximum of 31) via the data bus. RTs may be processors in their own right or may interface a number of subsystems (up to a maximum of 30) with the data bus. Data is transmitted at 1 MHz using a self-clocked Manchester bi-phase digital format. The transmission of data in true and complement form down a twisted screened pair, together with a message error correction capability offers a digital data link which is highly resistant to message corruption. Words may be formatted as data words, command words or status words as shown in Fig. 12.10. Data words encompass a 16-bit digital word while the command and status words are associated with the data bus transmission protocol. Command and status words are compartmented to include various address, sub-address and control functions as shown in the figure.

Fig. 12.10 *MIL-STD-1553B data bus word formats*

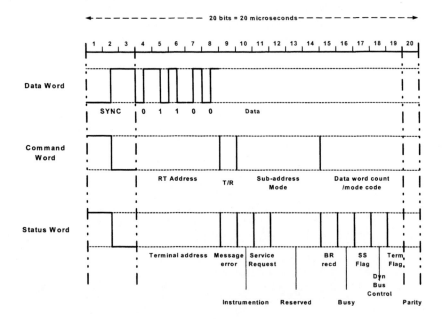

Fig. 12.11 *MIL-STD-1553B typical data transactions*

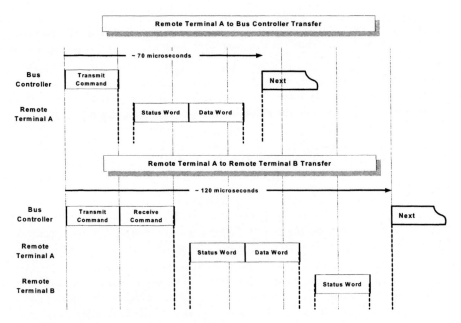

MIL-STD-1553B is a command-response system in which all transmissions are conducted under the control of the bus controller; although only one bus controller is shown in these examples a practical system will employ two bus controllers to provide control redundancy.

Two typical transactions are shown in Fig. 12.11. In a simple transfer of data from RT A to the BC, the BC sends a transmit to RT A, which replies after a short interval with a status word, followed immediately by one or more data words up to a maximum of 32 words. In the example shown in the upper part of Fig.12.11, transfer of one data word from

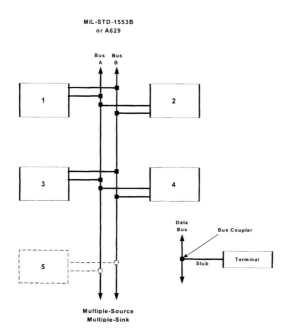

Fig. 12.12 *MIL-STD-1553 and A629 data bus topology*

the RT A to the BC will take an elapsed time of about 70 μsec. For the transfer of data between RTs as shown from RT A to RT B, the BC sends a receive word to RT B followed by a transmit word to RT A. RT A will send a status word plus a data word (up to a maximum of 32 words) to RT B which responds by sending a status word to the BC, thereby concluding the transaction. In the simple RT to RT transaction shown in Fig. 12.11 the total elapsed time is around 120 μsec for a single data word which appears to be rather expensive in overheads. However, if the maximum of words had been transferred the overheads would be the same, though now representing a much lower percentage of the overall message time. For further reading on MIL-STD-1553B see reference (5).

ARINC 629 data bus

ARINC 629 (A629), like MIL-STD-1553B, is a true data bus in that the bus operates as a multiple-source, multiple-sink system – see Fig. 12.12. That is, each terminal can transmit data to, and receive data from every other terminal on the data bus. This allows much more freedom in the exchange of data between units in the avionics system than the single-source, multiple-sink A429 topology. Furthermore the data rates are much higher than for A429 where the highest data rate is 100 kbits/sec. The A629 data bus operates at 2 Mbytes/sec or twenty times that of A429. The true data bus topology is much more flexible in that additional units can be fairly readily accepted physically on the data bus. A further attractive feature is the ability to accommodate up to a total of 131 terminals on a data bus, though in a realistic implementation data bus traffic would probably preclude the use of this large number of terminals. The protocol utilized by A629 is a time-based, collision-avoidance concept in which each terminal is allocated a particular time slot access to transmit data on to the bus. Each terminal autonomously decides when the appropriate time slot is available and transmits the necessary data. This protocol was the civil aircraft industry's response to the military MIL-STD-1553B data bus that utilizes a dedicated controller to decide what traffic passes down the data bus.

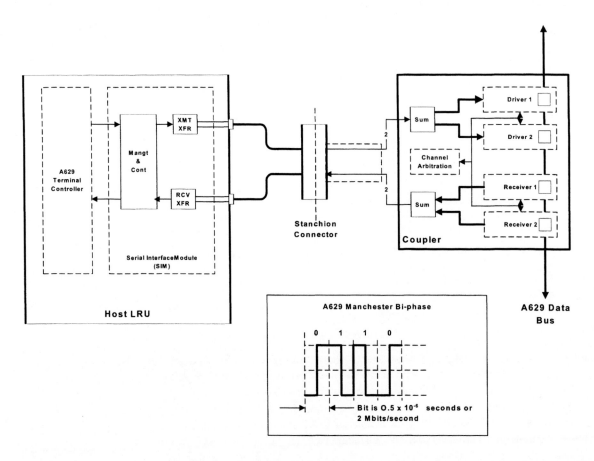

Fig. 12.13 A629 bus coupler interface and encoding format

Because of the higher data rates and higher technology baseline, the A629 bus coupler arrangement is slightly more involved than for A429. Figure 12.13 shows how the host LRU connects to the A629 data bus via the serial interface module (SIM), embedded in the LRU, and via a stanchion connector to the coupler itself. Due to the transmit/receive nature of the A629 protocol there are separate channels for transmit and receive. Transformer coupling is used due to concern that a single bus failure could bring down all the terminals associated with the data bus. The bus couplers are all grouped in a fairly low number of locations to ease the installation issues.

Also shown within the box in Fig. 12.13 is a simplified portrayal of the Manchester bi-phase encoding which the A629 data bus (and MIL-STD-1553B) uses. In this protocol a logic 0 is signified when there is a negative to positive change of signal; this change of state occurs midway during the particular bit duration. Similarly, logic 1 is denoted when there is a positive to negative change of signal during the bit period. This timing is aided by the fact that the first three bits in a particular data word act as a means of synchronization for the whole of the word. The data is said to be 'self-clocked' on a word by word basis and therefore these rapid changes of signal state may be accurately and consistently recognized with minimal risk of mis-reads.

Figure 12.14 shows the typical A629 20-bit data word format. The first three bits are related to word time synchronization as already described. The next 16 bits are data related and the final bit is a parity bit. The data words may have a variety of formats

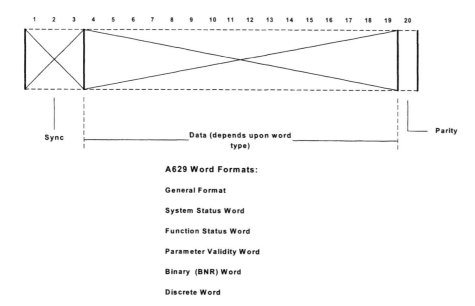

Fig. 12.14 *A629 digital word format*

depending upon the word function; there is provision for general formats, systems status, function status, parameter validity, binary and discrete data words. Therefore although the data format is simpler than A429, the system capabilities are more advanced as the bit rate is some twenty times faster than the fastest (100 kbit) option for A429.

The only aircraft utilizing A629 data buses so far is the Boeing 777. The widespread application of technology such as A629 is important as more widespread application drives component prices down and makes the technology more cost-effective. Certainly that has been the case with A429 and MIL-STD-1553B.

For more detail on A629 see references (6), (7) and (8).

Data bus examples – integration of aircraft systems

The increasing cost-effectiveness, which system integration using digital data buses and micro-electronic processing technologies offers, has led to a rapid migration of the technology into the control of aircraft systems. Three examples are shown below which highlight how all-embracing this process has become.

- MIL-STD-1553B - EAP utilities management system
- A429 - Airbus A330/340 aircraft systems
- A629 - Boeing 777 aircraft systems

Experimental Aircraft Programme (EAP)

The first aircraft to utilize MIL-STD-1553B for the integration of aircraft as opposed to avionics systems was the UK Experimental Aircraft Programme (EAP) which was a technology demonstrator forerunner to the Eurofighter. This aircraft first flew in August 1986 and was demonstrated at the Farnborough Air Show the same year, also being flown at the Paris Air Show the following year. This system is believed to be the first integrated system of its type; given purely to the integration of aircraft utility systems. The system encompassed the following functions.

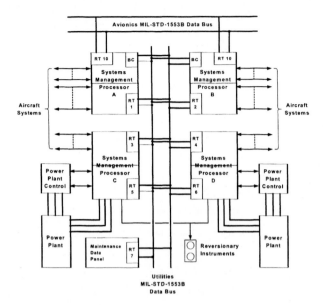

Fig. 12.15 EAP
utilities system
architecture

- Engine control and indication
- Fuel management and fuel gauging
- Hydraulic system control and indication, undercarriage indication and monitoring, wheel brakes
- Environmental control systems, cabin temperature control – and later an On-Board Oxygen Generating System (OBOGS)
- Secondary power system
- LOX contents, electrical generation and battery monitoring, probe heating, emergency power unit

The system comprised four LRUs – Systems Management Processors (SMPs) – which also housed the power switching devices associated with operating motorized valves, solenoid valves etc. These four units, comprised of a set of common modules or building blocks, replaced a total of 20 to 25 dedicated controllers and 6 power switching relay units which a conventional system would use. The system comprised several novel features; offering a level of integration which has not been equalled to-date. See Fig. 12.15.

The technology and techniques applied to aircraft utilities systems demonstrated on EAP have been used successfully on Eurofighter and Nimrod MRA4. The lessons learned on each aircraft have been passed on through the generations of utilities management systems, and are now being used on a number of aircraft projects under the heading of vehicle management systems.

Airbus A330/340

The two-engined A330 and four-engined A340 Airbus aircraft make extensive use of A429 data buses to integrated aircraft systems control units with each other and with the avionics and displays. Table 12.1 lists some of the major subsystems and control units.

Table 12.1 *A330/A340 typical aircraft system controllers*

Control unit	A330	A340	Remarks
Bleed air control	2	4	One per engine
Fuel control	2	2	
Landing-gear control	3	3	
Flight control:			
– Flight control primary computer	3	3	
– Flight control secondary computer	2	2	
– Flight control data concentrator	2	2	
– Slat/flap control computer	2	2	
Probe heat	3	3	
Zone controller	1	1	
Window heat control	2	2	
Cabin pressure control	2	2	
Pack controller	2	2	
Avionics ventilation computer	1	1	
Generator control unit	2	4	One per engine
Full-authority digital engine control (FADEC)	2	4	One per engine
Flight warning computer	2	2	
Central maintenance computer	2	2	
Hydraulic control	1	1	

Boeing 777

The Boeing 777 makes extensive use of the A629 digital data bus to integrate the avionics, flight controls and aircraft systems. Fig. 12.16 depicts a simplified version of the Boeing 777 aircraft systems which are integrated using A629 buses. Most equipment is connected to the left and right aircraft system buses but some are also connected to a centre bus. Exceptionally, the engine Electronic Engine Controllers (EECs) are connected to left, right, centre 1 and centre 2 buses to give true dual-dual interface to the engines. The systems so connected embrace the following.

- Fuel
- Electrical:
  - Electrical load management
  - Bus control
  - Generation control
- Landing gear
  - Brakes and anti-skid
  - Tyre pressure monitoring
  - Brake temperature monitoring
- APU and environmental control:

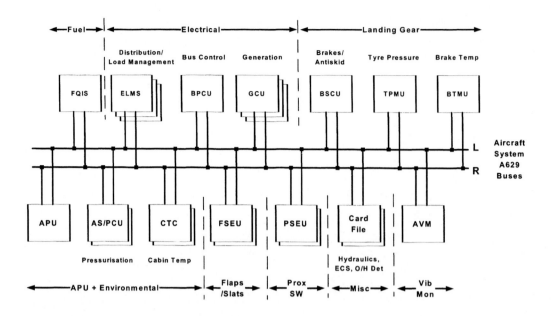

Fig. 12.16 *Boeing 777 aircraft systems integration using A629 data buses*

– APU controller
– Air pressure and pressurization control
– Cabin temperature control

- Flaps and slats
- Proximity switches
- Card files: Boeing produced modules used for the management of hydraulics, overheat detection, environmental and other functions
- Vibration monitoring

Regional aircraft/business jets

The foregoing examples relate to fighter and civil transport aircraft. The development of regional aircraft and business jet integrated avionics systems is rapidly expanding to include the aircraft utilities functions. The Honeywell EPIC system as being developed for the Hawker Horizon, Dornier 728 family and Embraer ERJ-170/190 is an example of how higher levels of system integration are being achieved. See Fig. 12.17.

This is a much more closed architecture than the ones already described which utilize open, internationally agreed standards. This architecture uses a proprietary Avionics Standard Communication Bus (ASCB) – Variant D data bus developed exclusively by Honeywell, originally for General Aviation (GA) applications. Previous users of ASCB have been Cessna Citation, Dassault Falcon 900, DeHavilland Dash 8 and Gulfstream GIV.

The key characteristics of ASCB D are:

- Dual data bus architecture
- 10 MHz bit rate – effectively a hundred times faster than the fastest A429 rate (100 kbits/sec)
- Up to 48 terminals may be supported
- The architecture has been certified for flight critical applications

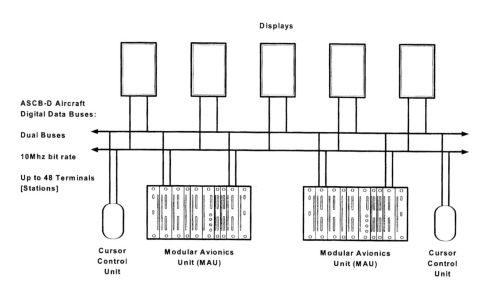

Fig. 12.17 *Honeywell EPIC system – typical*

This example shows two Modular Avionics Units (MAUs), however it is more typical to use four such units to host all the avionics and utilities functions. It can be seen from this example that the ambitions of Honeywell in wishing to maximize the return on their EPIC investment is driving the levels of system integration in the regional aircraft and business jets to higher levels than the major OEMs such as Boeing and Airbus. A publication which addresses ASCB and compares it with other data buses is at reference (9). This reference contains much useful data regarding the certification of data bus systems. For an example of a fuel system integrated into the EPIC system – see the paper at reference (10) which describes the integration of the Hawker Horizon fuel system into the Honeywell EPIC system.

Fibre-optic buses

The examples described thus far relate to electrically signalled data buses. Fibre-optic interconnections offer an alternative to the electrically signalled bus that is much faster and more robust in terms of ElectroMagnetic Interference (EMI). Fibre-optic techniques are widely used in the telecommunications industry and those used in cable networks serving domestic applications may typically operate at around 50–100 MHz.

A major problem with fibre-optic communication is that it is unidirectional. That is the signal may only pass in one direction and if bidirectional communication is required then two fibres are needed. There is also no 'T-junction' in fibre optics and communication networks have to be formed by 'Y-junctions' or ring topologies. An example of the ring topology is shown in Fig. 12.18 in which the bi-directional interconnection between four terminals requires a total of eight unidirectional fibres. This network does have the property that inter-unit communication is maintained should any terminal or fibre fail.

This particular topology is similar to that adopted by the Raytheon Control-By-Light™ (CBL™) system that has been demonstrated in flight controlling the engine and thrust reversers of a Raytheon business jet. In this application the data rate is a modest

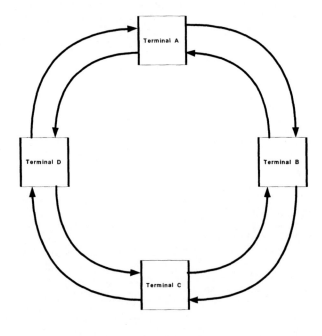

Fibre- Optic Ring Topology:

- **Each Link transmits Uni-directional Data**

- **Ring Topology allows Data to be passed from Terminal to Terminal**

- **Dual Ring Topology allows Data top be passed following a Terminal Failure**

- **Data transfer @ 1.25 MBits/sec**

Fig. 12.18 *Fibre-optic ring topology*

1.25Mbits/sec which is no real improvement over conventional buses such as MIL-STD-1553B and indeed is slower than A629. A fibre-optic bus does have the capability of operating at much higher data rates. It appears that the data rate in this case may have been limited by the protocol (control philosophy) which is an adaptation of a US personal computer/industrial Local Area Network (LAN) protocol widely used in the US.

Fibre-optic standards have been agreed and utilized on a small scale within the avionics community, usually for On-board Maintenance System (OMS) applications.

Avionics packaging – Line Replaceable Units (LRUs)

Line Replaceable Units (LRUs) were developed as a way of removing functional elements from an avionics system with minimum disruption. LRUs have logical functional boundaries associated with the task they perform in the aircraft. LRU formats were standardized to the following standards.

Air Transport Radio (ATR)

The origins of ATR standardization may be traced back to the 1930s when United Airlines and ARINC established a standard racking system called Air Transport Radio (ATR) unit case. ARINC 11 identified three sizes: ½ ATR, 1 ATR, 1½ ATR with the same height and length. In a similar time-scale, standard connector and pin sizes were specified for the wiring connections at the rear of the unit. The US military and the military authorities in the UK adopted these standards although to differing degrees and they are still in use in military parlance today. Over the period of usage ATR 'short' approximately 12.5 in length and ATR 'long' approximately 19.5 in length have also been derived. ARINC 404A developed the standard to the point where connector and cooling duct positioning were specified to give true inter-changeability between units from different suppliers. The relatively dense packaging of modern electronics means the ATR 'long' boxes are seldom used.

Modular Concept Unit (MCU)

The civil airline community developed the standardization argument further which was to develop the Modular concept unit (MCU). An 8 MCU box is virtually equivalent to 1 ATR and boxes are sized in MCU units. A typical small aircraft systems control unit today might be 2 MCU while a larger avionics unit, such as an Air Data and Inertial Reference System (ADIRS) combining the Inertial Reference System (IRS) with the air data computer function, may be 8 or 10 MCU. One 1MCU is roughly equivalent to 1¼ in width but the true method of sizing an MCU unit is given in Fig. 12.19 below.

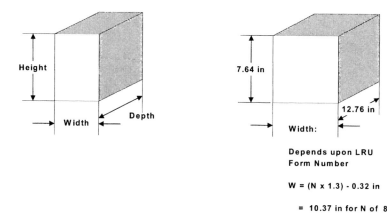

Fig. 12.19 *MCU sizing*

An 8 MCU box will therefore be 7.64 in high x 12.76 in deep x 10.37 in wide. The adoption of this concept was in conjunction with ARINC 600 which specifies connectors, cooling air inlets etc. in the same way that ARINC 404A did earlier.

Typical LRU architecture

The architecture of a typical avionics Line Replaceable Unit (LRU) is shown in Fig. 12.20. This shows the usual interfaces and component elements. The unit is powered by a power supply unit (PSU) which converts either 115 VAC or 28 VDC aircraft electrical power to low voltage DC levels (+5V and ±15V are typical) for the predominant micro-electronic devices. In some cases where commercially driven technology is used +3.3V may also be required. The processor/memory module communicates with the various I/O modules via the processor bus. The 'real world' to the left of the LRU interfaces with the processor bus via a variety of I/O devices which convert true analogue values to/from a digital format. The right portion of the LRU interfaces with other LRUs by means of digital data buses; in this example A429 is shown and it is certainly the most common data bus in use in civil avionics systems today.

One of the shortcomings exhibited by micro-electronics is their susceptibility to external voltage surges and static electricity. Extreme care must be taken when handling

Fig. 12.20 *Typical LRU architecture*

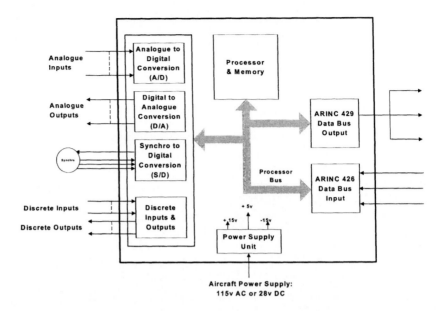

the devices outside the LRU as the release of static electricity can irrevocably damage the devices. The environment that the modern avionics LRU has to withstand and be tested to withstand is onerous as will be seen later.

The environmental and EMI challenges faced by the LRU in the aircraft can be quite severe, typically including the following.

- Electromagnetic interference
  - EMI produced by sources emission external to the aircraft; surveillance radars, high power radio stations and communications
  - Internal EMI: interference between aircraft equipment or by passenger carried laptops, gaming machines or mobile phones
  - Lightning effects

Mil-Std-461 and 462 are useful military references

- Physical effects due to one or more of the following:
  - Vibration: sinusoidal or random in three orthogonal axes
  - Temperature
  - Altitude
  - Temperature and altitude
  - Temperature, altitude and humidity
  - Salt fog
  - Dust
  - Sand
  - Fungi

Figure 12.21 shows the construction of a typical LRU; most avionics suppliers adopt this or similar techniques to meet the EMI requirements being mandated today. The EMI sensitive electronics is located in an enclosure on the left which effectively forms a 'Faraday cage'. This enclosed EMI 'clean' area is shielded from EMI effects such that the sensitive micro-electronics can operate in a protected environment. All signals

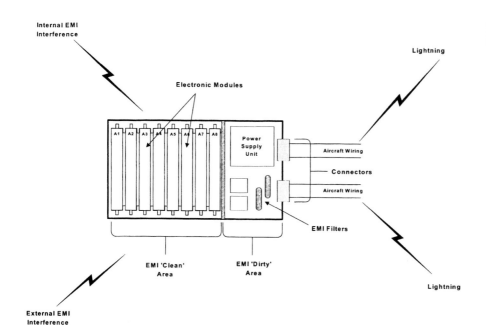

Fig. 12.21 LRU EMI hazards

entering this area are filtered to remove voltage spikes and surges. To the right of the EMI boundary are the EMI filters and other 'dirty' components such as the power supply unit (PSU). These components are more robust than the sensitive electronics and can successfully operate in this environment. Finally, in many cases the external wiring will be shielded and grounded to screen the wiring from external surges or interference induced by lightning – and more recently and perhaps more certainly – from emissions from passengers' laptop computers and hand-held computer games.

A typical test plan for modern avionics units will include many or all of the above tests as part of the LRU/system, as opposed to the aircraft certification process. Additionally, production units may be required to undergo an environmental stress screening (ESS) during production testing which typically includes 50 h of testing involving temperature cycling and/or vibration testing to detect 'infant mortality' prior to units entering full time service.

Integrated modular avionics

Integrated Modular Avionics (IMA) is a new packaging technique which could move electronic packaging beyond the ARINC 600 era. ARINC 600 as described earlier relates to the specification of LRUs in recent transport aircraft and this is the packaging technique used by many aircraft flying today. However, the move towards a more integrated solution is being sought as the avionics technology increasingly becomes smaller and the benefits to be attained by greater integration become very attractive. Therefore the advent of integrated modular avionics introduces an integrated cabinet approach where the conventional ARINC 600 LRUs are replaced by fewer units.

The IMA concept is shown in Fig. 12.22. The diagram depicts how the functionality of seven ARINC 600 LRUs (LRUs A through to J) may instead be installed in an integrated rack or cabinet as seven Line Replaceable Modules (LRMs) (LRMs A through to J). In fact the integration process is likely to be more aggressive than this, specifying common

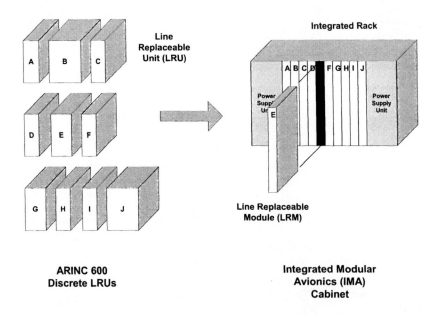

Fig. 12.22 *LRU and integrated modular cabinet comparison*

Line Replaceable Unit (LRU)

Integrated Rack

Power Supply Unit

Power Supply Unit

Line Replaceable Module (LRM)

ARINC 600 Discrete LRUs

Integrated Modular Avionics (IMA) Cabinet

modules and inter-leafing multiple processing tasks within common processor modules.

The US military were the first to implement modular avionics, starting with the Pave Pillar program and then applying the principles to the F-22 Raptor integrated avionics suite. In this implementation dynamic re-configuration is employed which enables the remaining computer resources to take over computational tasks should a computing module fail.

As the diagram suggests there are a number of obvious potential advantages to be realized by this integration:

- Volume and weight savings
- Sharing of resources, such as power supplies, across a number of functional modules
- More unified approach to equipment design
- LRMs are more reliable than LRUs

These advantages must be weighed against the disadvantages:

- Possibly more expensive overall to procure
- Possibly more risky
- May pose proprietary problems by having differing vendors working more closely together
- Segregation considerations (more eggs in one basket)
- Will an 'open' or 'closed' architecture prevail?
- What standards will apply – given the fact that a lot of effort has been invested in ARINC 600?
- Possibly more difficult to certify
- Who takes responsibility for systems integration?

Clearly there are some difficult issues to be answered. Also, critics might say: as the technology is becoming more reliable anyway; is the reliability increase due to the concept or the technology? Nevertheless this approach is gaining credence in both military and civil fields and the EPIC system described earlier has adopted this approach.

References

(1) **Lovell, Sir Bernard** (1991) *Echoes of War - The Story of H2 S Radar*, Adam Hilger, Bristol.

(2) **Middleton, D.M.** et al, (1989) *Avionics Systems*, Longman Scientific & Technical, Harlow.

(3) **Spitzer, Cary R.**, (1993) *Digital Avionics Systems - Principles & Practice*, McGraw-Hill.

(4) ARINC Specification 429: Mk 33 Digital Information transfer System, Aeronautical Radio, Inc., 1977.

(5) MIL-STD-1553B Digital Time Division Command/Response Multiplex data Bus, Notice 2, 8th September 1986.

(6) ARINC Characteristic 629, Multi-Transmitter Data Bus, Aeronautical Radio, Inc., November 1989.

(7) Boeing 777 ARINC 629 Data Bus - Principles, Development and Application, RAeS Conference - Advanced Avionics on the Airbus A330/A340 and the Boeing 777 Aircraft, November 1993.

(8) **Aplin, Newton & Warburton**, A Brief Overview of Databus Technology, RAeS Conference - The Design and Maintenance of Complex Systems on Modern Aircraft, April 1995.

(9) *Principles of Avionics Data Buses*, Avionics Communications Inc.

(10) **Tully, T.** Fuel Systems as an Aircraft Utility, International Conference – Civil Aerospace Technologies, FITEC '98, London, September 1998.

Index (misc)

Index of Aircraft

Index of Companies and Organizations

[Organizations such as CAA, FAA, etc. are also referenced below]

Index of Aircraft Engines